Quantifying Software

Quantifying Software
Global and Industry Perspectives

Capers Jones

CRC Press
Taylor & Francis Group
Boca Raton London New York

CRC Press is an imprint of the
Taylor & Francis Group, an **informa** business
AN AUERBACH BOOK

MIX
Paper from
responsible sources
FSC® C014174

CRC Press
Taylor & Francis Group
6000 Broken Sound Parkway NW, Suite 300
Boca Raton, FL 33487-2742

Printed on acid-free paper

International Standard Book Number-13: 978-1-138-03311-5 (Hardback)

Visit the Taylor & Francis Web site at
http://www.taylorandfrancis.com

and the CRC Press Web site at
http://www.crcpress.com
Printed and bound in the United States of America by Sheridan

Contents

Preface

The software industry is one of the most successful industries in history and has created numerous billionaires and millionaires as well as given employment to millions of technology workers.

Software and computers have changed the way the world operates. Every business and government agency now depends upon computers and software for all their major operations: payroll, human resources, finance, market analysis, research, and in many cases, actual computer-controlled manufacturing.

Probably on any given business day, out of the world population of more than 7.5 billion people, at least 1 billion people use computers as part of their daily jobs. About another two billion people use smartphones with embedded computers or own smart appliances with embedded computer technology.

Most forms of modern transportation such as automobiles, aircraft, trucks, and trains are heavily dependent on computers for many technical functions. We have all seen the news of driverless automobiles, which may be the wave of the future.

All large aircraft have computerized autopilot systems and most have computerized basic controls. Some modern military aircraft cannot even fly without computer controls because normal human reaction times are not fast enough.

Yet for all the benefits that software and computers have provided, software remains a troubling technology. Software cost and schedule overruns are endemic and occur for over 50% of software projects larger than 1,000 function points or 50,000 lines of source code. Canceled projects above 10,000 function points amount to about 35% of all projects and waste millions of dollars annually. Of the large systems that are not canceled, over 70% run late, exceed their budgets, and have marginal-to-poor quality.

The main reason for cancelation of large software projects is negative return on investment (ROI) due to massive schedule delays and cost overruns. The main reason for schedule delays and cost overruns is poor quality that stretches out test schedules. Unplanned and uncontrolled requirements creep add to the delays. Numerous software bugs and poor software quality are endemic. As of 2017, the industry average for defect removal efficiency (DRE) is only about 92.50% when it should be more than 99.50%.

Very few companies or government agencies actually know their software productivity and quality results for completed projects. Even fewer companies can estimate software quality, productivity, and schedules with reasonable accuracy before software projects start because the project managers keep using optimistic manual estimates instead of accurate parametric estimates. It is professionally embarrassing that so many companies and government groups have so little accurate quantified data available.

Inaccurate Software Metrics Are a Chronic Problem

Among the reasons for these software problems is a chronic lack of reliable quantified data. This book will provide quantified data from many countries and many industries based on about 26,000 projects developed using a variety of methodologies and team experience levels. The data has been gathered between 1970 and 2017, so interesting historical trends are available.

The book also examines a variety of software types and sizes. For example, there are big differences in software productivity and quality between Web applications, embedded software, information technology applications, open-source software, commercial software applications, and defense software.

Not many people know that defense software generates about three times the volume of paper documents than civilian software projects of the same size and type. In fact, for military software producing paper documents costs more than the source code. The huge volumes of defense software documents are due to arcane military standards and have little or no technical value.

The book will highlight results from samples of all modern software development methodologies such as Agile, containers, DevOps, spiral, waterfall, and many more. It will also highlight results from over 80 programming languages such as Go, Ruby, Swift, Objective-C, and also older languages such as C, Fortran, and COBOL. Since most modern applications use multiple languages such as Java and HTML, the book will also discuss productivity and quality results from common programming language combinations.

The book will also show the impact of the five levels of the Software Engineering Institute (SEI) Capability Maturity Model Integration (CMMI) for both quality and productivity. Although the CMMI has been useful for defense software, it is very sparsely used in the civilian sector. The author has met quite a few CIO and CFO executives who have never even heard of the SEI or the CMMI. About 95% of U.S. military and defense software producers use the CMMI, but <5% of civilian companies use it.

This book will show readers both current software cost drivers and also what might be possible in the future if current problems can be eliminated or at least greatly reduced.

Historically, software measurement practices and common software metrics are so bad that the software industry has very sparse reliable data on software schedules, costs, quality, and productivity.

Much of the available published data is incorrect, incomplete, or both. Few people know that government and defense software has paperwork volumes three times larger than civilian projects of the same size and spends more money on paperwork than on source code. Few people know that the #1 cost driver for the software industry is "finding and fixing bugs" and the #2 cost driver for the software industry is "producing paper documents." Actual coding in 2017 is only the #3 software cost driver. Few people know that software requirements grow and creep at about 1% per calendar month during development and keep growing after delivery at about 8% per year.

A key reason for this lack of accurate quantified data is the fact that the software industry has had the worst metrics and the worst measurement practices of any industry in human history. Two of the most widely used software metrics are hopelessly inaccurate, distort reality, and conceal actual progress in software engineering. The two most inaccurate software metrics are the traditional "lines of code" (LOC) metric and the "cost per defect" metric, both of which have been in continuous use since the late 1950s, with only a few people understanding their serious problems and how they distort reality and conceal true progress.

Quite a few other software metrics are weak or ambiguous, such as "technical debt" and "defect density," neither of which have standard definitions in 2017. Both story points and use case points lack standard definitions and vary widely from group to group.

Even though it has been used for more than 60 years, the LOC metric has never had a standard definition and varies wildely between counts of physical lines with blanks and comments and counts of actual code statements and data definitions. The literature on LOC metrics can vary by over 2000% between physical and logical LOC. No other industry has ever had a metric with such huge variations in apparent size.

The LOC metric has been in continuous use for software since the 1950s. But it penalizes modern high-level programming languages and cannot measure the effort or quality of requirements, design, and other noncode tasks. The LOC metric worked when assembly language was the only programming language, but it is a professional malpractice for measuring productivity and quality for modern high-level programming languages because LOC reverses true economic productivity and makes the assembly look better than Ruby or Java or Python.

The cost per defect metric has also been in use since the 1950s. But this metric penalizes quality and is cheapest for the buggiest software applications. Eventually, the software industry will achieve zero-defect software and of course cost per defect will be totally useless in this situation. The problem with the cost-per-defect metric is that as bugs decline, the cost per defect gets higher and higher due to the fixed costs of defect removal.

There are still large costs for writing and running test cases and inspecting source code even if there are zero bugs in the code. The main reason for the problems with both LOC metrics and cost-per-defect metrics is that they both ignore fixed costs and therefore neither metric actually measures true economic costs.

Software Engineering Fails to Use Standard Manufacturing Economic Measures

For more than 200 years, the standard economic definition of productivity has been, "Goods or services produced per unit of labor or expense." This definition is used in all industries, but has been hard to use in the software industry. For software, there is ambiguity in what constitutes our "goods or services."

The oldest unit for software "goods" was a LOC. More recently, software goods have been defined as "function points." Fortunately, function point metrics do provide accurate information on both quality and productivity.

Even more recent definitions of software goods include "story points," "use case points," and a newer metric called "SNAP points" for nonfunctional requirements. The pros and cons of these metric units will be discussed, but the main metric used in this book and indeed by almost all software benchmark organizations is the function point metric first developed by IBM in the mid-1970s.

This book uses function point metric defined by IFPUG version 4.3. Other kinds of function points such as COSMIC, FISMA, NESMA, etc. will be discussed and illustrated but not used for the main economic and quality analyses in this book. This is because the author has been collecting data with IFPUG function points since the metric was first developed in the 1970s. Indeed, the inventor of function point metrics, Allan Albrecht, worked with the author after he retired from IBM and were personal friends as well.

An important topic from manufacturing economics has a big impact on software productivity that is not well understood by the software industry even in 2017: the impact of fixed costs.

A basic law of manufacturing economics that is valid for all industries including software is the following: "When a development process has a high percentage of fixed costs, and there is a decline in the number of units produced, the cost per unit will go up."

When an LOC is selected as the manufacturing unit and there is a switch from a low-level programming language such as assembly to a higher level language such as Java or Objective-C, there will be a reduction in the number of "units" developed.

For example, developing a specific software function of two function points in size might take 100 code statements in assembly language, 50 code statements in the Java language, and only 25 code statements in the Objective-C language. But it is exactly the same function and creates the same outputs regardless of the programming language used to code it.

The assembly version might cost $3000 to produce, the Java version might cost $2000, and the Objective-C version might cost $1200. In terms of real economic productivity or producing goods for the lowest cost, Objective-C clearly has the best economic productivity and is only about one-third as expensive as assembly language.

But the cost per LOC for assembly is $30.00, the cost per LOC for Java is $40.00, and the cost per LOC for Objective-C is $48.00. As can be seen from this simple example, the cost per LOC reverses real economic productivity and penalizes modern programming languages. This problem is best seen when measuring the same kind of software. It was originally discovered by IBM when comparing productivity results for new IBM programming languages and looked at the productivity of various compilers, which are of course quite similar.

All three versions of the example were two function points in size since they were actually the same function. Using cost per function point instead of cost per LOC, assembly language cost $1500 per function point, Java was $1000, and Objective-C was $600 per function point. As can be seen, function points are a perfect fit with standard economics, while LOC reverses actual economic results and penalizes the two modern programming languages.

Yet, even in 2017, thousands of companies still try and measure software productivity using the invalid LOC metric and do not yet understand the problems of LOC metrics. This is professionally embarrassing for one of the richest industries in human history.

The noncode tasks of analysis, requirements, design, and documentation act like fixed costs for all three languages. Therefore, the "cost per LOC" will go up for the high-level languages compared to assembly language. This means that LOC is not a valid metric for measuring economic productivity.

Software Economic Analysis

For software, there are two definitions of productivity that match standard economic concepts, although one metric is inconsistent from country to country:

1. Producing a specific quantity of deliverable units for the lowest number of work hours
2. Producing the largest number of deliverable units in a standard work period such as an hour, month, or year

In definition 1, deliverable goods are constant and *work hours are variable.*

In definition 2, *deliverable goods are variable* and work periods are constant.

The common metric "work hours per function point" is a good example of productivity definition 1.

The metric "function points per month" is a good example of definition 2.

Definition 2 will encounter the fact that the number of work hours per month varies widely from country to country. For example, India works 190 h per month while the Netherlands work only 115 h per month. This means that productivity definitions 1 and 2 will not be the same.

Table 1 Global Work Hour Differences (1000 Function Points)

	Work Hours/ Month	*Overtime*	*Total Hours/ Month*	*Work Hours/ Function Point*	*Function Point/ Month*
China	186	8	194	15.00	12.93
Israel	159	8	167	15.00	11.13
The United States	132	10	142	15.00	9.47
Australia	127	0	127	15.00	8.47
Germany	116	0	116	15.00	7.73

A given number of work hours would take fewer calendar months in India than in the Netherlands due to the larger number of monthly work hours. Table 1 illustrates the differences between work hours per function point and function points per month for five countries. Table 1 assumes 1000 function points for all the applications and a fixed value of 15.00 work hours per function point for all applications in all five countries:

Note that work hours per function point is the same in every country for this example. However, because of the wide range of global work hours per month, the function points per month metric varies from country to country. For that matter, there are also industry variations and local company variations in the number of hours that software teams work each month, and also large variations in the amount of unpaid overtime used.

Thus, to understand the actual economic productivity of software development, the work hours per function point metric is the best choice available as of 2017. This metric will be used later in the book for both complete projects and also for specific activities such as requirements, design, and coding.

The new SNAP metric that separates nonfunctional requirements from functional requirements is too new to have substantial data as this book was being written. However, normal function point metrics have been used successfully since 1975 and are still effective software metrics for carrying out studies of software economic and quality results.

IBM Software Metrics Inventions Circa 1975

Fortunately, for the software industry, IBM developed a set of very effective software metrics in the early 1970s. Three of these effective IBM metrics are used in this book to show accurate quantified data on software productivity and quality.

The three effective software metrics developed by IBM are as follows:

1. Function point metrics
2. Defect potentials quantified with function point metrics
3. DRE

Function point metrics have formal definitions and ISO standards. Function point metrics have formal training and also certification examinations available. In general, function points are derived from counts of five key topics: inputs, outputs, inquiries, interfaces, and logical files. (The actual counting rules for function points in 2017 are about 100 pages in size and need to be studied carefully.)

Assuming 132 paid hours and 10 unpaid hours per month, 142 h in total, the 2017 average U.S. productivity rate is about 8.00 function points per month or 17.75 work hours per function point. The upper range is about 25.00 function points per month or 5.68 work hours per function point. The lowest results are 1.50 function points per month or 94.66 work hours per function point for large military software projects. The civilian low is about 3.00 function points per month or 47.33 work hours per month.

The IBM "defect potential" metric is based on IBM's huge collection of accurate software quality data. The defect potential of a software application is the sum total of probable bugs found in requirements, design, source code, documents, and "bad fixes." (A bad fix is a new bug accidentally created when fixing an older bug. The U.S. average for bad fixes is about 7% of all bug repairs, but high cyclomatic complexity more than about 25% can raise the percentage to more than 30% bad fixes.)

The current U.S. average for software defect potentials in 2017 is about 4.25 defects per function point, with a range from below 2.00 defects per function point to more than 6.50 defects per function point. All the bugs are not in the code, of course. The distribution of bugs is shown in Table 2.

Table 2 Average Software Defect Potentials circa 2017 for the United States

• Requirements: 0.70 defects per function point
• Architecture: 0.10 defects per function point
• Design: 0.95 defects per function point
• Code: 1.15 defects per function point
• Security code flaws: 0.25 defects per function point
• Documents: 0.45 defects per function point
• Bad fixes: 0.65 defects per function point
• Totals: 4.25 defects per function point

Although code defects individually are present in the largest numbers, they are only 27% of the overall defect potential. It is obvious why requirements and design inspections are also useful for software quality control.

The IBM DRE metric is the percentage of bugs found by any specific form of defect removal such as static analysis or unit testing, and also the sum total of defects found by all forms of defect removal including pretest activities such as static analysis and all forms of software testing.

The current U.S. average for total DRE during development is about 92.50%, with a range from a high of 99.80% down to a low about 78.00%. Most forms of testing are about 30% efficient or find one bug out of three. Static analysis is about 55% efficient, while formal inspections of requirements, design, and source code average about 85% in DRE. To top 99% in DRE, effective quality control needs a synergistic combination of pretest defect removal such as static analysis or inspections and about eight stages of formal testing, with formal test case design as well.

These three metrics developed originally by IBM are widely used by technology companies in 2017 and have proven to be valuable for both estimating software projects before they begin and for measuring software projects when they are complete.

U.S. Software Cost Drivers in 2017 and Projected for 2027

One of the historical problems of software engineering due to bad metrics and poor measurement practices is the fact that hardly anyone knows the major expenses or "cost drivers" for software applications in rank order. Table 3 shows 15 key software cost drivers in rank order as of 2017.

What is troubling about Table 3 is that many of the 15 cost drivers are due to poor quality and sloppy software engineering. We should not be spending more money on finding and fixing bugs than on anything else, but it is the #1 cost driver. Requirements changes should not be #4. Canceled projects should not be #5. Successful cyber-attacks should not be #6. For that matter, litigation for failures and disasters should not even be included, but it is #12 out of 15.

With better metrics and measurement practices combined with more sophisticated software engineering based on standard reusable components instead of custom designs and manual coding, it is theoretically possible to make major and beneficial changes to the pattern of software cost drivers. Table 4 shows a hypothetical pattern of software cost drivers for the year 2027, or 10 years from today.

In Table 4, all the high cost drivers due to poor quality control and sloppy software engineering have moved to the bottom of the table, and the #1 cost driver could theoretically be that of innovation and creating new kinds of software.

Assuming a nominal cost structure of $10,000 per staff month (salary, burden, insurance, etc.), the average 2017 cost for building software applications is roughly

Table 3 U.S. Software Cost Drivers circa 2017 in Rank Order

1. The cost of finding and fixing bugs
2. The cost of producing paper documents
3. The cost of programming or coding
4. The cost of requirements changes
5. The cost of canceled projects
6. The cost of successful cyber-attacks
7. The cost of post-release customer support
8. The cost of meetings and communication
9. The cost of project management
10. The cost of post-release renovation and migration
11. The cost of innovation and new kinds of software
12. The cost of litigation for failures and disasters
13. The cost of training and learning for users and staff
14. The cost of avoiding security flaws
15. The cost of constructing reusable components

$1,000 per function point, with a range from below $400 per function point to a high over about $3,500 per function point.

Assuming that better measures and metrics lead to quick software engineering progress, by calendar year 2027, the average cost for building software applications in 10 years might drop to $400 per function point, with a range from below $150 per function point to a high of perhaps $2,000 per function point, assuming $10,000 per month as the nominal expense rate.

What Is Needed to Improve Software Engineering Metrics and Performance

Improving software engineering costs, schedules, quality, and productivity requires a multifaceted set of technological and sociological changes. The set of changes required to achieve the theoretical 2027 software engineering results includes the following:

Table 4 U.S. Software Cost Drivers circa 2027 in Rank Order

1. The cost of innovation and new forms of software
2. The cost of renovation and migration
3. The cost of post-release customer support
4. The cost of constructing reusable components
5. The cost of meetings and communications
6. The cost of avoiding security flaws
7. The cost of training and learning for users and staff
8. The cost of project management
9. The cost of requirements changes
10. The cost of producing paper documents
11. The cost of programming or coding
12. The cost of finding and fixing bugs
13. The cost of successful cyber-attacks
14. The cost of canceled projects
15. The cost of litigation for failures and disasters

Software Metric and Measurement Changes

1. Stop using bad metrics such as LOC and cost per defect.
2. Stop using ambiguous metrics such as technical debt, story points, and use case points.
3. Revise the SNAP metric so it becomes additive and equivalent to function point metrics.
4. Provide conversion rules among all forms of function points (IFPUG, COSMIC, etc.).
5. Adopt function point metrics for both software cost estimates and software benchmarks.
6. Use work hours per function point for productivity as a standard metric.
7. Adjust function points for month for national and local work hour patterns.
8. Increase the number of governments that mandate function points to >100 countries.
9. Use defect potentials with function points for quantifying software bugs.
10. Use DRE for all software projects.

11. Stop using single points of data or phases for software benchmarks.
12. Adopt the concept of activity-based costs for all software benchmarks.
13. Measure software requirements creep for all projects >100 function points.
14. Improve function point sizing speed to <15 min per project.
15. Use formal parametric estimates for all software >100 function points in size.

Software Technology Stack Improvements

1. Increase the volume of standard reusable components from <15% to >85%.
2. Drop software defect potentials down <2.00 defects per function point.
3. Achieve >96% DRE for average projects and >99.50% DRE for critical projects.
4. Use static analysis on all projects for both new code and maintenance changes.
5. Use formal inspections of requirements, design, and source code on critical features.
6. Use mathematical test case design such as cause–effect graphs or design of experiments.
7. Use pattern-based development for all software projects with legacy antecedents.
8. Achieve test coverage of <90% for all projects and <97% for critical projects.
9. Eliminate "error-prone modules" in all software applications.
10. Decrease bad-fix injections from 7.00% to <0.50%.
11. Decrease bad test cases from 15.00% to <0.50%.
12. Decrease duplicate test cases from 10% to <0.50%.
13. Decrease requirements creep from >1.00% per month to <0.10% per month.
14. Keep average cyclomatic complexity for all modules <10.
15. Validate effectiveness of all methodologies before using; don't adopt due to popularity.

Software Sociological Improvements

1. Teach software function points in graduate and undergraduate software classes.
2. Teach standard economics including the impact of fixed costs to software managers.
3. Teach activity-based cost principles to software engineers.
4. Teach quality control using defect potential metrics.
5. Teach quality control using DRE metrics.
6. Establish effective measurement programs in every company and government agency.
7. Establish effective progress monitoring for all projects that includes quality and creep.

8. Measure activity-based costs for all projects >250 function points.
9. Measure defect potentials and DRE for all projects >250 function points.
10. Perform formal and early risk analysis for all projects >250 function points.
11. Measure the effectiveness of major software engineering methods such as agile and TSP.
12. Measure the effectiveness of software defect prevention such as requirements models.
13. Measure the effectiveness of pretest defect removal such as static analysis.
14. Measure the effectiveness of all forms of software testing.
15. Measure the accuracy of manual versus parametric software estimates.

The combination of these 15 metric changes, 15 technology changes, and 15 sociological changes, in theory, could reduce software development schedules by >50% compared to 2017, lower software costs by >60%, and reduce delivered software defects by >85% compared to 2017 norms. The technologies and measurement principles for improvement are clear, but social resistance to progress due to cognitive dissonance is a strong barrier that needs to be overcome.

These beneficial changes also have a large sociological component, and it is not easy to get an entire industry to move in new directions, even if all data supports the fact that the current direction is hazardous and expensive while the new direction would be safe and cheap.

(Many effective technologies were initially resisted or rejected including Copernican astronomy, sterile surgical procedures, vaccination, continental drift, evolution, self-leveling naval cannons, and Samuel Colt's revolver. For that matter, function point metrics have been resisted by many companies even though all empirical data and all controlled studies prove that they are the best available software metric for both productivity and quality, while the older LOC metric has been mathematically proven to distort reality and conceal software engineering productivity and quality gains.)

In summary, this book will provide current data on average software schedules, effort, costs, and quality for several industries and countries. Because productivity and quality vary by technology and size, the book will show quantitative results for applications between 100 function points and 100,000 function points in size using powers of 10, i.e., size plateaus of 100, 1,000, 10,000, and 100,000 function points.

The book will also show quality results using defect potential and DRE metrics. This is because the #1 cost driver for software is finding and fixing bugs. The book will show data on cost of quality for software projects. It will discuss technical debt but that metric is not standardized and has no accurate benchmarks as of 2017.

The book will include some data on 3 years of software maintenance and enhancements, and also some data on total cost of ownership. However, the main focus of the book is on software development and software quality control.

Since current average software productivity and quality results in 2017 are suboptimal, the book will focus on "best in class" results and show not only quantified quality and productivity data from best-in-class organizations, but also the technology stacks used to achieve best-in-class results. The overall goal of this book is to encourage readers to adopt best-in-class software metrics and best-in-class technology stacks.

Readers of this book are encouraged to also read *The Social Transformation of American Medicine* by Paul Starr. This book was published in 1982 and won both a Pulitzer Prize in 1984 and also a Bancroft Prize. It is not a software book, but it is a revealing history of how to transform a major industry for the better.

The reason for suggesting Paul Starr's book is because at one time medicine was even more chaotic and disorganized than what software engineering is today. Medical schools had only 2 years of training and most did not require college or even high school graduation to be enrolled. Medical malpractice was common and there were no medical licenses or board certification. Paul Starr's excellent book explains how medical practice evolved from a poorly organized and minimally trained craft into a top learned profession in about 100 years.

Hopefully software can also evolve from a poorly trained craft into a top learned profession. Using Paul Starr's book as a guideline, this might be accomplished in perhaps 10–20 years rather than the 100 years required for medical practice evolution.

Software is one of the most important products in human history and is widely used by all industries and all countries. It is also one of the most expensive and labor-intensive products in human history. Software also has very poor quality that has caused many major disasters and wasted many millions of dollars. Software is also the target of frequent and increasingly serious cyber-attacks.

The goal of all of us in software engineering should be to optimize software engineering methods and prove the success of these optimal methods by using accurate and effective economic metrics such as function points, defect potentials, and DRE. Software needs to eliminate cost overruns, schedule delays, expensive failures, poor quality, and cyber-attacks. It is technically possible to achieve major software engineering improvements if we can overcome cognitive dissonance and sociological resistance to change.

MATLAB® is a registered trademark of The MathWorks, Inc. For product information, please contact:

The MathWorks, Inc.
3 Apple Hill Drive
Natick, MA 01760-2098 USA
Tel: 508-647-7000
Fax: 508-647-7001
E-mail: info@mathworks.com
Web: www.mathworks.com

Acknowledgments

This is my 18th software book and the 20th book overall. (I also have written two history books; one is a history of Narragansett Bay from before the last ice age, and the second is mainly a history of ancient civil engineering from 3500 BCE through about 1000 CE).

This will perhaps be my final software book. It is part of a series of three new books completed for CRC Press. The first book discussed how to measure software; the second was a quantified comparison of 60 software development methodologies; and this third book has large volumes of quantitative data from many industries and countries.

My first software book was in 1979 while working at IBM. I also have about a hundred journal articles, although I tend to lose track of these since there have been so many.

As always, many thanks to my wife, Eileen Jones, for making this book possible. Thanks for her patience when I get involved in writing and disappear for several hours. Also, thanks for her patience on holidays and vacations when I take my portable computer and write early in the morning.

Thanks to my neighbor and business partner Ted Maroney, who handles contracts and the business side of Namcook Analytics LLC, which frees up my time for books and technical work. Thanks also to Aruna Sankaranarayanan for her excellent work with our Software Risk Master (SRM) estimation tool and our website. Thanks to Bob Heffner for marketing plans. Thanks also to Gary Gack and Jitendra Subramanyam for their work with us at Namcook.

Thanks to other metrics and measurement research colleagues who also attempt to bring order into the chaos of software development: Special thanks to the late Allan Albrecht, the inventor of Function Points, for his invaluable contribution to the industry and for his outstanding work. Without Allan's pioneering work in function points, the ability to create accurate baselines and benchmarks would probably not exist today in 2017.

The new Software NonFunctional Assessment Process (SNAP) team from International Function Point Users Group (IFPUG) also deserves thanks: Talmon Ben-Canaan, Carol Dekkers, and Daniel French.

Thanks also to Dr. Alain Abran, Mauricio Aguiar, Dr. Victor Basili, Dr. Barry Boehm, Dr. Fred Brooks, Manfred Bundschuh, Tom DeMarco, Dr. Reiner Dumke, Christof Ebert, Gary Gack, Tom Gilb, Scott Goldfarb, Peter Hill, Dr. Steven Kan, Dr. Leon Kappelman, Dr. Tom McCabe, Don Reifer, Dr. Howard Rubin, Dr. Akira Sakakibara, Manfred Seufort, Paul Strassman, Dr. Gerald Weinberg, Cornelius Wille, the late Ed Yourdon, and the late Dr. Harlan Mills for their own solid research and for the excellence and clarity with which they communicated ideas about software. The software industry is fortunate to have researchers and authors such as these.

Appreciation is also due to various corporate executives who supported the technical side of measurement and metrics by providing time and funding. From IBM, the late Ted Climis and the late Jim Frame both supported my measurement work and in fact commissioned several studies of productivity and quality inside IBM as well as funding IBM's first parametric estimation tool in 1973. Rand Araskog and Dr. Charles Herzfeld at ITT also provided funds for metrics studies, as did Jim Frame who became the first ITT VP of software.

Appreciation is also due to the officers and employees of the IFPUG. This organization started 30 years ago in 1987 and has grown to become the largest software measurement association in the history of software. When the affiliates in other countries are included, the community of function point users is the largest measurement association in the world.

There are other function point associations such as COSMIC, FISMA, and NESMA, but all 18 of my prior software books (2 of my books are history books and don't deal with software) have used IFPUG function points. This is in part because Al Albrecht and I worked together at IBM and later at Software Productivity Research (SPR). Al and Jean Albrecht were both good family friends and we enjoyed our visits with them.

Author

Capers Jones is currently vice president and chief technology officer of Namcook Analytics LLC (www.Namcook.com). Namcook Analytics LLC designs leading edge risk, cost, and quality estimation and measurement tools. Software Risk Master (SRM) is the company's advanced estimation tool with a patent-pending early sizing feature that allows sizing before requirements via pattern matching. Namcook Analytics also collects software benchmark data and engages in longer range software process improvement, quality, and risk assessment studies. These Namcook studies are global and involve major corporations and some government agencies in many countries in Europe, Asia, and South America.

Chapter 1

Introduction to Quantifying Software Results

In order to quantify software productivity and software quality with high precision, quite a few variable factors need to be measured and included in the overall measurement suite because they impact software results. This chapter illustrates 50 factors that influence software productivity and quality. For those who might prefer to skip directly to a specific factor, the 50 are in the following sequence:

1	Metrics used for accurate quantification
2	Metrics to avoid for accurate quantification
3	Metrics that are ambiguous in 2017
4	Twenty criteria for metrics selection
5	Project taxonomy
6	Software industry codes
7	Early and rapid software application size metrics
8	Size metrics with and without conversion rules
9	Certified reusable components
10	Development team, project management, and client experience

11	Team work hours per month and unpaid overtime
12	Costs of canceled projects
13	Requirements creep
14	Activity-based costs
15	Assignment scopes and production rates
16	Software schedule measures
17	Software quality measures
18	Cost of quality (COQ)
19	Gaps and omissions in software quality data
20	Software cyber-attack defense and recovery
21	Leakage from software project historical data
22	Software risk measures
23	Software documentation costs
24	Software occupation groups
25	Software methodologies
26	Software programming languages
27	Software Non-Functional Assessment Process (SNAP) nonfunctional requirements
28	Software user costs for internal applications
29	Software breach of contract litigation costs
30	Capability maturity model integrated (CMMI)
31	Software wasted workdays
32	Tools, standards, and certification
33	Software technology stack evaluation
34	Software meeting and communication costs
35	Software outsource contract estimates and benchmarks
36	Venture capital and software start-up estimation
37	Software maintenance, enhancement, and customer support
38	Software total cost of ownership (TCO)

39	Life expectancy of software benchmark data
40	Executive interest levels in software benchmark types
41	Available benchmarks circa 2017
42	Variations by application size
43	Mix of new development and modification projects
44	Portfolio benchmarks
45	Problem, code, and data complexity
46	Deferred features
47	Multinational development
48	Special costs that are hard to estimate and measure
49	Software value and return on investment (ROI)
50	Summary overview of software measurements

The overall goals of the book are twofold: One is to bring software measurement and metric practices into conformance with standard manufacturing economic practices; the second is to show ranges of productivity and quality from various industries and countries.

For too long, the software industry has used metrics such as "lines of code" (LOC) and "cost per defect" that distort reality and conceal true economic progress. Further, few people have known that state and federal government software tends to lag corporate software.

This introductory chapter discusses key metric and measurement topics, in summary form, in order to illustrate the major measurements and metrics needed to evaluate software economic productivity and software quality with high precision. The goal is to be able to measure software projects with about 1% precision and predict them with about 5% precision.

Unfortunately, historical data on productivity circa 2017 is only about 37% complete and quality data is even worse. These "leaks" include the work of part-time specialists for productivity and failing to measure bugs before function test for quality.

Of course, to measure productivity and quality you need to know what those words actually mean in an economic sense. The standard economic definition of productivity for over 200 years, which is used across all industries, is "Goods or services produced per unit of labor or expense."

Another important aspect of productivity is a basic law of manufacturing economics, which is poorly understood (if at all) by the software community: "If a manufacturing process has a high percentage of fixed costs and there is a decline in the number of units produced, the cost per unit will go up."

For software, if you select "a line of code" as a manufacturing unit and you switch from a low-level language such as assembly to a high-level language such as Java, you have reduced the number of "units" produced. But requirements, design, and non-coding work act like fixed costs and hence drive up the "cost per line of code."

Function points are a synthetic metric that define the features that users want in a software application. If you reduce function points, you will also reduce costs; so, function points can be used for economic analysis but LOC distort reality and penalize high-level languages and conceal the true economic value of high-level languages.

As for quality, the software industry needs a metric that can be predicted before projects start and then measured when they are finished. Subjective metrics such as "fitness for use" or "meeting customer requirements" are impossible to predict and difficult to measure.

A workable definition for software quality is "The prevention or removal of defects that might cause a software application to either stop completely or to produce incorrect results."

This definition can be predicted before projects start because "defect potentials" and "defect removal efficiency (DRE)" have been included in parametric estimation tools since the 1980s. This definition can also be used with requirements and design defects as well as with code defects, since software defects originate in many sources.

When a software application goes through a testing cycle, fewer and fewer bugs will be found, but the costs of writing and running test cases act like fixed costs and hence drive up "cost per defect." Cost per defect does not measure the economics of software quality and indeed penalizes software quality. Cost per defect is cheapest for the buggiest software, even though defect repair costs may be astronomical.

However, the metric "defect removal cost per function point" declines in later test stages and does measure the economic value of higher quality. Defect removal cost per function point is cheapest for high-quality software and hence demonstrates the economic value of effective software quality control.

The key factors that need to be evaluated and included in overall software measurements include the 50 factors in summary form. The 50 factors in the text are listed in approximate order of when they are needed; i.e., the factors needed to get started with a project are at the top of the list. Factors such as maintenance and ROI that can't be measured until several years after release are at the end of the list.

Since many readers will be interested in the importance of these factors, the 25 most important factors for software estimation and measurement precision are listed in Table 1.0.

All 50 factors are significant and should be understood. But as can be seen from the ranking, a formal taxonomy, application size in function points, and an activity-based cost structure are the top three.

Table 1.0 Twenty-Five Software Measurement Factors in Order of Importance

1	Project taxonomy
2	Early and rapid application size metrics
3	Activity-based costs
4	Assignment scopes and production rates
5	Certified reusable components
6	Development, management, and client experience
7	Team work hours per month
8	Requirements creep
9	Software quality measures
10	Software schedule measures
11	Software risk measures
12	Software documentation costs
13	Software occupation groups
14	Software programming languages
15	Software methodologies
16	Software value and ROI
17	Software cyber-attack defense and recovery
18	Problem, code, and data complexity
19	Software TCO
20	COQ
21	CMMI
22	Software technology stack evaluation
23	Software user costs for internal projects
24	Software meeting and communication costs
25	SNAP nonfunctional requirements

Factor 1—Metrics used for accurate quantification: There are over 100 possible software metrics available in 2017, but only a few are effective in quantifying software productivity and quality. The metrics used in this book include (1) function point metrics as defined by the International Function Point Users Group (IFPUG) but predicted by the author's Software Risk Master (SRM) tool; (2) defect potentials or the sum total of possible bugs originating in requirements, design, code,

documents, and "bad fixes" or new bugs in bug repairs themselves; and (3) DRE or the percentage of bugs found via specific defect removal operations and also cumulative DRE for all defect removal operations. The industry average in 2017 for cumulative DRE is only about 92.5%, but available technologies such as static analysis and inspections plus formal testing and mathematical test case design can raise cumulative DRE to over 99.75% and save time and money as well.

Factor 2—Metrics to avoid for accurate quantification: It is an unfortunate fact that many common software metrics being used in 2017 are inaccurate and indeed some distort reality and reverse true economic productivity. Among these hazardous metrics that should be avoided are LOC, cost per defect, story points, and use case points.

None of these metrics have standard definitions and all vary widely. LOC and cost per defect distort reality and conceal true progress and true economic results. LOC metrics penalize high-level programming languages. Cost per defect metrics penalize quality. Story points are highly subjective and have no standard counting rules, and these statements are true for use case points too.

Factor 3—Metrics that are ambiguous in 2017: Some popular metrics are ambiguous and poorly defined. These metrics include technical debt, goal-question metrics, defect density, Agile velocity, Agile burnup, and several recent function point variations such as engineering function points and unadjusted function points.

The function point metrics with ISO standards and certification examinations are better choices. These include COSMIC, FISMA, IFPUG, and Netherlands Software Metrics Association (NESMA) function point metrics. The new SNAP metric for nonfunctional requirements is ambiguous at the time of writing this book but is in the process of being standardized.

Factor 4—Twenty criteria for software metrics selection: Software metrics are created by ad hoc methods, often by amateurs and are broadcasted to the world with little or no validation or empirical results. This set of 20 criteria shows the features that effective software metrics should have as attributes before being adopted by the software community.

Currently, IFPUG function point metrics meet 19 of these 20 criteria. Function points are somewhat slow and costly, so criterion 5 is not fully met.

Other function point variations such as COSMIC, NESMA, FISMA, unadjusted, engineering function points, feature points, etc. vary in the number of criteria they meet, but most meet more than 15 of the 20 criteria.

Automated function points using tools developed by CAST Software and Relativity Technologies meet criteria 1 through 5, which are important, but don't meet any of the other 20 criteria.

The new SNAP metric for nonfunctional requirements meets criteria 1 through 4 and 6 through 8. It is not easy to count and does not support all kinds of software nor does it support enhancements to legacy software.

The older "LOC" metric meets only criterion 5 and none of the others. LOC metrics are fast and cheap but otherwise fail to meet the other 19 criteria. The LOC metric makes requirements and design invisible and penalizes high-level languages.

The "cost per defect" metric does not actually meet any of the 20 criteria and also does not address the value of high quality in achieving shorter schedules and lower costs.

The "technical debt" metric does not currently meet any of the 20 criteria, although it is such a new metric that it probably will be able to meet some of the criteria in the future. It has a large and growing literature but does not actually meet criterion 9 because the literature resembles the blind men and the elephant, with various authors using different definitions for technical debt. Technical debt comes close to meeting criteria 14 and 15.

The "story point" metric for Agile projects seems to meet five criteria, i.e., numbers 6, 14, 16, 17, 18, but varies so widely and is so inconsistent that it cannot be used across companies and certainly can't be used without user stories.

The "use case" metric seems to meet criteria 5, 6, 9, 11, 14, and 15, but can't be used to compare data from projects that don't utilize use cases.

This set of metric criteria is a useful guide for selecting metrics that are likely to produce results that match standard economics and do not distort reality, as do so many current software metrics.

Factor 5—Project taxonomy: A basic factor for all software projects is "What kind of software is being built?" It is obviously invalid to compare productivity of a small Web application against productivity for a large military weapons system. The starting point of all effective software benchmarks, and also software estimates, is to use a formal taxonomy to identify the specific size and kind of software project under study.

The author developed a formal taxonomy for use in both project estimation and project benchmark measures with his SRM tool. The taxonomy includes the factors of project nature, scope, class, type, and complexity. Samples of this taxonomy are shown in Table 1.2.

The information provided by this taxonomy is useful for both estimating projects before they begin and measuring completed projects after they are delivered. It is also used for sizing since applications with the same taxonomy patterns are usually of almost the same size in function points, but of course not in LOC.

Only the final results of the author's taxonomy are shown in Table 1.1. In day-to-day use, the taxonomy topics are selected from multiple-choice lists, some of which have more than 25 possible selections. For example, "scope" can range from an individual module through stand-alone programs, through component of large systems, up to the largest choice, which is "global system."

Factor 6—Software industry codes: Note that the taxonomy shown in the previous section includes "industry" as a factor. This is an important factor for both productivity and quality economic analysis. The author uses industry codes provided by the U.S. Census Bureau in the "North American Industry Classification" (NAIC) tables as shown in Table 1.3.

NAIC codes have several formats ranging from two digits to four digits. The author uses the three-digit codes for benchmarks. Recording industries are what

Table 1.1 Twenty Criteria for Software Metrics Adoption

1	Metrics should be validated before release to the world
2	Metrics should be standardized, preferably by ISO
3	Metrics should be unambiguous
4	Metrics should be consistent from project to project
5	Metrics should be cost-effective and have automated support
6	Metrics should be useful for both predictions and measurements
7	Metrics should have formal training for new practitioners
8	Metrics should have a formal user association
9	Metrics should have ample and accurate published data
10	Metrics should have conversion rules for other metrics
11	Metrics should support both development and maintenance
12	Metrics should support all activities (requirements, design, code, test, etc.)
13	Metrics should support all software deliverables (documents, code, tests, etc.)
14	Metrics should support all sizes of software from small changes through major systems
15	Metrics should support all classes and types of software (embedded, systems, Web, etc.)
16	Metrics should support both quality and productivity measures and estimates
17	Metrics should support measuring requirements creep over time
18	Metrics should support consumption and usage of software as well as construction
19	Metrics should support new projects, enhancement projects, and maintenance projects
20	Metrics should support new technologies (new languages, cloud, methodologies, etc.)

allowed regression analysis to show that medical devices have the highest software quality; that defense projects produce the largest volumes of software documents; and that state governments lag in both software quality and software productivity. Also, state and federal government software packages have higher volumes of non-functional requirements than corporate software packages.

Table 1.2 Taxonomy Differences Leading to Project Differences

Project Taxonomy	*Project 1*	*Project 2*
Nature	New project	New project
Scope	PC program	Corporate system
Class	Internal	Commercial
Type	Web application	Big data application
Hardware platform	Personal computer	Mainframe computer
Software platform	Windows	IBM operating system
Problem complexity	Low	High
Code complexity	Average	Average
Data complexity	Low	High
Languages	Ruby on Rails	PHP and SQL
Industry	Banking	Telecommunications
Country	United States	Japan
Cities	Cambridge, MA	Tokyo
Methodologies	Agile/Scrum	TSP/PSP
Team experience	Above average	Above average
Management experience	Above average	Above average
Client experience	Above average	Above average

The full set of industry codes total to more than 300 industries. The example used here only includes a small sample of industries with significant software construction and usage.

Factor 7—Early and rapid software application size metrics: As all readers know, software productivity and quality are inversely related to application size. The bigger the application in both function points and LOC, the lower the productivity rates and the higher the defect potentials become.

Because application size has such a strong impact on project results, it is desirable to know application size early before requirements are fully defined. Unfortunately, LOC metrics can't be evaluated until code is written.

Function points can't be counted until requirements are complete. Also function point analysis is somewhat slow and expensive, averaging only about 500 function points per day for a certified counter. This means that application size may not be known for quite a few months after the project has started. For major

Table 1.3 North American Industry Classification

Codes	*Selected Industry Sectors*
11	Agriculture
21	Mining
211	Oil and gas
212	Mining
213	Support
23	Construction
31–33	Manufacturing
311	Food
312	Beverages and tobacco
324	Petroleum
325	Chemicals
333	Machinery
334	Computers
42	Trade/wholesale
44–45	Trade/retail
48–49	Transportation
481	Air transportation
482	Rail transportation
483	Ship transportation
484	Truck transportation
51	Information
511	Publishing
512	Motion pictures
515	Broadcasting
516	Internet
517	Telecommunications

(*Continued*)

Table 1.3 (*Continued*) North American Industry Classification

Codes	*Selected Industry Sectors*
518	Data processing/hosting
52	Finance
521	Banks
523	Securities
524	Insurance
54	Scientific/management consulting
92	Public administration

Note: Full NAIC codes available from U.S. Census Bureau: www.census.gov.

applications, this uncertainty in size creates uncertainty in cost and schedule estimates and of course in risk analysis.

The author has developed a high-speed early sizing method based on pattern matching. Using the standard taxonomy discussed above, size is derived by comparing the project's taxonomy to historical projects in the author's knowledge base of some 26,000 projects. It happens that projects with the same taxonomy patterns are usually the same size in function points, although LOC size varies widely based on which programming languages are used.

This method is embedded in the author's SRM estimation tool. Sizing via pattern matching using SRM is quick and averages only about 1.8 min to size an application. Manual function point counting, on the other hand, averages perhaps 500 function points per day. This is why manual function points are seldom used on large systems above 10,000 function points. The author's high-speed method has no problem with sizing massive applications such as SAP and Oracle in the 250,000-function point size range.

Table 1.4 shows a sample of 100 software applications sized using SRM in about a 3 h time span.

The author's sizing tool is calibrated for IFPUG function points and also supports LOC metrics. (In fact, it can actually produce size data in 23 metrics.) It was first created in 2011 and has been used on many applications in many countries. The new SNAP metric was added in 2016 as an experiment but has since become a new feature of IFPUG function points.

Early and rapid application sizing opens up a new technical window that also permits early risk analysis, early selection of optimal methodologies that match size

Table 1.4 Samples of Application Sizes Using Pattern Matching

	Applications	*Size in Function Points IFPUG 4.3*	*SNAP Nonfunction Points IFPUG*	*Size in Logical Code Statements*
1	IBM Future Systemss FS/1 (circa 1985 not completed)	515,323	108,218	68,022,636
2	Star Wars missile defense	352,330	68,800	32,212,992
3	Worldwide military command and control system (WWMCCS)	307,328	65,000	28,098,560
4	U.S. air traffic control	306,324	70,133	65,349,222
5	Israeli air defense system	300,655	63,137	24,052,367
6	North Korean border defenses	273,961	50,957	25,047,859
7	Iran's air defense system	260,100	46,558	23,780,557
8	SAP	253,500	32,070	18,480,000
9	Aegis destroyer C&C	253,088	49,352	20,247,020
10	Oracle	229,434	29,826	18,354,720
11	Windows 10 (all features)	198,050	21,786	12,675,200
12	Obamacare website (all features)	107,350	33,450	12,345,250
13	Microsoft Office Professional 2010	93,498	10,285	5,983,891

(Continued)

Table 1.4 (*Continued*) Samples of Application Sizes Using Pattern Matching

	Applications	*Size in Function Points IFPUG 4.3*	*SNAP Nonfunction Points IFPUG*	*Size in Logical Code Statements*
14	Airline reservation system	38,392	8,900	6,142,689
15	North Korean long-range missile controls	37,235	4,468	5,101,195
16	NSA code decryption	35,897	3,590	3,829,056
17	FBI Carnivore	31,111	2,800	3,318,515
18	FBI fingerprint analysis	25,075	3,260	2,674,637
19	NASA space shuttle	23,153	3,010	2,116,878
20	Veteran's Administration (VA) Patient monitoring	23,109	6,500	4,929,910
21	Data warehouse	21,895	2,846	1,077,896
22	NASA Hubble controls	21,632	2,163	1,977,754
23	Skype	21,202	3,392	1,130,759
24	Shipboard gun controls	21,199	4,240	1,938,227
25	American Express billing	20,141	4,950	1,432,238
26	M1 Abrams battle tank operations	19,569	3,131	1,789,133
27	Apple I Phone v6 operations	19,366	2,518	516,432

(Continued)

Table 1.4 (*Continued*) Samples of Application Sizes Using Pattern Matching

	Applications	*Size in Function Points IFPUG 4.3*	*SNAP Nonfunction Points IFPUG*	*Size in Logical Code Statements*
28	IRS income tax analysis	19,013	5,537	1,352,068
29	Cruise ship navigation	18,896	2,456	1,343,713
30	MRI medical imaging	18,785	2,442	1,335,837
31	Google search engine	18,640	2,423	1,192,958
32	Amazon website	18,080	2,350	482,126
33	Statewide child support	17,850	4,125	952,000
34	Linux	17,505	2,276	700,205
35	FEDEX shipping controls	17,378	4,500	926,802
36	Tomahawk cruise missile	17,311	2,250	1,582,694
37	Denver Airport baggage (original)	17,002	2,166	1,554,497
38	Inventory management	16,661	2,111	1,332,869
39	eBay transaction controls	16,390	2,110	1,498,554
40	Patriot missile controls	16,239	2,001	1,484,683
41	IBM IMS database	15,392	1,939	1,407,279
42	Toyota robotic manufacturing	14,912	1,822	3,181,283

(Continued)

Table 1.4 (*Continued*) Samples of Application Sizes Using Pattern Matching

	Applications	*Size in Function Points IFPUG 4.3*	*SNAP Nonfunction Points IFPUG*	*Size in Logical Code Statements*
43	Android operating system	14,019	1,749	690,152
44	Quicken 2015	13,811	1,599	679,939
45	State transportation ticketing	12,300	1,461	656,000
46	State motor vehicle registrations	11,240	3,450	599,467
47	Insurance claims handling	11,033	2,567	252,191
48	SAS statistical package	10,927	1,349	999,065
49	Oracle CRM features	10,491	836	745,995
50	DNA Analysis	10,380	808	511,017
51	EZPass vehicle controls	4,751	1,300	253,400
52	CAT Scan medical device	4,575	585	244,000
53	Chinese submarine sonar	4,500	522	197,500
54	Microsoft Excel 2007	4,429	516	404,914
55	Citizens Bank Online	4,017	1,240	367,224
56	MapQuest	3,969	493	254,006
57	Bank ATM controls	3,917	571	208,927

(Continued)

Table 1.4 (*Continued*) Samples of Application Sizes Using Pattern Matching

	Applications	*Size in Function Points IFPUG 4.3*	*SNAP Nonfunction Points IFPUG*	*Size in Logical Code Statements*
58	NVIDIA graphics card	3,793	464	151,709
59	Lasik surgery (wave guide)	3,625	456	178,484
60	Sun DTrace utility	3,505	430	373,832
61	Microsoft Outlook	3,450	416	157,714
62	Microsoft Word 2007	3,309	388	176,501
63	Adobe Illustrator	2,507	280	178,250
64	Spy Sweeper antispyware	2,227	274	109,647
65	Norton antivirus software	2,151	369	152,942
66	Microsoft Project 2007	2,108	255	192,757
67	Microsoft Visual Basic	2,068	247	110,300
68	All-in-one printer	1,963	231	125,631
69	AutoCAD	1,900	230	121,631
70	Garmin handheld GPS	1,858	218	118,900
71	Intel Math function library	1,768	211	141,405
72	Private Branch Exchange (PBX) switching system	1,658	207	132,670

(Continued)

Table 1.4 (*Continued*) Samples of Application Sizes Using Pattern Matching

	Applications	*Size in Function Points IFPUG 4.3*	*SNAP Nonfunction Points IFPUG*	*Size in Logical Code Statements*
73	Motorola cell phone contact list	1,579	196	144,403
74	Seismic analysis	1,564	194	83,393
75	Sidewinder missile controls	1,518	188	60,730
76	Apple iPod	1,507	183	80,347
77	Property tax assessments	1,492	457	136,438
78	Mozilla Firefox (original)	1,450	174	132,564
79	Google Gmail	1,379	170	98,037
80	Digital camera controls	1,344	167	286,709
81	Individual Retirement Account (IRA) account management	1,340	167	71,463
82	Consumer credit report	1,332	345	53,288
83	Sun Java compiler	1,310	163	119,772
84	All-in-one printer driver	1,306	163	52,232
85	Laser printer driver	1,285	162	82,243
86	Microsoft C# compiler	1,281	162	91,096

(*Continued*)

Table 1.4 (*Continued*) Samples of Application Sizes Using Pattern Matching

	Applications	*Size in Function Points IFPUG 4.3*	*SNAP Nonfunction Points IFPUG*	*Size in Logical Code Statements*
87	Smart bomb targeting	1,267	150	67,595
88	Wikipedia	1,257	148	67,040
89	Cochlear implant (embedded)	1,250	135	66,667
90	Casio atomic watch with compass, tides	1,250	129	66,667
91	APAR analysis and routing	1,248	113	159,695
92	Computer BIOS	1,215	111	86,400
93	Automobile fuel injection	1,202	109	85,505
94	Antilock brake controls	1,185	107	63,186
95	CCleaner utility	1,154	103	73,864
96	Hearing aid (multi program)	1,142	102	30,448
97	Logitech cordless mouse	1,134	96	90,736
98	Instant messaging	1,093	89	77,705
99	Twitter (original circa 2009)	1,002	77	53,455
100	Denial of service virus	866	–	79,197
	Averages	42,682	7,739	4,250,002

Note: SRM sizing takes about 1.8 min per application for sizing.

and taxonomy patterns, and early planning for a software quality control including inspections and static analysis.

Early sizing may also lead to business decisions not to build selected applications because risks are too high. But it is best if these business decisions can be made early before expending large amounts of money on applications that will later be canceled or will run so late that ROI turns negative.

Factor 8—Size metrics with and without conversion rules: For sociological reasons, the software industry has more size metrics than any other industry in human history. There is no valid technical reason for this. Many size metrics are pushed into service without any validation or proof of effectiveness. The software industry also lacks published conversion rules among many of these diverse metrics.

As a convenience to users, the author's SRM estimating tool predicts size in a total of 23 metrics simultaneously, and will probably add new metrics such as data envelopment analysis in 2017. SRM is calibrated to match IFPUG function points version 4.3.

Table 1.5 shows the current metrics supported by SRM for an application of 2500 IFPUG function points coded in the Java programming language.

Factor 9—Certified reusable components: Custom designs and manual coding are intrinsically expensive and error prone, no matter what methodologies are used and no matter what programming languages are used. The future progress of effective software engineering should concentrate on moving from labor-intensive and error-prone manual development toward construction of software from certified materials.

Software reuse includes not only source code but also reusable requirements, architectures, designs, test materials, test cases, training materials, documents, and other key items. Table 1.2 illustrates the impacts of various quantities of certified reusable materials on both software productivity and software quality. Table 1.6 assumes 1000 function points and the Java programming language.

Of all the possible technologies that can benefit software engineering productivity and quality, increasing the availability of certified reusable components will have the greatest impact. Note that the key word is "certified." Casual reuse of aging code segments is not likely to be effective due to latent bugs. The reusable materials need to approach zero defects in order to be cost-effective. That means, rigorous use of static analysis on all reusable code segments and inspections of reusable requirements and design.

The current U.S. average for software reuse in 2017 hovers around 20% for code reuse but below 10% for requirements, design, and test material reuse, i.e., somewhere in the range of 15% overall reuse. It would be desirable for certified reuse to top 85% across the board, and 95% for common applications such as compilers and telephone switching systems.

Factor 10—Development team, project management, and client experience: It is obvious that development team experience is a major factor for project outcomes. Not so obvious but also important, project management experience is also a

Table 1.5 SRM Alternate Size Metrics (Size at Delivery)

	Metrics	*Size*	*IFPUG (%)*
1	IFPUG 4.3	2,500	100.00
2	Automated code-based function points	2,675	107.00
3	Automated Unified Modeling Language (UML)-based function points	2,575	103.00
4	Backfired function points	2,375	95.00
5	COSMIC function points	2,857	114.29
6	Fast function points	2,425	97.00
7	Feature points	2,500	100.00
8	Finnish Software Metrics Association (FISMA) function points	2,550	102.00
9	Full function points	2,925	117.00
10	Function points light	2,413	96.50
11	IntegraNova function points	2,725	109.00
12	Mark II function points	2,650	106.00
13	NESMA function points	2,600	104.00
14	Reports, Interfaces, Conversions, Enhancements (RICE) objects	11,786	471.43
15	Software Common Controls, Queries, Interfaces (SCCQI) function points	7,571	302.86
16	Simple function points	2,438	102.63
17	SNAP nonfunctional size metrics	325	13.00
18	SRM pattern matching function points	2,500	100.00
19	Story points	1,389	55.56
20	Unadjusted function points	2,225	89.00
21	Use case points	833	33.33
		Source Code	**LOC per FP**
22	Logical code statements	133,325	53.33
23	Physical lines of code (with blanks, comments)	538,854	215.54

Table 1.6 Impact of Software Reuse on Productivity and Quality

Reuse (%)	*Work Hours per FP*	*FP per Month*	*Defect Potential per FP*	*Defect Removal Percentage (%)*	*Delivered Defects per FP*
95	2.07	63.63	1.25	99.75	0.003
85	2.70	48.94	1.68	98.25	0.029
75	3.51	37.65	2.10	96.78	0.068
65	4.56	28.96	2.53	95.33	0.118
55	5.93	22.28	2.95	93.90	0.180
45	7.70	17.14	3.38	92.49	0.253
35	10.01	13.18	3.80	91.10	0.338
25	13.02	10.14	4.23	89.74	0.434
15	16.92	7.80	4.65	88.39	0.540
5	22.00	6.00	5.08	87.06	0.656
0	33.00	4.00	5.50	85.76	0.783

critical factor. From working as an expert witness in a number of software lawsuits, project managers seem to cause more failures than developers. Common software management failings include short-cutting quality control to "make schedules," status tracking that omits or even conceals serious problems, and grossly inaccurate and optimistic cost and schedule estimates.

Client experience is often important too. Novice clients are much more likely to cause excessive requirement changes than experienced clients. Table 1.7 shows approximate results for software experience levels on project productivity for a project of 1000 function points in the Java programming language.

Because experts are much more likely than novices to have high productivity and high quality at the same time, it is tempting to consider staffing big projects with 100% expert teams. The problem with this idea is that true experts are fairly scarce and seldom top 15% of the total population of software engineers and project managers in large corporations, and even less in government software organizations. (As demonstrated by the frequent delays and cancellations of state government benefits and motor vehicle software systems.) Some wealthy and profitable companies such as Apple, IBM, Amazon, and Microsoft have higher ratios of experts, but overall true expertise is a rare commodity.

The author evaluates experience on a scale of 1 to 5, with 1 being experts, 3 average, and 5 novices. We look at several flavors of experience:

Table 1.7 Impact of Experience Levels on Software Results

	Expert	*Average*	*Novice*
Percentage of staff (%)	14.00	63.00	23.00
Monthly costs	$12,500	$10,000	$8,000
Project staffing	5.50	6.50	7.50
Project schedule (months)	12.45	15.85	19.50
Project effort (months)	68.45	103.02	146.24
Function points per month	14.61	9.71	6.84
Work hours per function point	9.04	13.60	19.30
Lines of code per month	730	485	342
Project cost	$855.604	$1,030,181	$1,169,907
Cost per function point	$855.60	$1,030.18	$1,169.91
Defects per function point	2.25	4.00	6.50
Defect potential (all defect sources)	2,250	4,000	6,500
DRE (%)	99.50	93.50	87.50
Delivered defects per function point	0.01	0.26	0.81
Delivered defects	11	260	813
High-severity defects	1	42	154
Delivered security flaws	0	4	20

Software Experience Factors
Development team experience
Test team experience
Management experience
Client experience
Methodology experience
Language experience

Average experience: Experience is an important factor. However, experts are fairly rare, while novices are fairly common. Average personnel are in the majority, as might be expected.

Factor 11—Team work hours per month and unpaid overtime: Software is a global commodity that is developed in every industrialized country in the worlds. This means that both estimates and benchmarks need to include local work hours, which vary widely.

The author uses published work hour tables from the Organization for Economic Cooperation and Development (OECD) as the basis for work hours in the SRM tool. Table 1.8 shows local work hours per month for 52 countries.

Variations in monthly work hours is why the metric "function points per month" varies from country to country. The metric "work hours per function point" on the other hand does not vary by country, although it does vary by application size, experience, and technology stacks.

There are also large differences in U.S. work hour patterns by industry sector. Start-up technology companies tend to work long hours that can top 200 h per month. State and federal software workers tend to work shorter months of below 130 h per month. The U.S. average is about 142 regular hours and 7 unpaid overtime hours; 149 h per month.

Unpaid overtime has a large impact on software costs and can also have a large impact on apparent productivity, if the unpaid overtime is invisible and not tracked.

Assume a project of 1000 function points cost $1000 per function point or $1,000,000 dollars and had zero unpaid overtime. Now assume a similar project of 1000 function points had 20% untracked unpaid overtime. This project would only cost $800,000 or $800 per function point. But it really took the same amount of effort.

Assume a project of 1000 function points had a productivity rate of 10 function points per month, but zero unpaid overtime. Now assume a similar project had 20% untracked unpaid overtime. The productivity would seem to be 12 function points per month. Here too the amount of effort is the same for both projects, but one project used untracked unpaid overtime. This is a common phenomenon and it distorts software benchmark results.

Many European countries and Australia also have large quantities of paid overtime due in part to having unionized software workers. Paid overtime is rare in the United States. However, since paid overtime usually has a premium of perhaps 50% additional costs, it would definitely make software projects more expensive.

Factor 12—Costs of canceled projects: About 35% of software projects >10,000 function points are canceled without being completed. These canceled projects are almost never measured or at least never have published data. This is because companies are embarrassed by cancellations and don't want information about them made available.

From an economic standpoint, canceled projects have zero productivity because they have no delivered "goods or services." However, they still have known sizes in terms of function points or LOC, and they have accumulated cost data.

Table 1.8 Software Work Hours per Month in Selected Countries

	Countries	*Software Work Hours per Month*	*Software Unpaid Overtime per Month*	*Software Total Hours per Month*	*U.S. Total Hours per Month (%)*
1	India	190.00	12.00	202.00	146.38
2	Taiwan	188.00	10.00	198.00	143.48
3	Mexico	185.50	12.00	197.50	143.12
4	China	186.00	8.00	194.00	140.58
5	Peru	184.00	6.00	190.00	137.68
6	Colombia	176.00	6.00	182.00	131.88
7	Pakistan	176.00	6.00	182.00	131.88
8	Hong Kong	168.15	12.00	180.15	130.54
9	Thailand	168.00	8.00	176.00	127.54
10	Malaysia	169.92	6.00	175.92	127.48
11	Greece	169.50	6.00	175.50	127.17
12	South Africa	168.00	6.00	174.00	126.09
13	Israel	159.17	8.00	167.17	121.14
14	Vietnam	160.00	6.00	166.00	120.29
15	Philippines	160.00	4.00	164.00	118.84
16	Singapore	155.76	8.00	163.76	118.67
17	Hungary	163.00	6.00	163.00	118.12
18	Poland	160.75	2.00	162.75	117.93
19	Turkey	156.42	4.00	160.42	116.24
20	Brazil	155.76	4.00	159.76	115.77
21	Panama	155.76	4.00	159.76	115.77
22	Chile	149.64	8.00	157.64	114.23
23	Estonia	157.42	0.00	157.42	114.07
24	Japan	145.42	12.00	157.42	114.07
25	Switzerland	148.68	8.00	156.68	113.54
26	Czech Republic	150.00	0.00	150.00	108.70
27	Russia	145.51	4.00	149.51	108.34

(*Continued*)

Table 1.8 (*Continued*) Software Work Hours per Month in Selected Countries

	Countries	*Software Work Hours per Month*	*Software Unpaid Overtime per Month*	*Software Total Hours per Month*	*U.S. Total Hours per Month (%)*
28	Argentina	148.68	0.00	148.68	107.74
29	South Korea	138.00	6.00	144.00	104.35
30	United States	132.00	6.00	138.00	100.00
31	Saudi Arabia	141.60	0.00	141.60	102.61
32	Portugal	140.92	0.00	140.92	102.11
33	United Kingdom	137.83	2.00	139.83	101.33
34	Finland	139.33	0.00	139.33	100.97
35	Ukraine	138.06	0.00	138.06	100.04
36	Venezuela	134.52	2.00	136.52	98.93
37	Austria	134.08	0.00	134.08	97.16
38	Luxembourg	134.08	0.00	134.08	97.16
39	Italy	129.21	2.00	131.21	95.08
40	Belgium	131.17	0.00	131.17	95.05
41	New Zealand	128.25	2.00	130.25	94.38
42	Denmark	128.83	0.00	128.83	93.36
43	Canada	126.11	2.00	128.11	92.84
44	Australia	127.44	0.00	127.44	92.35
45	Ireland	127.42	0.00	127.42	92.33
46	Spain	124.34	2.00	126.34	91.55
47	France	123.25	0.00	123.25	89.31
48	Iceland	120.00	0.00	120.00	86.96
49	Sweden	119.55	0.00	119.55	86.63
50	Norway	118.33	0.00	118.33	85.75
51	Germany	116.42	0.00	116.42	84.36
52	Netherlands	115.08	0.00	115.08	83.39
	Average	148.21	3.85	151.94	110.10

From working as an expert witness in lawsuits on canceled projects and from studying data on canceled projects in client organizations, some interesting facts have emerged:

- An average canceled project is about 12 calendar months late when terminated.
- An average canceled project cost about 20% more than similar delivered projects.
- An average canceled project is about 8700 function points in size.
- Canceled projects cause C-level executives to distrust software organizations.
- An average canceled project had over 20% nonfunctional requirements.
- A majority of canceled projects had poor quality control.
- A majority of canceled projects did not use static analysis.
- A large majority of canceled projects (>99%) did not use inspections.
- A majority of canceled projects did not use parametric estimation tools.
- A majority of canceled projects had zero risk analysis before starting.
- A majority of canceled projects did not track schedules or effort with accuracy.
- Some canceled projects concealed problems until too late to fix them.
- About 75% of the outsourced canceled projects end up in litigation.
- State governments have the highest frequency of canceled projects.
- The defense industry is number 2 in frequency of canceled projects.
- Large commercial software groups (Microsoft, IBM, etc.) have fewest canceled projects.

Because of the high frequency and high costs of canceled software projects, there is a need for more study of this topic. Table 1.9 shows three comparisons of successfully completed projects versus canceled projects of the same size and type. Assume all three were coded in Java. Assume the successes were developed by above-average managers and personnel, and the canceled projects were developed by below-average managers and personnel.

Not only are canceled projects more expensive than successful projects, but the ratio of costs also increases with application size. If you include the probable litigation costs for the canceled projects, the overall difference can top 300% higher costs for canceled projects than for successful projects of the same size and type.

Table 1.9 Successful versus Canceled Projects

Size in Function Points	*Successful $ per FP*	*Canceled $ per FP*	*Percentage (%)*
1,000	$1,000	$1,180	118.00
10,000	$1,700	$2,142	126.00
100,000	$2,300	$3,358	146.00

Factor 13—Requirements creep: Measuring software productivity requires knowing the size of the application. The author and his colleagues use application size on the day of delivery to customers as the baseline. We also measure and estimate size at the end of the requirements phase, and then size at the end of each calendar or fiscal year after delivery.

Software requirements creep and change at about 1%–2% per calendar month during development and about 8%–12% per calendar year after release. Commercial software applications grow more rapidly after release than internal software.

Also about every 4 years, commercial applications tend to have something called a "mid-life kicker" or a large quantity of new features in order to stay ahead of competitors. Table 1.10 shows a 10-year illustration of software growth patterns during development and after release for commercial software.

Note that not all size changes are positive. Some 4800 function points were deferred from the first release due to attempts to make a specific delivery date.

Table 1.10 Software Multiyear Size Growth

	Nominal Application Size in IFPUG Function Points	*10,000 Function Points*	*1,369 SNAP Points*	*Mid-Life Kickers*
1	Size at the end of requirements	10,000	1,389	
2	Size of requirement creep	2,000	278	
3	Size of planned delivery	12,000	1,667	
4	Size of deferred functions	–4,800	(667)	
5	Size of actual delivery	7,200	1,000	
6	Year 1	12,000	1,667	
7	Year 2	13,000	1,806	
8	Year 3	14,000	1,945	
9	Year 4	17,000	2,361	Kicker
10	Year 5	18,000	2,500	
11	Year 6	19,000	2,639	
12	Year 7	20,000	2,778	
13	Year 8	23,000	3,195	
14	Year 9	24,000	3,334	Kicker
15	Year 10	25,000	3,473	

Because of constant size changes, software productivity results need to be recalibrated from time to time. The author uses "size at delivery" as the basis for cost estimates and initial benchmarks. After 3 or 4 years, productivity rates need to be updated due to size growth.

Table 1.10 is based on commercial software application packages. These tend to add "mid-life kickers" or large functional additions about every 4 years to stay competitive. Years 9 and 13 illustrate these large bursts of functional growth.

Size is not a constant value either before release or afterward. So long as there are active users, applications grow continuously. During development, the measured rate is 1%–2% per calendar month; after release, the measured rate is 8%–12% per calendar year. A typical post-release growth pattern might resemble the following:

Over a 10-year period, a typical mission-critical departmental system starting at 15,000 function points might have:

Two major system releases of:	2,000 function points
Two minor system releases of:	500 function points
Four major enhancements of:	250 function points
Ten minor enhancements of:	50 function points
Total growth for 10 years:	6,500 function points
System size after 10 years:	21,500 function points
10-year total growth percentage:	43%

Over long time periods, SNAP nonfunctional requirements may not change at the same rate as true user requirements. However, SNAP does not have a long enough history to actually measure results over a 10-year period.

As can be seen, software applications are never static if they have active users. This continuous growth is important to predict before starting and to measure at the end of each calendar or fiscal year.

The cumulative information on original development, maintenance, and enhancement is called "total cost of ownership" or TCO. Predicting TCO is a standard estimation feature of SRM, which also predicts growth rates before and after release.

From time to time, software productivity and quality results need to be renormalized based on growth patterns. It is not accurate to use size at delivery as a fixed value over long time periods. It is even worse to use size at requirements end, when function points are first calculated, as a fixed value for the rest of development or after release.

Factor 14—Activity-based costs: In order to understand the economics of software, it is important to know the productivity results of specific activities. Knowing only gross overall results for a project is insufficient because overall results are hard to validate. You need to know the specific productivity rates of all activities used including business analysis, requirements, architecture, design, coding, testing, integration, quality assurance, technical writing, project management, and others as well. Activity-based costs are best measured using the metric "work hours per function point."

Measuring projects with a single point or by phase are not accurate because they cannot be validated. Phase-based metrics are not accurate either because too many activities such as quality assurance and project management span multiple phases. Activity-based costs can be validated because they show specific work patterns.

Table 1.11 illustrates a small version of activity-based costs with seven activities. Table 1.6 assumes 1000 function points, the Java programming language, average experience, 132 work hours per month, CMMI level 3, and a burdened cost of $10,000 per calendar month.

Table 1.11 only shows a small sample of seven activities. For large applications in the size range of 10,000 function points, as many as 40 activities might be performed. Some of these activities include architecture, business analysis, configuration control, function point analysis, and risk analysis.

For defense software applications, two activities are mandated by Department of Defense standards that seldom occur for civilian projects: independent verification and validation (IV&V) and independent testing. Military projects may also use earned-value analysis, which is also rare in the civilian software sector. Commercial software projects often perform market analysis and competitive analysis. The bottom line is that activity-based cost analysis is a critical factor for a true understanding of software engineering economics.

Factor 15—Assignment scopes and production rates: An assignment scope is the amount of a product that can be typically handled by one person. A production rate is the amount of product development a knowledge worker can perform in a fixed

Table 1.11 Example of Activity-Based Costs

Activities	*Staff*	*Effort Months*	*Schedule Months*	*Project Costs*
Requirements	2.45	10.90	4.46	$109,036
Design	3.49	15.45	4.42	$154,468
Coding	7.01	46.68	6.66	$466,773
Testing	6.25	34.31	5.49	$343,094
Documentation	1.47	3.37	2.29	$33,702
Quality assurance	1.28	4.21	3.30	$42,128
Management	1.38	14.83	17.78	$148,289
Totals	7.30	129.75	17.78	$1,297,490

time period such as a month. Assignment scopes and production rates are useful corollaries to activity-based cost analysis.

Metrics for assignment scopes and production rates can be natural metrics such as pages of a book or LOC, or synthetic metrics such as function points. Suppose a company wants to produce a new users' guide for a software application. The book will be 200 pages in size. If the assignment scope for this book is 200 pages, then one author can produce it. Assume that the production rate for this book is 10 pages per day. Obviously, the author would need 20 work days to write the book.

The assignment scope and production rate metrics make it easy to estimate software projects. Experts have larger assignment scopes and higher production rates than novices. Some programming languages such as Ruby and Python have larger assignment scopes and production rates than older languages such as assembly. Empirical data from historical projects are needed to adjust assignment scopes and production rates.

Table 1.12 shows typical values for assignment scopes and production rates for 40 activities for a large system of 10,000 function points coded in Java. All values are expressed in terms of IFPUG function point metrics. Also shown are work hours per function point, schedules by activity, and staff by activity, which are derived from assignment scopes and production rates.

As can be seen, the logic of assignment scopes and production rates can be applied to all software activities. For example, staffing for an activity can be calculated by dividing function point size by assignment scope. Effort months can be calculated by dividing function point size by production rates. Schedules can be calculated by dividing effort months by staff.

Of course, overlap is not shown by means of assignment scopes and production rates, so net schedules require additional calculations based on average overlaps between adjacent activities as shown in the next section on software schedule measures.

The usefulness of assignment scopes and production rates is that they allow easy and accurate adjustments to match software personnel experience levels. For example, the assignment scope might be 5000 LOC for an expert programmer in Java; 3000 LOC for an average programmer; and 1500 LOC for a novice Java programmer. Monthly production rate might be 700 LOC for the expert, 350 for an average programmer, and 200 for a novice programmer.

Factor 16—Software schedule measures: Software schedules are probably the most critical variables of interest to C-level executives and to project management. This is partly due to the fact that large software projects have a very high incidence of schedule slips: about 70% of software projects >10,000 function points run late by several months and some by more than a year.

What makes schedules difficult to measure is the "overlap" or the fact that new activities start before existing activities are finished. Although the "waterfall" project more or less assumes that activities are sequential and one starts when a predecessor ends, design in fact starts before requirements are finished and so does coding.

Table 1.12 Samples of Assignment Scopes and Production Rates

	Development Activities	*Staff Function Points Assignment Scope ASCOPE*	*Monthly Function Points Production Rate PRATE*	*Work Hours per Function Point*	*Schedule Months*	*Staff*
1	Business analysis	5,000	25,000.00	0.01	0.20	2.00
2	Risk analysis/sizing	20,000	75,000.00	0.00	0.27	0.50
3	Risk solution planning	25,000	50,000.00	0.00	0.50	0.40
4	Requirements	2,000	450.00	0.29	4.44	5.00
5	Requirements inspection	2,000	550.00	0.24	2.22	5.00
6	Prototyping	10,000	350.00	0.38	2.86	1.00
7	Architecture	10,000	2,500.00	0.05	4.00	1.00
8	Architecture inspections	10,000	3,500.00	0.04	0.95	3.00
9	Project plans/estimates	10,000	3,500.00	0.04	2.86	1.00
10	Initial design	1,000	200.00	0.66	5.00	10.00
11	Detail design	750	150.00	0.88	5.00	13.33
12	Design inspections	1,000	250.00	0.53	4.00	10.00
13	Coding	150	20.00	6.60	7.50	66.67
14	Code inspections	150	40.00	3.30	3.75	66.67

(Continued)

Table 1.12 (*Continued*) Samples of Assignment Scopes and Production Rates

	Development Activities	*Staff Function Points Assignment Scope ASCOPE*	*Monthly Function Points Production Rate PRATE*	*Work Hours per Function Point*	*Schedule Months*	*Staff*
15	Reuse acquisition	20,000	75,000.00	0.00	0.27	0.50
16	Static analysis	150	15,000.00	0.01	0.67	66.67
17	COTS package purchase	10,000	25,000.00	0.01	0.40	1.00
18	Open-source acquisition	50,000	50,000.00	0.00	1.00	0.20
19	Code security audit	33,000	2,000.00	0.07	16.50	0.30
20	Independent Verification and Validation (IV&V)	10,000	10,000.00	0.01	1.00	1.00
21	Configuration control	15,000	5,000.00	0.03	3.00	0.67
22	Integration	10,000	6,000.00	0.02	1.67	1.00
23	User documentation	10,000	500.00	0.26	20.00	1.00
24	Unit testing	150	125.00	1.06	1.20	66.67
25	Function testing	250	140.00	0.94	1.79	40.00
26	Regression testing	350	90.00	1.47	3.89	28.57
27	Integration testing	450	125.00	1.06	3.60	22.22

(*Continued*)

Table 1.12 (*Continued*) Samples of Assignment Scopes and Production Rates

	Development Activities	*Staff Function Points Assignment Scope ASCOPE*	*Monthly Function Points Production Rate PRATE*	*Work Hours per Function Point*	*Schedule Months*	*Staff*
28	Performance testing	1,200	500.00	0.26	2.40	8.33
29	Security testing	1,250	350.00	0.38	3.57	8.00
30	Usability testing	1,500	600.00	0.22	2.50	6.67
31	System testing	300	175.00	0.75	1.71	33.33
32	Cloud testing	3,000	2,400.00	0.06	1.25	3.33
33	Field (Beta) testing	600	4,000.00	0.03	0.15	16.67
34	Acceptance testing	600	5,000.00	0.03	0.12	16.67
35	Independent testing	5,000	6,000.00	0.02	0.83	2.00
36	Quality assurance	10,000	750.00	0.18	13.33	1.00
37	Installation/training	10,000	4,000.00	0.03	2.50	1.00
38	Project measurement	25,000	9,500.00	0.01	2.63	0.40
39	Project office	10,000	550.00	0.24	18.18	1.00
40	Project management	2,500	75.00	1.76	26.67	4.00
	Cumulative results	343	6.02	21.91	34.68	29.12

If requirements take 3 months, design takes 3 months, coding takes 3 months, and testing takes 3 months, it would seem to be a 12-month schedule. But design starts at about month 2 of requirements, coding starts at about month 2 of design, and so forth. Instead of a 12-month schedule, the actual schedule will be about 8 calendar months, even though all four activities individually took 3 months.

The best way to illustrate the concept of schedule overlap is to use a traditional Gantt chart as shown in Figure 1.1.

The actual delivery schedule for this sample project is 16 calendar months. However, the sum total of the activity schedules is 52 calendar months. The difference is because of overlap, which averages about 50%. That is, an activity such as requirements is about 50% complete when design starts. Design is about 50% complete when coding starts, and so on. The sum of the activity schedules is shown in the following table.

SNO	Activity	Months
1	Requirements	4
2	Design	6
3	Coding	6
4	Testing	6
5	Documentation	6
6	Quality assurance	8
7	Project management	16
	Total	52

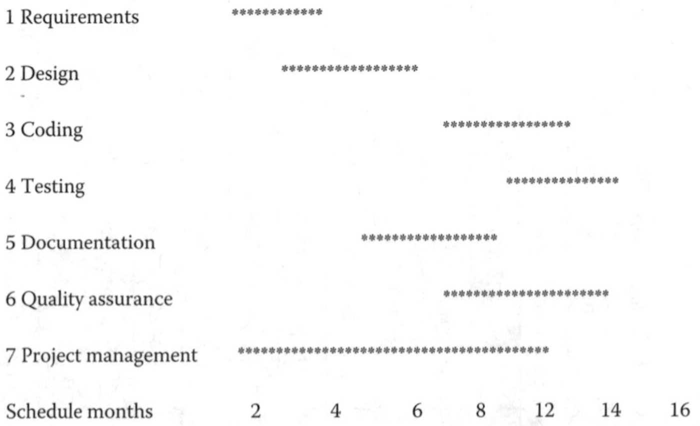

Figure 1.1 Gantt chart example of schedule overlap for 1000 function points.

Measuring software project schedules is complicated by the fact that software development activities always overlap and are never a true waterfall. Schedule overlap is a natural adjunct to activity-based cost analysis. Some parametric estimation tools such as the author's SRM tool use both activity-based costs and schedule overlap calculations.

Factor 17—Software quality measures: Because finding and fixing bugs is the #1 cost driver of the entire software industry, it is not possible to understand software engineering economics without understanding the COQ. COQ is the sum total of defect removal costs during development and after release. The author uses a fixed period of 3 years after release for COQ calculations.

The metrics the author uses for COQ quality calculations include defect potentials and DRE. Also included are delivered defects and delivered security flaws. Supplemental quality measures include numbers of test cases, test coverage, and cyclomatic complexity. (Although the author's SRM estimating tool predicts technical debt, we caution users that there are no standard definitions for that metric and everybody calculates it differently.) Table 1.13 shows a sample of key software quality metrics and measurement practices. Table 1.13 assumes 1000 function points, the Java programming language, and average team experience.

Table 1.13 illustrates the quality predictions produced by the author's SRM estimation tool. The most important topic shown by Table 1.6 is that of DRE. Due to the software industry's chronic poor measurement practices, very few people realize that most kinds of testing are below 30% in DRE or only find about one bug out of three.

A synergistic combination of defect prevention, pretest defect removal, and formal testing can raise DRE values above 99%, which is where they should be for all important software projects. The current observed upper limit of DRE is 99.75%. It is true that a few applications receive zero bug reports, but that may just mean that all latent bugs have not been found and reported.

About 7% of U.S. software bug repairs have new bugs in the repairs themselves. These are called "bad fixes." Bad-fix injections were first studied by IBM circa 1969 and are a key quality factor that is not widely understood. Bad-fix predictions are a standard feature of the author's SRM estimation tool. For projects with high levels of cyclomatic complexity above about 25, bad-fix injections can top 30% of all bug repairs!

Factor 18— COQ: The COQ metric is much older than software and was first made popular by the 1951 book entitled *Juran's QC Handbook* by the well-known manufacturing quality guru Joseph Juran. Phil Crosby's later book, *Quality is Free*, also added to the literature.

COQ is not well named because it really focuses on the cost of poor quality rather than the cost of high quality. In its general form, COQ includes prevention, appraisal, failure costs, and total costs.

When used for software, the author of this paper modifies these terms for a software context: defect prevention, pretest defect removal, test defect removal,

Table 1.13 Examples of Software Quality Measurements

	Defect Potentials		**Number**		
	Requirements defect potential		746		
	Design defect potential		777		
	Code defect potential		1,110		
	Document defect potential		139		
	Total defect potential		2,773		
	Defect Prevention	**Efficiency (%)**	**Remainder**	**Bad Fixes**	**Costs**
1	Joint application design (JAD)	22.50	2,149	19	$52,906
2	Quality function deployment (QFD)	26.00	1,609	16	$72,352
3	Prototype	20.00	1,300	9	$26,089
4	Models	62.00	498	24	$91,587
	Subtotal	81.19	522	68	$242,933
	Pretest Removal	**Efficiency (%)**	**Remainder**	**Bad Fixes**	**Costs**
1	Desk check	25.00	391	12	$21,573
2	Pair programming	14.73	344	10	$27,061
3	Static analysis	59.00	145	4	$16,268
4	Inspections	85.00	22	1	$117,130
	Subtotal	95.57	23	27	$182,031
	Test Removal	**Efficiency (%)**	**Remainder**	**Bad Fixes**	**Costs**
1	Unit	30.0	16	0	$34,384
2	Function	33.0	11	1	$62,781
3	Regression	12.0	10	0	$79,073
4	Component	30.0	7	0	$88,923
5	Performance	10.0	7	0	$51,107

(Continued)

Table 1.13 (*Continued*) Examples of Software Quality Measurements

	Defect Potentials		Number		
6	System	34.0	5	0	$98,194
7	Acceptance	15.0	4	0	$23,589
	Subtotal	81.6	4	2	$438,052
	Defects delivered		4		
	High severity (sev 1+sev 2)		1		
	Security flaws		0		
	Cumulative defect removal efficiency		99.84%		$863,016

post-release defect removal. The author also includes several topics that are not part of standard COQ analysis: cost of projects canceled due to poor quality; cost of consequential damages or harm to customers from poor quality; cost of litigation and damage awards due to poor quality, and costs of cyber-attacks.

Table 1.14 illustrates a typical COQ calculation for an application of 1000 function points, coded in Java by an average team, and with a user population of 1000 people.

Table 1.14 Software COQ

Basic COQ Costs	*Costs*	*Cost per FP*	*Cost per KLOC*
Defect prevention	$95,000	$95.00	$1,900
Pretest defect removal	$125,000	$125.00	$2,500
Testing (all test stages)	$325,000	$325.00	$6,500
Post-release defect repairs	$175,000	$175.00	$3,500
Total	$720,000	$720.00	$14,400
Expanded COQ costs			
Cyber-attacks	$300,000	$300	$6,000
Litigation for poor quality	$550,000	$550	$11,000
Total	$850,000	$850	$17,000
Grand total	$1,570,000	$1,570	$31,400

Many companies only measure testing costs that are only 45% of the development COQ and only 20% of the expanded COQ, which includes both cyber-attack costs and litigation for poor quality.

Not included because it is seldom reported would be consequential damages or financial costs to users of software due to poor quality.

Ward Cunningham's new metric of "technical debt" has attracted interest and is popular, but is highly erratic from company to company. Whenever this metric is discussed, one of the first questions is "How do you calculate technical debt?"

Factor 19—Gaps and omissions in software quality data: When we look at software quality data from companies that collect their own data, we see major leakages and woefully incomplete data. Many companies do not track any bugs before release. Only sophisticated companies such as IBM, Raytheon, Motorola, and the like track pretest bugs.

At IBM there were even volunteers who recorded bugs found during desk check sessions, debugging, and unit testing, just to provide enough data for statistical analysis. Table 1.15 shows the pattern of missing data for software defect measurements.

Out of 25 total forms of defect removal, data are only collected for 13 of these under normal conditions. The apparent defect density of the measured defects is less than one-third of the true volume of software defects. In other words, true defect potentials would be about five defects per function point, but due to gaps in the measurements apparent defect potentials would seem to be only about two defects per function point.

For the software industry as a whole, the costs of finding and fixing bugs is the #1 cost driver. It is professionally embarrassing for the industry to be so lax about measuring the most expensive kind of work since software began.

Factor 20—Software cyber-attack defenses and recovery: In 2017, cyber-attacks have become daily occurrence for the software industry and frequent occurrences for major corporations and many important software applications dealing with finances, personnel information, and even manufacturing production.

Cyber-attacks have costs that involve defenses and preparations before attacks occur, short-term defense costs during an active cyber-attack, and then recovery costs and sometimes reparations to clients after a successful cyber-attack.

Several estimating tools such as SRM can predict cyber-attack defense and recovery costs for various threats such as denial of service, phishing, data theft, etc. Table 1.7 illustrates cyber-attack defense and recovery costs. Table 1.16 illustrates a 10,000-function point financial application with 10,000 customers and a monthly average cost for personnel and equipment of $10,000.

Table 1.16 illustrates cyber-attack costs as predicted by the author's SRM estimation tool. As can be seen, cyber-attacks are expensive and are also becoming increasingly common for the software industry. This is a chronic problem that may well get worse before it gets better.

Table 1.15 Measured versus Actual Quality

	Development Projects	*Defects Removed*	*Defects Measured*
1	Requirements inspection	75	
2	Requirements changes	10	
3	Architecture inspection	15	
4	Initial design inspection	50	
5	Detail design inspection	100	
6	Design changes	20	
7	Code inspections	350	
8	Code changes	75	
9	User document editing	25	
10	User document changes	10	
11	Static analysis	235	
12	Unit test	50	
13	Function testing	50	50
14	Regression testing	25	25
15	Integration testing	75	75
16	Performance testing	25	25
17	Security testing	10	10
18	Usability testing	20	20
19	System testing	50	50
20	Cloud testing	10	10
21	Field (Beta) testing	20	20
22	Acceptance testing	15	15
23	Independent testing	10	10
24	Quality assurance	25	25
25	Customer bug reports	150	150
	Cumulative results	1,500	485
	Percentage of total defects (%)	100.00	32.33

Table 1.16 Software Cyber-Attack Defense and Recovery Cost Example

Cyber Attack Annual Insurance				*Costs $15,000.00*
Cyber Attack Deterrence	*Staff*	*Schedule*	*Effort*	*Costs*
Security static analysis before release	3.57	0.66	2.36	$23,628
Security inspections	10.13	7.62	77.13	$771,330
Security testing	6.39	7.45	47.59	$475,889
Ethical hackers	3.40	6.46	21.92	$219,223
Security-hardened hardware				$350,000
Security-hardened software				$87,500
Security-hardened offices				$217,143
Total deterrence	23.48	22.19	149.01	$2,144,713
Cyber-attacks and recovery	**Staff**	**Schedule**	**Effort**	**Costs**
Hacking, data theft	8.34	8.94	74.60	$745,988
Denial-of-service attack	7.48	6.62	49.54	$495,420
Virus, worm, botneck attacks	4.89	6.95	34.01	$340,125
Phishing/whale phishing	4.32	5.96	25.72	$257,237
Cyber blackmail (locked files)	3.45	4.64	16.01	$160,059
Infrastructure/equipment attacks	8.46	9.11	77.03	$770,282
Total	36.94	42.22	276.91	$2,769,111
Cyber Security Consultants	**Staff**	**Schedule**	**Effort**	**Costs**
Software recovery specialists	10.16	11.59	117.74	$2,354,864
Hardware recovery specialists	10.82	8.94	96.73	$2,128,158
Total	20.98	20.53	214.48	$4,483.022
	Staff	Schedule	Effort	Costs
Total	57.92	62.75	491.39	$7,252,132

Note: Effective cyber deterrence can lower attack odds and recovery costs.

In any case every major company and every government agency needs both internal software security experts and also contracts with external security experts. They also need state-of-the-art defenses such as firewalls and antivirus packages. Really critical software applications may also need physical security defenses such as Faraday cages and biometric sensors on terminals.

Factor 21—Leakage from software project historical data: A chronic problem of the software industry for more than 60 years has been "leakage" of historical data. The term "leakage" implies leaving out important activities such as business analysis, architecture, quality assurance, and project management. Unpaid overtime is also commonly omitted. (See the discussion of unpaid overtime earlier in this chapter.) Many companies only measure "design, code, and unit test" (DCUT), which collectively are only about 40% of the total software effort.

Software quality data also "leak" and usually omit unit test, static analysis, and informal walk-throughs.

Among the companies that collect their own software productivity and quality data, as opposed to having the data collected by a professional benchmark group, the average for software project effort is only 37% of the actual total efforts. Software quality data only averages about 25% of actual bugs that are discovered.

These leaks are professionally embarrassing and should not occur. Every company that produces significant quantities of software applications should measure 100% of the effort and 100% of the bugs found on all major software applications >500 function points in size (Table 1.17).

As can be seen, gaps and leaks in software measurement data make it impossible to understand the true economics of significant software projects. It is professionally embarrassing to have errors of this magnitude as common occurrence in majority of companies and government agencies.

Factor 22—Software risk measures: Software projects are among the riskiest of any manufactured products in human history. The author's master list of software risks contains a total of 210 software risks including security risks, quality risks, financial risks, and many others. The author recommends that all important software projects >250 function points in size carry out a formal risk analysis concurrently with software requirements analysis, or even as a standard part of software requirements analysis.

Software risks increase dramatically as application size goes up in terms of function points. Small projects <250 function points have few risks, but large systems >10,000 function points face over 50% probabilities of cancellation, cost overruns, schedule delays, poor quality, and cyber-attacks.

The author's SRM tool predicts the most likely risks for software projects based on factors such as size in function points, team experience levels, methodologies used, project type, quality control methods used, and programming languages. Table 1.18 illustrates a sample of 15 risks predicted by SRM.

The percentages shown in Table 1.18 are based on analysis of thousands of applications of approximately the same size plateaus. Since these risks can be predicted

Table 1.17 Measured versus Actual Effort

	For Development Projects	*Percent of Activities*	*Measured Results (%)*
1	Business analysis	1.25	
2	Risk analysis/sizing	0.26	
3	Risk solution planning	0.25	
4	Requirements	4.25	
5	Requirements inspection	1.50	
6	Prototyping	2.00	
7	Architecture	0.50	
8	Architecture inspection	0.25	
9	Project plans/estimates	0.25	
10	Initial design	5.00	
11	Detail design	7.50	7.50
12	Design inspections	2.50	
13	Coding	22.50	22.50
14	Code inspections	20.00	
15	Reuse acquisition	0.03	
16	Static analysis	0.25	
17	COTS package purchase	0.03	
18	Open-source acquisition	0.03	
19	Code security audit	0.25	
20	IV&V	1.00	
21	Configuration control	1.00	
22	Integration	0.75	
23	User documentation	2.00	
24	Unit testing	0.75	0.75
25	Function testing	1.25	1.25

(*Continued*)

Table 1.17 (*Continued*) Measured versus Actual Effort

	For Development Projects	*Percentage of Total (%)*	*Measured Results (%)*
26	Regression testing	1.50	1.50
27	Integration testing	1.00	1.00
28	Performance testing	0.50	
29	Security testing	0.50	
30	Usability testing	0.75	
31	System testing	2.50	2.50
32	Cloud testing	0.50	
33	Field (Beta) testing	0.75	
34	Acceptance testing	1.00	
35	Independent testing	1.50	
36	Quality assurance	2.00	
37	Installation/training	0.65	
38	Project measurement	0.50	
39	Project office	1.00	
40	Project management	10.00	
	Cumulative results	100.00	37.00
	Unpaid overtime	7.50	

before requirements via pattern matching, there should be adequate time for proactive risk reduction approaches. For example, the risk of inaccurate cost estimates can be minimized by switching from optimistic manual estimates to accurate parametric estimates. The risks of poor quality and high warranty repairs can be minimized by using static analysis before testing on all software, and by using formal testing with mathematical test case design. The essential message of the SRM risk prediction feature is "forewarned is forearmed."

Factor 23—Software documentation costs: Due to the software industry's poor measurement practices, the #2 cost driver for large software applications is almost invisible and almost never studied. In fact, it is often not even included in software cost estimates! The #2 cost driver for the software industry is producing paper documents such as requirements, designs, test plans, user manuals, etc.

Table 1.18 Overview of Software Risks for Selected Size Levels

		Size in FP	*Size in FP*	*Size in FP*
		100	*1,000*	*10,000*
	Project Risks	*Risk (%)*	*Risk (%)*	*Risk (%)*
1	Optimistic manual cost estimates	9.30	22.50	45.70
2	Poor status tracking	12.50	18.00	55.00
3	Significant requirements creep (>10%)	3.00	15.50	35.60
4	Feature bloat (>15% not used)	2.00	11.00	27.00
5	Cancellation	8.80	14.23	26.47
6	Negative ROI	11.15	18.02	33.53
7	Cost overruns	9.68	15.65	34.00
8	Schedule slip	10.74	18.97	38.00
9	Unhappy customers	7.04	11.38	34.00
10	Litigation (breach of contract)	3.87	6.26	11.65
11	Poor quality and high COQ	5.00	16.00	26.21
12	Cyber-attacks	7.00	9.75	15.30
13	Financial risk	9.00	21.00	41.00
14	High warranty repairs	6.00	14.75	32.00
15	Poor maintainability	2.00	11.00	21.00
	Risk average	7.14	14.94	31.76

The author has studied software paperwork costs since the 1970s, and documentation cost estimation is a standard feature of the author's SRM estimation tool. Document costs are especially important for defense software projects, since military standards tend to create document volumes about three times larger than civilian software projects of the same size.

Although the word "documentation" implies deliberately planned documents such as requirements, design, and user manuals, in fact over a full development cycle, more pages will be created for change requests and bug reports than for any other form of documentation. Table 1.19 shows a sample of 13 typical documents for a software application of 1000 function points in size.

Table 1.19 illustrates a common failing of software documentation. It is never 100% complete. Overall document sizes go up rapidly as function points increase, but completeness declines rapidly as well.

Table 1.19 Document Sizes and Completeness

	Documents Types	*Pages*	*Words*	*Percentage Complete (%)*
1	Requirements	275	110,088	91.59
2	Architecture	76	30,238	91.07
3	Initial design	325	130,039	87.76
4	Detail design	574	229,796	89.74
5	Test plans	145	57,867	88.05
6	Development plans	55	22,000	89.07
7	Cost estimates	76	30,238	92.07
8	User manuals	267	106,723	93.39
9	HELP text	191	76,313	93.64
10	Courses	145	58,000	93.27
11	Status reports	116	46,424	91.07
12	Change requests	212	84,800	96.59
13	Bug reports	963	385,191	90.87
	Total	3,420	1,367,717	91.40

In some industries such as banking and insurance, front-end documents such as requirements and design may be sparse. But for military and defense projects, many documents are enormous and expensive due to Department of Defense standards. The defense sector spends more money producing paper documents than producing source code.

In all industries, change requests and bug reports tend to be numerous. Even though most of the costs for bug reports and change requests are not associated with document costs per se, they still create large volumes of paper that need to be read and understood by the people who process the change requests and bug reports.

Factor 24—Software occupation groups: Several years ago, AT&T funded an interesting study to see how many different kinds of occupation groups worked in large companies. Among the participants of the study were AT&T, IBM, Texas Instruments, the U.S. Navy, and a dozen other companies. The study found a total of 126 different software occupations. No single company employed more than about 50 of these occupations. Companies building large and complex software applications such as Microsoft and IBM employed many more specialists than traditional companies such as banks, insurance, and manufacturing.

One piece of bad news uncovered by the study was that not even one human resource department in any of the organizations participating in the study had accurate records of occupation groups or even numbers of software personnel. The lack of data on software personnel was due in part to the use of ambiguous job titles such as "member of the technical staff," which includes both software and hardware personnel. In order to get accurate data on both occupation groups and the number of software engineering employees, the author's team of consultants had to interview unit software managers.

The failure of human resource departments in keeping accurate records of software engineering occupations and personnel makes government statistics on software engineering numbers highly suspect. The government agencies that produce such statistics such as the Department of Commerce get them from HR groups, and if the HR groups don't have good data then the government can't publish good statistics on software engineering populations.

Table 1.20 shows a sample of 20 common software occupation groups for a large application of 10,000 function points.

Software staffing is a complex situation due to the fact that over half of the occupation groups are part-time specialists that only work for short periods. For example, business analysts usually only work for the first 3 months or so of applications whose total schedules may top 36 months. Testers work for the last half of software projects but are not heavily involved in the first half.

The result of these part-time specialists is that there is a disconnect between actual live workers and average staff. For example, suppose that business analysts worked 2 months on a 12-calendar month software project, designers worked 3 months, quality assurance worked 3 months, programmers worked 6 months, testers worked 6 months, technical writers worked 4 months, and the project manager worked 12 months for a total of 36 months of effort.

If you divide 36 months of effort by the 12-calendar month schedule, the result is an average staff of three workers per calendar month. But the project had a total of seven different occupation groups:

1. Business analysts
2. Designers
3. Programmers
4. Testers
5. Quality assurance
6. Technical writers
7. Project management

To understand software economics, both occupation groups and total efforts need to be measured and understood.

Although there are more programmers than any other single occupation group, for large systems in the 10,000-function point size range, the effort for programming is only about 30% of the total effort for the whole project.

Table 1.20 Occupation Groups and Part-Time Specialists

		Normal Staff	*Peak Staff*
1	Programmers	32	47
2	Testers	28	42
3	Designers	19	31
4	Business analysts	19	29
5	Technical writers	7	10
6	Quality assurance	6	10
7	First-line managers	6	8
8	Database administration	3	4
9	Project office staff	3	4
10	Administrative support	3	4
11	Configuration control	2	2
12	Project librarians	1	2
13	Second-line managers	1	2
14	Estimating specialists	1	1
15	Architects	1	1
16	Security specialists	0	1
17	Performance specialists	0	1
18	Function point counters	0	1
19	Human factors specialists	0	1
20	Third-line managers	0	0
	Totals	132	201

To understand software engineering economics, it is important to measure the contributions and accumulated effort of all occupation groups, and not just programming and test personnel.

Factor 25—Software methodologies: A prior book by the author discussed 60 software development methodologies. This book only shows results from a sample of more common methodologies. All methodologies have ways of dealing with requirements and design. Some include effective quality control techniques and some do not.

Table 1.21 shows a sample of 20 modern software development methodologies sorted by quality results.

Methodologies are not a panacea in spite of the growing claims some receive such as Agile. Each methodology has its own strengths and weaknesses. For example, Agile seems to be most effective for small projects <500 function points but not effective for large systems above 5000 function points, although the newer

Table 1.21 Best Quality Results by Methodology

	Selected Methodologies	*Defect Potential per FP (2017)*	*DRE (2017) (%)*	*Delivered Defects per FP (2017)*
1	Team software process (TSP+PSP)	2.35	98.50	0.035
2	Feature Driven Development (FDD)	2.45	98.50	0.037
3	Kaizen development	2.00	97.00	0.060
4	Micro service development	1.35	95.50	0.061
5	RUP from IBM	3.40	98.00	0.068
6	Container development (65% reuse)	2.35	97.00	0.071
7	Disciplined Agile delivery (DAD)	2.50	96.00	0.100
8	Product Line engineering	2.85	96.24	0.107
9	Mashup development	2.40	95.16	0.116
10	Information engineering (IE)	2.65	95.50	0.119
11	Extreme programming (XP)	2.90	95.50	0.131
12	Microsoft solutions	3.00	95.50	0.135
13	Spiral development	3.20	95.50	0.144
14	GIT development	2.75	94.70	0.146
15	Iterative development	3.30	95.50	0.149
16	Hybrid (Agile+waterfall)	3.05	94.70	0.162
17	DevOps development	2.85	92.60	0.211
18	Agile+scrum	3.15	92.50	0.236
19	Waterfall development	3.75	90.00	0.375
20	Cowboy development	5.00	88.00	0.600
	Averages	2.86	95.09	0.153

disciplined Agile delivery (DAD) is effective for large systems. Team software process (TSP) is effective for large systems above 10,000 function points but is somewhat heavy for small applications.

A basic problem of the software industry is that methodologies are usually chosen on the basis of popularity and not on the basis of technical merits or empirical data. Due to the fact that no method is universally effective, there is a continuous stream of new methodologies being developed at a rate of about one new methodology every 14 months. This will probably continue to be the pattern indefinitely.

As with the following discussion of programming languages, selecting a software methodology is more like joining a religious cult based on faith than it is in making an informed technical decision based on empirical data of success and usefulness.

Factor 26—Software programming languages: For sociological rather than technical reasons as of 2017, the software industry has over 3000 programming languages, and new languages are being developed at the rate of more than two languages per year.

None of these languages have been measured well for either productivity or quality results because of poor software measurement practices. As with methodology selection, choosing a programming language is more like joining a religious cult based on faith rather than making a rational technical decision.

Modern software applications generally use several programming languages, and a few may use over a dozen languages in the same application. An average application circa 2017 uses about 2.5 programming languages such as Java, SQL, and HTML. The common practice of using several programming languages is a proof that no language is universal and that it fits all aspects of software engineering.

The existence of thousands of programming languages and the frequent release of new languages is a challenge for software estimation tool developers. For example, as of 2017, the author's SRM tool supports 84 common programming languages, and it can estimate combinations of up to three programming languages for the same application. We add new languages every year. But a problem with adding new languages to estimating tools is that until the languages have been used on software projects that are measured and benchmarked, there is no data available to judge the effectiveness of the language.

Table 1.22 shows the sets of programming languages used in the author's SRM tool as of early 2017. But new languages will be added before the end of 2017, as they are added every year.

When used for software project estimation, we like to show clients results in terms of both function points and LOC (even though LOC distort reality). This is the reason we show LOC per function point.

The column labeled "language level" was first developed at IBM circa 1970. Languages had been called "low level" or "high level," but no mathematical value was associated with those terms. IBM developed a "level" metric based on basic assembly code. The "level" of a programming language was defined as being equal

Table 1.22 Software Languages Used for Project Estimation

Language Levels	*Programming Languages*	*Statements per Function Point*
4.00	ABAP	80.00
6.50	Ada 95	49.23
3.00	Algol	106.67
10.00	APL	32.00
19.00	APS	16.84
13.00	ASP NET	24.62
5.00	Basic (interpreted)	64.00
1.00	Basic Assembly	320.00
3.00	Bliss	106.67
45.00	BPM	7.11
2.50	C	128.00
6.25	C#	51.20
6.00	C++	53.33
6.50	Ceylon	49.23
3.00	Chill	106.67
7.00	CICS	45.71
3.00	COBOL	106.67
7.50	ColdFusion	42.67
3.00	Coral	106.67
6.75	Dart	47.41
8.00	DB2	40.00
11.00	Delphi	29.09
7.00	DTABL	45.71
14.00	Eiffel	22.86
8.50	Elixir	37.65
0.10	English text	3,200.00

(*Continued*)

Table 1.22 (*Continued*) Software Languages Used for Project Estimation

Language Levels	*Programming Languages*	*Statements per Function Point*
7.50	Erlang	42.67
4.50	ESPL/I	71.11
50.00	Excel	6.40
7.00	F#	45.71
6.50	Fantom	49.23
18.00	Forte	17.78
5.00	Forth	64.00
3.00	Fortran	106.67
45.00	Generators	7.11
6.00	Go	53.33
7.00	Groovy	45.71
3.25	GW Basic	98.46
8.50	Haskell	37.65
5.00	Haxe	64.00
2.00	HTML	160.00
16.00	IBM ADF	20.00
60.00	IntegraNova	5.33
6.00	Java	53.33
4.50	JavaScript	71.11
1.45	JCL	220.69
3.00	Jovial	106.67
9.00	Julia	35.56
5.00	Lisp	64.00
8.00	LiveScript	40.00
9.00	M	35.56

(*Continued*)

Table 1.22 (*Continued*) Software Languages Used for Project Estimation

Language Levels	*Programming Languages*	*Statements per Function Point*
0.50	Machine language	640.00
1.50	Macro Assembly	213.33
35.00	Mathematica 10	9.14
25.00	Mathematica9	12.80
12.00	MATLAB®	26.67
8.50	Mixed Languages	37.65
4.00	Modula	80.00
17.00	MUMPS	18.82
12.00	Objective C	26.67
9.00	OPA	35.56
8.00	Oracle	40.00
3.50	Pascal	91.43
9.00	Perl	35.56
6.00	PHP	53.33
4.00	PL/I	80.00
3.50	PL/S	91.43
5.00	Prolog	64.00
6.00	Python	53.33
25.00	QBE	12.80
5.25	Quick Basic	60.95
8.00	R	40.00
6.75	RPG III	47.41
7.00	Ruby	45.71
8.50	Ruby on Rails	37.65
5.00	SH (shell scripts)	64.00

(*Continued*)

Table 1.22 (*Continued*) Software Languages Used for Project Estimation

Language Levels	*Programming Languages*	*Statements per Function Point*
7.00	Simula	45.71
15.00	Smalltalk	21.33
25.00	SQL	12.80
12.00	Swift	26.67
20.00	TELON	16.00
25.00	TranscriptSQL	12.80
12.00	Visual Basic	26.67
13.00	Visual J++	24.62
14.00	WebDNA	22.86
25.00	X	12.80
6.25	X10	51.20
2.50	XML	128.00
5.50	Zimbu	58.18
10.42	Average	99.10

to the number of statements in basic assembly language needed to produce the functionality of one statement in the target language; i.e., COBOL was a "level 3" language because it took three assembly statements to create the functionality of one COBOL statement.

Factor 27—SNAP nonfunctional requirements: SNAP metrics are a new variation on function points introduced by IFPUG in 2012. The term SNAP is an awkward acronym for "software nonfunctional assessment process."

The basic idea of the SNAP metric is that software requirements have two flavors: (1) functional requirements needed by users; and (2) nonfunctional requirements due to laws, mandates, or physical factors such as storage limits or performance criteria.

As an example of a "nonfunctional requirement," consider these situations in building a home in the author's state of Rhode Island. If a new home is to be built within a half mile of the ocean, state law mandates hurricane-proof windows at additional costs that may top \$15,000–\$50,000 based on the size of the house. If a new home is to be built within 1000 yards of an aquifer or stream, state law mandates a special kind of septic tank at an added cost of about $25,000. These are not things that users would necessarily want to include when building a home, but since they are state mandates they must be included.

Software applications also have nonfunctional requirements or mandates by government agencies such as the Food and Drug Administration (FDA), Federal Aviation Administration (FAA), Securities and Exchange Commission (SEC), IRS, etc.

The SNAP committee view is that these nonfunctional requirements should be sized, estimated, and measured separately from function point metrics. Thus, SNAP and function point metrics are not additive, although they could have been.

Having two separate metrics for economic studies is awkward at best and inconsistent with other industries. For that matter, it seems inconsistent with standard economic analysis in every industry. Almost every industry has a single normalizing metric such as "cost per square foot" or "cost per square meter" for home construction; "cost per gallon" for gasoline and diesel oil.

As of 2017 none of the parametric estimation tools have fully integrated SNAP and they might not do so since the costs of adding SNAP are painfully expensive. As a rule of thumb, nonfunctional requirements are equal to about 15% of functional requirements, although the range is very wide. The effort for building nonfunctional requirements may be even larger at perhaps 18% of the effort for function point features.

The author's view of SNAP, and the way SNAP is calculated in the author's SRM tool, matches standard economic assumptions but don't necessarily agree with the IFPUG SNAP committee.

Our assumption is that the true purpose for building software is to satisfy user requirements, which of course are measured by function point metrics. Nonfunctional requirements would not occur at all unless the user requirements are satisfied, so they are a secondary topic whose costs can be subsumed into cost per function point.

Suppose we are building an application of 1000 function points at a cost of $1000 per function point or $1,000,000 for meeting user requirements. Now suppose that the application also has 150 SNAP points that are necessary to support the functionality of the application.

Assume these 150 function points were built at a cost of $1200 per SNAP point or $180,000. The sum of the costs for functional requirements and the costs for SNAP nonfunctional requirements is $1,180,000. However, with the current SNAP rules there is no longer a single metric such as "cost per function point" because SNAP points and function points are not additive.

Our solution to this is to show clients both costs separately. However, for economic analysis of the total application we add the SNAP costs to the functional costs and use "cost per function point" for the combination. Thus, cost per function point for user functionality is $1000 per function point, and nonfunctional SNAP costs are $180,000. But we aggregate both user features and nonfunctional requirements that sum up to $1,180,000. We then say that the total application cost $1180 per function point.

In other words, we tack on the nonfunctional costs on to the cost per function point on the grounds that nonfunctional costs are secondary topics and would not occur at all without the system that satisfies user requirements. Basically, they are just a small extra cost appended to the basic cost of software construction using function points.

The same thing is done in Rhode Island with home construction costs. A home far away from both the ocean and an aquifer might cost $500 per square foot. But if the same house is built close to the ocean and an aquifer, the costs might be $550 per square foot due to the added costs of state mandates such as hurricane-proof windows, special septic systems, hurricane-proof roofs, leak-proof oil tanks, and other nonfunctional mandates by state and local governments.

Our method allows current data collected in 2017 to be compared against older software projects created perhaps 25 years ago. Using the current SNAP guidelines there is no easy way to compare new applications against legacy applications.

Factor 28—Software user costs for internal applications: For internal software applications created for use by companies and government agencies, user involvement is fairly expensive and sometimes approaches 90% of software development costs, although 35% is the norm. For example, the Agile method assumes that one or more users will be embedded in the development team. Users also participate in focus groups, phase reviews, and they participate in software requirements, in software user testing, and sometimes in software inspections.

A very common omission from software economic studies is the costs and defects contributed by the users themselves. Of course, for commercial and embedded software, users are not active participants, other than focus groups or field trials.

However, for internal information technology projects, users contribute a majority of effort for requirements, and significant amounts of effort later in activities such as change control and acceptance testing. Indeed, for Agile projects, at least one future user is assigned full time to the development teams and therefore comprises perhaps 20% of the development effort.

For ordinary information systems applications developed internally for a company's own use, user effort is a large but invisible cost element. Table 1.23 shows the major activities where users participate and the approximate effort is utilized.

As can be seen, the total quantity of user efforts on typical internal applications of about 1000 function points in size is a bit more than one-third of the total efforts. However, user costs are seldom included in software cost estimates or budgets, and are seldom included in corporate cost tracking systems too. For some applications such as state government motor vehicle and welfare applications, user costs can approach 90% of development costs. This cost data should be included in software benchmarks and economic studies, but usually it is ignored.

There are a total of 10 activities in which users are active participants during internal software development projects.

	Common User Activities
1	User requirements team
2	User architecture team
3	User planning/estimating team
4	User prototype team
5	User design review team
6	User change control team
7	User governance team
8	User document review team
9	User acceptance test team
10	User installation team

Table 1.23 User and Development Team Efforts

	Software Development Projects	*Team Percentage of Total (%)*	*User Percentage of Total (%)*
1	Business analysis	1.25	3.75
2	Risk analysis/sizing	0.26	
3	Risk solution planning	0.25	
4	Requirements	4.25	5.31
5	Requirements inspection	1.50	1.50
6	Prototyping	2.00	0.60
7	Architecture	0.50	
8	Architecture inspection	0.25	
9	Project plans/estimates	0.25	
10	Initial design	5.00	
11	Detail design	7.50	
12	Design inspections	2.50	
13	Coding	22.50	
14	Code inspections	20.00	

(*Continued*)

Table 1.23 (*Continued*) User and Development Team Efforts

	Software Development Projects	*Team Percentage of Total (%)*	*User Percentage of Total (%)*
15	Reuse acquisition	0.03	
16	Static analysis	0.25	
17	COTS package purchase	0.03	1.00
18	Open-source acquisition	0.03	
19	Code security audit.	0.25	
20	IV&V	1.00	
21	Configuration control	1.00	
22	Integration	0.75	
23	User documentation	2.00	1.00
24	Unit testing	0.75	
25	Function testing	1.25	
26	Regression testing	1.50	
27	Integration testing	1.00	
28	Performance testing	0.50	
29	Security testing	0.50	
30	Usability testing	0.75	
31	System testing	2.50	
32	Cloud testing	0.50	
33	Field (Beta) testing	0.75	9.00
34	Acceptance testing	1.00	4.00
35	Independent testing	1.50	
36	Quality assurance	2.00	
37	Installation/training	0.65	9.75
38	Project measurement	0.50	
39	Project office	1.00	
40	Project management	10.00	
	Cumulative results	100.00	35.91
	Unpaid overtime	5.00	5.00

You can expect user effort to exceed 5 h per function point on normal waterfall projects of 1000 function points in size. For Agile projects of 1000 function points in size, user effort might top 7 h per function point.

Factor 29—Software breach of contract litigation costs: The author has worked as an expert witness in over a dozen lawsuits for breach of contract between clients and outsource vendors. Usually the suits involve either canceled projects or software that has such poor quality that it can't be used for its intended purpose. Samples of such lawsuits include the State of California vs Lockheed and Heidtman Steel vs Computer Associates (CA). Another case involved the State of South Carolina vs Computer Sciences Corporation (CSC). There were several cases involving the Canadian government vs U.S. outsource companies.

It is interesting that there seem to be about twice as many breach of contract cases involving state governments (perhaps 10% vs 5%) than those involving corporations. Apparently, state governments are lax in project tracking and governance.

There will probably be a lawsuit between the author's State of Rhode Island and Deloitte due to the huge cost and schedule overruns for a state welfare software upgrade.

(There are already lawsuits involving the failed Studio 38 project and company in Rhode Island, although these were not breach of contract cases. The State performed zero due diligence prior to awarding over $75,000,000 to Curt Schilling's game company in order to move it to Rhode Island. The author's SRM tool was run retroactively after Studio 38 went bankrupt and it predicted a 88% probability of application failure and cancellation.)

The Rhode Island welfare benefit system could not go online when planned, and as a result, many low-income Rhode Island citizens did not receive welfare checks or food stamps. Worse, the state had laid off a number of welfare workers in the naïve assumption that manual checks would not be needed due to the new software upgrade—a big mistake.

A number of Rhode Island (RI) executives have been fired due to the problems with this software mishap. Table 1.24 shows approximate litigation expenses for a sizable application of 10,000 function points.

(All of the lawsuits where the author has been an expert witness have been in the 10,000-function point size range.)

The author's SRM tool predicts the odds of breach of contract litigation occurring for every project estimated, and also the probable costs of litigation for the plaintiff and defendant. This is an optional feature but useful for outsource contracts.

Since these risk and litigation cost estimates can be carried out before requirements are complete and before contracts are signed, they provide very early warnings of possible problems that might be remedied before getting underway.

It is interesting that discovery documents and depositions in most breach of contract showed that project managers were the chief source of problems leading to litigation. They either did not track costs and quality well enough to see the

Table 1.24 Software Breach of Contract Costs for Plaintiff and Defendant

Size in IFPUG function points at requirements	8,095
Size in function points at delivery	10,000
Requirements creep in function points	1,905
Percentage of requirements creep	23.53%
Size of SNAP points at delivery	1,389
Programming language	Java
Language level	6.00
LOC per function point	53.33
LOC delivered	533,333
Planned application schedule months	33.11
Actual application schedule months	47.86
Months of schedule slippage	14.75
Percentage of schedule slippage	44.54%
Team burdened cost per month	$15,000
Team staff	66.67
Effort in staff months	3,190.87
Projects cost	$47,863,009
Cost per function point	$4,786.30
Client experience level	Inexperienced
Development team experience level	Inexperienced
Methodology	Waterfall
Pretest inspections used?	No
Pretest static analysis used?	No
Formal test case design used?	No
Test coverage tools used?	No
Cyclomatic complexity measures used?	No
Certified test personnel used?	No

(Continued)

Table 1.24 (*Continued*) Software Breach of Contract Costs for Plaintiff and Defendant

Contract cover out-of-scope changes?	No
Contract cover quality and defect repairs?	No
Contract includes penalties for late delivery?	No
Contract includes rewards for early delivery?	No
Defect potential	22,387
Defect removal efficiency	93.08%
Delivered defects	1,549
High-severity delivered defects	201
Security flaws in delivered software	25
Delivered defects per function point	0.155
Delivered defects per KLOC	15.947
High-severity delivered defects per FP	0.020
Security flaws per FP	0.003
Planned project cost	$33,113,112
Project cost at delivery/termination	$47,863,009
Cost overrun	$14,749,897
Percentage of overrun	44.54%
Probability of litigation	18.18%
Month litigation filed from project start date	51.86
Probable trial duration months	32.41
Odds of out-of-court settlement	90.75%
Plaintiff Litigation Expenses	
Plaintiff claims: late delivery	
Plaintiff claims: poor quality	
Plaintiff claims: consequential damages	
Plaintiff claims: failure to report problems	

(*Continued*)

Table 1.24 (*Continued*) Software Breach of Contract Costs for Plaintiff and Defendant

Plaintiff executive hourly costs	$500
Plaintiff staff hourly costs	$100
Plaintiff expert witness hourly costs	$475
Plaintiff attorney hourly costs	$450
Plaintiff paralegal hourly costs	$200
Plaintiff consequential damages	$18,500,000
Plaintiff executives involved in litigation	10
Plaintiff technical staff involved in litigation	18
Plaintiff expert witnesses involved in litigation	3
Plaintiff attorneys involved in litigation	5
Plaintiff paralegals involved in litigation	7
Plaintiff executive hours during litigation	2,139
Plaintiff technical staff hours during litigation	3,744
Plaintiff expert witness hours during litigation	333
Plaintiff attorneys' hours during litigation	1,637
Plaintiff paralegal hours during litigation	1,078
Plaintiff executive costs	$1,069,675
Plaintiff staff costs	$374,386
Plaintiff expert witness costs	$158,297
Plaintiff attorney costs	$736,471
Plaintiff paralegal costs	$215,646
Plaintiff total	$2,554,475
Total per function point	$255.45
Defendant litigation expenses	
Defendant claims: excessive scope creep	
Defendant claims: unprepared client	

(*Continued*)

Table 1.24 (*Continued*) Software Breach of Contract Costs for Plaintiff and Defendant

Defendant claims: lack of client cooperation	
Defendant claims: client approved deliverables	
Defendant executive hourly costs	$500
Defendant staff hourly costs	$100
Defendant expert witness hourly costs	$475
Defendant attorney hourly costs	$450
Defendant paralegal hourly costs	$200
Defendant executives involved in litigation	11
Defendant technical staff involved in litigation	18
Defendant expert witnesses involved in litigation	3
Defendant attorneys involved in litigation	5
Defendant paralegals involved in litigation	7
Defendant executive costs	$1,187,339
Defendant staff costs	$423,056
Defendant expert witness costs	$176,501
Defendant attorney costs	$826,320
Defendant paralegal costs	$243,680
Defendant Total Cost	$2,856,897
Total per function point	$285.69
Total litigation costs	$5,411,372
Litigation cost per function points	$3,234.11
Consequential damages (unpredictable)	$18,500,000
Possible damage award (unpredictable)	$8,429,767
Total litigation costs (both sides)	$32,341,139
Total litigation costs per function point	$3,234.11

problems, or in some cases they deliberately concealed the problems from executives and clients.

Accurate parametric estimates and risk analysis before starting can reduce litigation odds. Accurate progress tracking using tools such as the Automated Project Office (APO) of Computer Aid can also reduce the odds of litigation occurring.

Factor 30— CMMI: The author's company received a contract from the U.S. Air Force to evaluate the benefits of ascending the ladder of CMMI levels from 1 to 5. The participants included various defense contractors and projects provided by the Air Force for the study.

The study found that higher CMMI levels had better quality and productivity than lower CMMI levels. We also found that some civilian companies that don't use the CMMI at all had quality levels as good as CMMI level 5, since they routinely used static analysis and formal inspections as well as formal testing. Table 1.25 shows overall results for the five levels of the CMMI.

Note that Table 1.25 integrates normal function points and SNAP points. Having two size metrics that are not compatible and additive is a poor technical decision by IFPUG. C-level executives and standard economics need aggregate results and not piecemeal results.

This study was done before 2012 when the new SNAP metric for nonfunctional requirements was first introduced. As an experiment, the author calculated SNAP points and added them to the overall results. This table shows the author's approach of aggregating SNAP defects and regular defects to show clients overall results in a single table.

Factor 31—Software wasted workdays: The topic of "wastage" is not widely studied for software engineering, but it should be. A "wasted day" is a day spent on activities that have no positive value for software projects. Three common sources of wasted workdays are (1) working on projects that are canceled and never delivered, (2) working on finding and fixing bugs that should not have occurred, and (3) bug repairs on bad fixes or new bugs in bug repairs themselves.

The author carried out a 6-month study in a major Fortune 500 corporation. At a fixed time each day, the author checked with a number of colleagues to find out what kind of work they were doing at that moment. Table 1.26 shows the aggregate results.

Because finding and fixing bugs is the #1 cost driver for software in 2017, and because about 35% of large systems are canceled without completion, the software industry has an alarming total of wasted days.

As it happens, projects that have effective software quality control have many fewer wasted days than projects that are lax in quality control. Static analysis and formal inspections of key modules can greatly reduce wastage. Also, running static analysis on all bug repairs can reduce wastage.

If wastage could be eliminated or reduced, software productivity rates would double and software schedules would be shortened by over 35%. Wastage is a big software engineering problem that has essentially been ignored.

Table 1.25 CMMI for 2500 Function Points

CMMI Level	*Defect Potentials*	*Defect Potential per Function Point*	*Defect Removal Efficiency (%)*	*Delivered Defects per Function Point*	*Delivered Defects*
SEI CMMI 1	11,250	4.50	87.00	0.59	1,463
SEI CMMI 2	9,625	3.85	90.00	0.39	963
SEI CMMI 3	7,500	3.00	96.00	0.12	300
SEI CMMI 4	6,250	2.50	97.50	0.06	156
SEI CMMI 5	5,625	2.25	99.00	0.02	56
CMMI Level	*Defect Potentials*	*Defect Potential per SNAP Point*	*Defect Removal Efficiency (%)*	*Delivered Defects per SNAP Point*	*Delivered Defects*
SEI CMMI 1	2,500	1.00	80.00	0.20	500
SEI CMMI 2	2,125	0.85	83.00	0.14	361
SEI CMMI 3	1,875	0.75	92.00	0.06	150
SEI CMMI 4	1,625	0.65	95.00	0.03	81
SEI CMMI 5	1,125	0.45	97.00	0.01	34
CMMI Level	*Defect Potentials*		*Defect Removal Efficiency (%)*		*Delivered Defects*
SEI CMMI 1	13,750		85.72		1,963
SEI CMMI 2	11,750		88.73		1,324
SEI CMMI 3	9,375		95.20		450
SEI CMMI 4	7,875		96.99		237
SEI CMMI 5	6,750		98.67		90

Factor 32—Tools, standards, and certification: Software benchmarks should collect enough information to support multiple regression analysis to identify factors of greatest influence. Although neither tools nor standards nor certification are major factors with as much impact as reuse or experience, they are not trivial either. In benchmark studies, the author collects data on tools, standards, and certification, as shown in Table 1.27.

Table 1.26 Software Waste Patterns by Application Size

Function Points	*Poor Quality*		*High Quality*	
	Waste Days	*Productive Days*	*Waste Days*	*Productive Days*
10	80	100	20	152
100	99	81	45	134
1,000	111	61	49	131
10,000	129	50	55	124
100,000	160	25	66	114
Average	116	63.4	47	131

The information on tools, standards, and certification is significant for regression studies and for fine-tuning software estimates.

Factor 33—Software technology stack evaluation: When carrying out benchmarks, one of the topics evaluated is the technology stacks used for the project. Due to poor measurement practices, quite a few software projects make bad technology choices. Table 1.28 shows possible technologies ranked from best practices to worst practices.

The sample project scored a total of 226.50 using the weighted column. Projects scoring >200 have a majority of "best practices." Projects scoring <100 are likely to get in trouble and have cost and schedule overruns. Projects scoring <50 have a good chance of being canceled and ending up in court. This technology stack evaluation can be carried out early during or even before requirements.

Some defendants in breach of contract litigation had negative scores below 0. It is hard to see why a contract was given to a company with so little technical expertise. Of course, the client probably had low technical expertise as well.

Factor 34—Software meeting and communication costs: Software is a team activity with very intensive needs to cooperate. Communication with clients is important, and communication among team members is also important. In fact, the daily scrum sessions of Agile development push communication to the front row in terms of how information is transmitted. Communication costs are higher for internal projects whose users may be in the same building than they are for commercial software with invisible users scattered over perhaps hundreds of companies.

The author's SRM tool predicts communications effort and costs for live meetings, video conferences, and conference calls. Table 1.29 shows a small sample of communication and meetings effort for an application of 1000 function points.

As can be seen, meetings and communications are a significant percentage of monthly work hours. This is especially true with Agile projects that use daily scrum sessions.

Table 1.27 Tools, Standards, and Certification

	Tasks	*Tools Utilized*
1	Architecture	QEMU
2	Automated test	HP QuickTest Professional
3	Benchmarks	ISBSG, Namcook SRM, Davids, Q/P Management
4	Coding	Eclipse, SlickEdit
5	Configuration	Perforce
6	Cost estimate	SRM, SLIM, SEER, COCOMO
7	Cost tracking	Automated project office (APO), Microsoft Project
8	Cyclomatic	BattleMap
9	Debugging	GHS probe
10	Defect tracking	Bugzilla
11	Design	Projects Unlimited, Visio
12	Earned value	Deltek Cobra
13	ERP	Microsoft Dynamics
14	Function points 1	SRM
15	Function points 2	Function point workbench
16	Function points 3	CAST automated function points
17	Graphics design	Visio
18	Inspections	SlickEdit
19	Integration	Apache Camel
20	ISO tools	ISOXpress
21	Maintenance	Mpulse
22	Manual test	DevTest
23	Milestone track	KIDASA Software Milestone Professional
24	Progress track	Jira

(*Continued*)

Table 1.27 (*Continued*) Tools, Standards, and Certification

	Tasks	*Tools Utilized*
25	Project management	APO
26	Quality estimate	SRM
27	Requirements	Rational Doors
28	Risk analysis	SRM
29	Source code size 1	SRM
30	Source code size 2	Unified code counter
31	SQA	NASA Goddard ARM tool
32	Static analysis	OptimMyth Kiuwin
33	Support	Zendesk
34	Test coverage	Software Verify suite
35	Test library	DevTest
36	Value analysis	Excel and Value Stream Tracking
	Standards Used on Project	
	IEEE 610.12–1990	Software engineering terminology
	IEEE 730–1999	Software assurance
	IEEE 12207	Software process tree
	ISO/IEC 9001	Software quality
	ISO/IEC 9003	Software quality
	ISO/IEC 12207	Software engineering
	ISO/IEC 25010	Software quality
	ISO/IEC 29119	Software testing
	ISO/IEC 27034	Software security
	ISO/IEC 20926	Function point counting
	OMG Corba	Common Object Request Broker Architecture
	OMG Models	Meta models for software

(*Continued*)

Table 1.27 (*Continued*) Tools, Standards, and Certification

	Tasks	*Tools Utilized*
	OMG function points	Automated function points (legacy applications)
	UNICODE	Globalization and internationalization
	Professional certifications used on project	
	Certification used for project=1	
	Certification not used for project=0	
	Note: Some team members have multiple certifications.	
	Certification—Apple	0
	Certification—Computer Aid, Inc.	0
	Certification—Computer Associates	0
	Certification—FAA	0
	Certification—FDA	0
	Certification—Hewlett Packard	0
	Certification—IBM	1
	Certification—Microsoft	1
	Certification—Oracle	0
	Certification—PMI	1
	Certification—QAI	0
	Certification—Red Hat	0
	Certification—RSBC	0
	Certification—SAP	0
	Certification—Sarbanes–Oxley	0
	Certification—SEI	0
	Certification—Sun	0

(*Continued*)

Table 1.27 (*Continued*) Tools, Standards, and Certification

	Tasks	*Tools Utilized*
	Certification—Symantec	0
	Certification—TickIT	0
	Certified computing professionals	0
	Certified configuration management specialist	1
	Certified function point analyst	1
	Certified project managers	1
	Certified requirements engineers	0
	Certified scrum master	1
	Certified secure software lifecycle professional	0
	Certified security engineer	0
	Certified SEI appraiser	1
	Certified software architect	0
	Certified software business analyst	1
	Certified software development professional	0
	Certified software engineers	0
	Certified software quality assurance	1
	Certified test managers	0
	Certified testers	1
	Certified webmaster	1
	Certified software auditor	1
	Total	13

Factor 35—Software outsource contract estimates and benchmarks: A study by the author of major outsource vendors such as CA, Lockheed, IBM, and CSC found that productivity was about 5% higher than productivity of client companies for large applications over 1000 function points. Quality was slightly better, averaging about 4.10 defects per function point and 93.5% DRE against U.S. norms of 4.25 defects per function point and 92.5% DRE.

Table 1.28 Software Technology Stack Evaluation

		Value Scores	*Project Scores*	*Weighted Scores*
1	Benchmarks (validated data from similar projects)	10.00	1.00	10.00
2	Defect potential <2.5 per function point	10.00	1.00	10.00
3	DRE >99%	10.00	1.00	10.00
4	Estimates: parametric estimation tools	10.00	1.00	10.00
5	Estimates: TCO cost estimates	10.00	1.00	10.00
6	Formal and early quality predictions	10.00	1.00	10.00
7	Formal and early risk abatement	10.00	1.00	10.00
8	Inspection of all critical deliverables	10.00	1.00	10.00
9	Metrics: defect potential measures	10.00	1.00	10.00
10	Metrics: DRE measures	10.00	1.00	10.00
11	Static analysis of all source code	10.00	1.00	10.00
12	Metrics: IFPUG function points	10.00	1.00	10.00
13	Metrics: COSMIC function points	10.00	0.00	0.00
14	Estimates: activity-based cost estimates	9.00	1.00	9.00
15	Estimates: COQ estimates	9.00	1.00	9.00
16	Test coverage >96%	9.00	1.00	0.00
17	Test coverage tools used	9.00	1.00	9.00
18	Accurate cost tracking	8.50	1.00	8.50
19	Accurate defect tracking	8.50	1.00	8.50
20	Accurate status tracking	8.50	1.00	8.50
21	Methods: TSP/PSP	8.50	1.00	8.50
22	Methods: hybrid: (Agile/TSP)	8.50	0.00	0.00
23	Requirements change tracking	8.50	0.00	0.00
24	Reusable designs	8.50	0.00	0.00
25	Reusable requirements	8.50	0.00	0.00

(*Continued*)

Table 1.28 (*Continued*) Software Technology Stack Evaluation

		Value Scores	*Project Scores*	*Weighted Scores*
26	Reusable source code	8.50	1.00	8.50
27	Automated requirements modeling	8.00	0.00	0.00
28	Mathematical test case design	8.00	1.00	8.00
29	Methods: RUP	8.00	0.00	0.00
30	Metrics: cyclomatic complexity tools	8.00	1.00	8.00
31	Requirements change control board	8.00	1.00	8.00
32	CMMI 5	7.50	0.00	0.00
33	Methods: Agile <1000 function points	7.50	0.00	0.00
34	Methods: correctness proofs-automated	7.50	0.00	0.00
35	Methods: hybrid (waterfall/Agile)	7.50	0.00	0.00
36	Methods: QFD	7.50	0.00	0.00
37	Automated testing	7.00	0.00	0.00
38	Methods: JAD	7.00	0.00	0.00
39	Methods: DevOps	6.00	0.00	0.00
40	Metrics: automated function points	6.00	0.00	0.00
41	Metrics: earned value analysis (EVA)	6.00	0.00	0.00
42	Metrics: FOG/Flesch readability scores	6.00	0.00	0.00
43	Methods: extreme programming	5.50	0.00	0.00
44	Methods: service-oriented models	5.50	0.00	0.00
45	CMMI 3	5.00	0.00	0.00
46	Methods: continuous development	5.00	0.00	0.00
47	Methods: kanban/kaizen	5.00	0.00	0.00
48	Methods: iterative	4.50	0.00	0.00
49	Methods: iterative development	4.50	0.00	0.00
50	Methods: spiral development	4.50	0.00	0.00
51	Metrics: technical debt measures	4.50	0.00	0.00

(*Continued*)

Table 1.28 (*Continued*) Software Technology Stack Evaluation

		Value Scores	*Project Scores*	*Weighted Scores*
52	Certified quality assurance personnel	4.00	0.00	0.00
53	Certified test personnel	4.00	0.00	0.00
54	Defect potential 2.5–4.9 per function point	4.00	0.00	0.00
55	ISO risk standards	4.00	1.00	4.00
56	Maintenance: ITIL	4.00	1.00	4.00
57	ISO quality standards	3.00	1.00	3.00
58	Requirements modeling manual	3.00	0.00	0.00
59	Estimates: phase-based cost estimates	2.00	0.00	0.00
60	Test coverage <90%	2.00	0.00	2.00
61	Testing by untrained developers	2.00	0.00	0.00
62	CMMI 0 (not used)	0.00	0.00	0.00
63	Metrics: story point metrics	–1.00	0.00	0.00
64	Metrics: use case metrics	–1.00	0.00	0,00
65	Methods: correctness proofs manual	–1.00	0.00	0.00
66	Benchmarks (unvalidated self-reported benchmarks)	–2.00	0.00	0.00
67	CMMI 1	–2.00	0.00	0.00
68	Methods: Agile >5000 function points	–3.00	0.00	0.00
69	Cyclomatic complexity >20	–4.00	0.00	0.00
70	Methods: waterfall development	–4.50	0.00	0.00
71	Test coverage not used	–4.50	0.00	0.00
72	Methods: pair programming	–5.00	0.00	0.00
73	Metrics: cost per defect metrics	–5.00	0.00	0.00
74	Inaccurate cost tracking	–5.50	0.00	0.00
75	Estimates: manual estimation >250 function points	–6.00	0.00	0.00

(*Continued*)

Table 1.28 (*Continued*) Software Technology Stack Evaluation

		Value Scores	*Project Scores*	*Weighted Scores*
76	Inaccurate defect tracking	–6.00	0.00	0.00
77	Inaccurate status tracking	–7.00	0.00	0.00
78	Metrics: no function point measures	–7.00	0.00	0.00
79	DRE <85%	–7.50	0.00	0.00
80	Defect potential >5 per function point	–8.00	0.00	0.00
81	Metrics: no productivity measures	–8.50	0.00	0.00
82	Methods: Cowboy development	–10.00	0.00	0.00
83	Metrics: LOC for economic study	–10.00	0.00	0.00
84	Metrics: no quality measures	–10.00	0.00	0.00
85	No static analysis of source code	–10.00	0.00	0.00
	Scoring sum	307.5	27.00	226.50

Table 1.29 Software Meeting Hours per Month

Meeting Types	*Hours/Month*
Status meetings/scrum	12
Individual technical meetings	6
Client/user meetings	5
Project phase reviews	4
Conference/video calls	3
Meetings with SQA	2
Total	32
Work hours per month	132
Meeting percentage of work month (%)	22.73

However, the study found that about 15% of outsource contracts were in trouble and the client was considering termination. About 5% of outsource contracts had been terminated and the vendor was involved in breach of contract litigation with the client. In fact, the author often works as an expert witness in these cases.

From participating in a number of such assessments and lawsuits, it is obvious that most cases are remarkably similar. The clients charge that the contractor breached the agreement by delivering the software late, by not delivering it at all, or by delivering the software in an inoperable condition or with excessive errors.

The contractors, in turn, charge that the clients unilaterally changed the terms of the agreement by expanding the scope of the project far beyond the intent of the original agreement. The contractors also charge some kind of nonperformance by the clients, such as failure to define requirements or failure to review delivered material in a timely manner.

The fundamental root causes of the disagreements between clients and contractors can be traced to two problems:

1. Ambiguity and misunderstandings in the contract itself.
2. The historical failure of the software industry to quantify the dimensions of software projects before beginning them.

Although litigation potentials vary from client to client and contractor to contractor, the overall results of outsourcing within the United States approximates the following distribution of results after about 24 months of operations, as derived from observations among our clients (Table 1.30).

From process assessments performed within several large outsource companies, and analysis of projects produced by outsource vendors, our data indicate better than average quality control approaches when compared to the companies and industries who engaged the outsource vendors.

However, our data are still based on samples of only about 1600 outsource projects, as of 2017. Our main commissioned research in the outsource community has been with clients of the largest outsource vendors in the United States such as Accenture, CSC, EDS, IBM, Keane, Lockheed, and others in this class.

Table 1.30 Approximate Distribution of U.S. Outsource Results after 24 Months

Results	*Percentage of Outsource Arrangement (%)*
Both parties generally satisfied	70
Some dissatisfaction by client or vendor	15
Dissolution of agreement planned	10
Litigation between client and contractor probable	4
Litigation between client and contractor in progress	1

There are many other smaller outsource vendors and contractors where we have encountered only a few projects, or sometimes none at all since our clients have not utilized their services.

Software estimation, contracting, and assessment methodologies have advanced enough so that the root causes of software outsource contracts can now be overcome. Software estimation is now sophisticated enough so that a formal estimate using one or more of the commercial software parametric estimation tools in conjunction with software project management tools can minimize or eliminate unpleasant surprises later due to schedule slippages or cost overruns.

Indeed, old-fashioned purely manual cost and schedule estimates for major software contracts should probably be considered an example of professional malpractice. Manual estimates are certainly inadequate for software contracts or outsource agreements whose value is larger than about $500,000.

A new form of software contract based on the use of function point metrics is clarifying the initial agreement and putting the agreement in quantitative, unambiguous terms. This new form of contract can also deal with the impact of creeping user requirements in a way that is agreeable to both parties.

Function points have proven to be so successful for outsource contracts that a number of national governments now mandate function point metrics for all government software projects. These countries include Brazil, Italy, Japan, South Korea, and Malaysia. Poland may be the next country to mandate function point metrics for government software contracts.

Table 1.31 shows the overall results of software projects in terms of being on time, late, or terminated and never delivered at all.

As can be seen, delay and failure rates go up rapidly with application size in function points.

Table 1.31 Software Project Outcomes by Size of Project

	Probability of Selected Outcomes				
	Early (%)	*On Time (%)*	*Delayed (%)*	*Canceled (%)*	*Sum (%)*
1 FP	14.68	83.16	1.92	0.25	100.00
10 FP	11.08	81.25	5.67	2.00	100.00
100 FP	6.06	74.77	11.83	7.33	100.00
1,000 FP	1.24	60.76	17.67	20.33	100.00
10,000 FP	0.14	28.03	23.83	48.00	100.00
100,000 FP	0.00	13.67	21.33	65.00	100.00
Average	5.53	56.94	13.71	23.82	100.00

Factor 36—Venture capital and software start-up estimation: It is common knowledge that venture capital investments into start-up software companies is risky. The $100,000,000 loss by the State of Rhode Island for a poorly monitored investment in Curt Schilling's Studio 38 game company is a lesson in venture risks. The State performed zero due diligence prior to the investment, so the later bankruptcy of Studio 30 left the state with huge financial obligations, or so the governor claimed. (In fact, the debt was from "moral obligation bonds" that had no legal mandate for repayment.)

To ease the problems of venture capital for software start-ups, the author's SRM tool has a start-up estimating feature. This feature predicts odds of success or failures, number of rounds of venture investment probably needed prior to profitability, and also dilution of the ownership of the original entrepreneurs.

On average, software start-ups burn through three rounds of venture funding that total to more than $10,000,000 during which the entrepreneur's equity is diluted by over 60%. Since start-ups fail and go bankrupt at least 75% of the time, this is a hazardous situation for both sides.

The SRM venture feature was run retroactively on the Studio 38 investment and predicted a failure probability of 88%.

It is not clear if the State of Rhode Island planned for post-release investments. So far data on post-release maintenance and enhancement costs have not been published in local papers, but the predictions from my tool showed post-release maintenance costs of about $67,000,000 for a 5-year period.

These costs are over and above the initial investment by the State of $75,000,000. It is not clear whether these post-release costs would be funded by revenues from Studio 38 or would require additional rounds of investment.

Normally for start-ups, several rounds of investment are needed. Maintenance costs plus initial development bring the financial exposure for Studio 38 up to $142,000,000.

The final prediction was for the total amount of venture money that might be needed for development, maintenance, support, marketing, advertising, global sales, and other corporate expenses. My tool predicted a total investment of about $206,000,000 and three rounds of financing over about a 10-year period when all expense elements were aggregated. It is not clear from newspaper reports whether the State planned for continuing investments after the initial outlay, but no doubt the State investigation will find out.

Even without a formal risk analysis, the success rate of venture-funded software start-ups is <25%. This statement is made in the context of professional venture capitalists who know what they are doing.

The success rate of software start-ups funded by well-meaning but inexperienced State economic advisors is unknown because most states don't jump into such risky ventures. It will be interesting to see what the State Police investigation uncovers about the background of the Studio 38 investment or which come to the surface during the discovery and depositions of the current Studio 38 litigation.

Factor 37—Software maintenance, enhancement, and customer support: Post-release software costs are more complex to estimate and measure than development costs. In part, this is because no fewer than 25 different kinds of work are all called "maintenance." This makes maintenance benchmarks very difficult to perform at all, and even more difficult to perform with accuracy. The various kinds of work subsumed under the word maintenance include:

1. Major enhancements (new features of >20 function points)
2. Minor enhancements (new features of <5 function points)
3. Maintenance (repairing defects for good will)
4. Warranty repairs (repairing defects under formal contract)
5. Customer support (responding to client phone calls or problem reports)
6. Error-prone module removal (eliminating very troublesome code segments)
7. Mandatory changes (required or statutory changes)
8. Complexity or structural analysis (charting control flow plus complexity metrics)
9. Code restructuring (reducing cyclomatic and essential complexity)
10. Optimization (increasing performance or throughput)
11. Migration (moving software from one platform to another)
12. Conversion (changing the interface or file structure)
13. Reverse engineering (extracting latent design information from code)
14. Reengineering (transforming legacy application to modern forms)
15. Dead code removal (removing segments no longer utilized)
16. Dormant application elimination (archiving unused software)
17. Nationalization (modifying software for international use)
18. Mass updates such as Euro or Year 2000 Repairs
19. Refactoring, or reprogramming applications to improve clarity
20. Retirement (withdrawing an application from active service)
21. Field service (sending maintenance members to client locations)
22. Reporting bugs or defects to software vendors
23. Installing updates received from software vendors
24. Processing invalid defect reports
25. Processing duplicate defect reports

Post-release software costs are heavily impacted by the number of users of software applications. Table 1.32 shows an example of an application of 1000 function points coded in Java for 3 years of maintenance and support.

The number of users are shown in powers of 10: 10, 100, 1000, 10,000, 100,000, and 1,000,000. Applications with over a million users include many Microsoft projects and also many smartphone operating systems and some applications.

Table 1.32 shows cost per function point for development, enhancements, maintenance, and support for applications of 1000 function points that have between 10 and 1,000,000 users. Applications with over 1,000,000 users include Windows, Android, Office, and a few popular smartphone applications.

Table 1.32 Software Post-release Costs for Three Calendar Years

Software Users	*Develop $*	*Enhance $*	*Maintain $*	*Support $*	*Total $*
10	$1,000.00	$100.00	$100.00	$50.00	$1,250.00
100	$1,000.00	$150.00	$110.00	$112.50	$1,372.50
1,000	$1,000.00	$225.00	$165.00	$253.13	$1,643.13
10,000	$1,000.00	$337.50	$247.50	$569.53	$2,154.53
100,000	$1,000.00	$506.25	$371.25	$1,281.45	$3,158.95
1,000,000	$1,000.00	$759.38	$556.88	$2,883.25	$5,199.51

As can be seen, increases in software usage drive up enhancement and maintenance (bug repairs) and make very large increases in customer support costs. Bug repairs increase because more users mean a higher probability for finding latent bugs. Enhancements increase because more users mean a more diverse set of user requirements.

The standard 3-year predictions by SRM for maintenance and support allow users to specify the numbers of expected users and the annual rates of increase in user populations. The default value is 100 users and a 20% annual increase in user population.

Other variable factors that affect maintenance costs include "incidents" or the specific reasons why clients contact software vendors. There are many sources of incidents and the set used by the author include the following:

Software Incidents
Customer technical help requests
Customer complaints
Customer billing queries
Valid enhancement bug reports
Valid original bug reports
High severity bug reports
Security flaws discovered
Bad fixes (new bugs in bug repairs)
Duplicate bug reports
Invalid bug reports or user errors
Abeyant defects that can be fixed
Total incidents

Help requests are the most frequently occurring incidents, followed by bug reports. Overall, software maintenance is a complex issue and is more difficult to estimate and measure than software development due to the large impacts of user populations that are outside the control of the software development teams.

Factor 38—Software TCO: C-level executives and corporate officers are more interested in TCO than in development. In real life, some applications such as air traffic control are over 30 years old and quite a few major applications, especially government applications, are over 20 years old. Obviously, predictions of TCO are needed up front because C-level executives can't wait 20 years or more.

The author's SRM tool uses a fixed 3-year window for calculating TCO, with the understanding that changes will need to occur after the 3-year window ends.

Sample TCO values are shown in Table 1.33 for an application of 5000 function points coded in Java by an average team using iterative development. The application had 500 users.

The full TCO calculations by SRM also include work hours, function points per month, work hours per function point, staffing, and schedules. But these data items are too big for ordinary book pages in portrait mode.

Factor 39—Life expectancy of software benchmark data: Once benchmark data are collected and entered into a database, how long will they be useful? In other words, is data collected in 1995 still relevant in 2017? As it happens, the value of

Table 1.33 Example of Total Cost of Ownership

TCO Cost Elements	*Effort Months*	*Costs*	*Percentage (%)*
Development	170.45	$1,917,614	33.63
User effort	101.52	$1,370,455	24.03
Enhancements—3 years	78.00	$877,500	15.39
Maintenance—3 years	95.67	$932,750	16.36
Support—3 years	114.00	$538,650	9.45
Litigation—plaintiff (none)	0.00	$0	0.00
Litigation—defendant (none)	0.00	$0	0.00
Cyber-attack: prevention (included in maintenance)	0.00	$0	0.00
Cyber-attack: recovery (none)	0.00	$0	0.00
Other costs		$65,000	1.14
Total for project	559.64	$5,701,569	100.00

historical benchmark data is just about as good as the value of historical medical records. Historical data allow longitudinal studies that can show improvements over long time spans.

It is interesting that waterfall projects of 1000 function points developed in 1985, 1990, 1995, 2000, 2005, 2010, and 2015 are remarkably similar in their productivity and quality rates and resemble waterfall projects delivered in early 2017. On the other hand, projects using Agile, Rational Unified Process (RUP), and TSP all show higher productivity rates and better quality levels than waterfall at any year.

Thus, historical data can reveal interesting trends about the effectiveness of new methods by allowing them to be compared against historical records for similar classes and types of applications.

When comparing modern projects developed in 2017, it is possible to limit the years for comparative projects and have a cutoff point of perhaps 2010. However, calendar time is less significant than methodology. If you want to compare a new Agile project against others, you only need to search on "Agile" because there are no old Agile projects as yet. The database itself would probably contain no Agile entries prior to 2000.

Once benchmark data are collected, it will always have value. But benchmark data collections need to be refreshed with new projects on a monthly basis. To keep track of new tools, languages, and methods, monthly benchmark data for the United States should increase by about 500 projects per month in round numbers.

For specialized benchmarks such as those that deal with special topics such as embedded software or Agile development, benchmark volumes should grow at about 5–15 projects per month in round numbers.

The software industry does change and therefore benchmarks need to be identified by the year of project completion. Table 1.34 illustrates approximate software quality averages from the 1960s through today with projections to 2030.

As can be seen from Table 1.34, there are continuous changes in software results over time. However, the software industry resembles a drunkard's walk, with about as many companies going backward and getting worse as there are companies improving and getting better.

In fact, if every company got better every year, assessments and benchmarks would probably not be needed. It is the combination of regressions and advances that make benchmarks valuable for companies and government agencies.

Factor 40—Executive interest levels in software benchmark types: This topic primarily deals with individual project benchmarks for software project productivity and quality levels. There are many other kinds of benchmarks as well. Table 1.35 shows the levels of interest in all types of benchmarks based on interviews with executives in client companies.

As can be seen, all levels and types of executives are interested in benchmarks. However, chief technology officers (CTO) have the widest range of interest, as might be expected from the nature of CTO work.

Table 1.34 Software Quality Levels by Decade

Era	*Languages*	*Methods*	*Defect Potential*	*Defect Removal (%)*	*Delivered Defects*
1960s	Assembly	Cowboy	6.00	83	1.02
1970s	COBOL/ FORTRAN	Waterfall	5.50	85	0.83
1980s	C, Ada	Structured	5.00	85	0.75
1990s	COBOL/ FORTRAN	Object Oriented	4.50	87	0.59
2000s	Java, C#	Agile/RUP/TSP	4.25	90	0.43
2010s	PHP/mySQL/ JavaScript	Agile/RUP/TSP/ Mashups	4.00	95	0.20
2020s	Ruby on Rails	Mashups/75% reuse	3.00	98	0.06
2030s	New	SEMAT/90% reuse	2.50	99	0.03

Table 1.35 Executive Interest in Software Benchmark Types

	Software Benchmark Types	*CEO Interest*	*CFO Interest*	*CIO Interest*	*CTO Interest*
1	Competitive practices within industry	10	10	10	10
2	Project failure rates (size, methods)	10	10	10	10
3	Development schedules	10	10	10	10
4	Outsource contract success/failure	10	10	10	10
5	Project risks	10	10	10	10
6	CMMI assessments	9	9	9	10
7	Security attacks (number, type)	10	10	10	10
8	ROI	10	10	10	10
9	TCO	10	10	10	10
10	Customer satisfaction	10	10	10	10
11	Employee morale	10	9	10	10
12	Data quality	10	10	10	10

(Continued)

Table 1.35 (*Continued*) Executive Interest in Software Benchmark Types

	Software Benchmark Types	*CEO Interest*	*CFO Interest*	*CIO Interest*	*CTO Interest*
13	Customer support benchmarks	9	9	10	10
14	COQ	10	10	10	10
15	Best practices—requirements	6	9	9	10
16	Development costs	10	10	10	10
17	Team attrition rates	7	8	10	10
18	ISO standards certification	9	9	9	9
19	Best practices—maintenance	8	9	10	10
20	Best practices—test efficiency	8	7	9	10
21	Best practices—defect prevention	5	6	8	10
22	Best practices—pretest defects	6	6	9	10
23	Technical debt	8	8	9	9
24	Productivity—project	8	8	9	10
25	Coding speed	7	8	8	8
26	Code quality (code-nothing else)	7	7	10	10
27	Cost per defect (caution: unreliable)	6	7	10	10
28	Application sizes by type	8	8	8	9
29	Enhancement costs	8	10	10	8
30	Maintenance costs (annual)	8	9	9	9
31	Team morale	8	8	8	8
32	Litigation—patent infringements	10	10	10	10
33	Best Practices—design	5	5	9	10
34	Litigation—intellectual property	10	10	10	10
35	Portfolio maintenance costs	10	10	10	10
36	Team compensation level	10	9	9	7
37	Litigation—breach of contract	10	10	10	10

(*Continued*)

Table 1.35 (*Continued*) Executive Interest in Software Benchmark Types

	Software Benchmark Types	*CEO Interest*	*CFO Interest*	*CIO Interest*	*CTO Interest*
38	Earned value (EVA)	5	7	7	10
39	Methodology comparisons	5	6	9	10
40	Test coverage benchmarks	5	6	9	8
41	Productivity-activity	4	6	8	10
42	Application types	7	7	7	9
43	Litigation—employment contracts	7	8	8	8
44	Tool suites used	4	4	7	10
45	CMMI levels within industries	6	5	5	9
46	Standards benchmarks	5	5	6	8
47	Country productivity	8	5	8	9
48	Database size	9	8	8	9
49	Industry productivity	10	7	6	9
50	Metrics used	3	3	7	9
51	Programming Languages	3	3	6	8
52	Application class by taxonomy	4	5	6	8
53	Certification benchmarks	3	3	6	7
54	Cyclomatic complexity benchmarks	1	1	5	7
55	SNAP nonfunctional size metrics	3	4	5	8
	Totals	412	421	475	513
	Averages	7.49	7.65	8.64	9.33

CEOs are more interested in competitive and large-scale benchmarks than in specific project benchmarks, with the important exception of very high interest levels in applications that might be of strategic importance or cost more than $10,000,000. CEOs also care about schedule delays and cost overruns, and especially so when the delays top 12 months and the cost overruns top $1,000,000, which happen a lot, and especially so for state government software projects.

Factor 41—Available benchmarks circa 2017: The total number of software benchmarks circa 2017 is over 115,000. Unfortunately, these are fragmented among at least 20 different metrics, most of which have no conversion rules to other benchmark metrics. Table 1.36 shows the author's approximation of software benchmarks by metric.

The software benchmark situation is not as good as it should be. Many benchmarks show only development data. Indeed, quite a few don't even show complete development cycles but only "design, code, and unit test," which collectively are less than 40% of total project effort and costs.

Table 1.36 Estimated Benchmarks by Metric circa 2017

	Benchmark Metrics	*Average Size*	*Percentage of IFPUG size (%)*	*Benchmarks as of 2017*
1	Backfired IFPUG function points	1,000	100.00	50,000
2	IFPUG function points	1,000	100.00	45,000
3	Physical LOC	85,000	8500.00	7,500
4	COSMIC function points	1,143	114.00	5,500
5	Logical code statements	50,000	5000.00	5,000
6	NESMA function points	1,040	104.00	1200
7	Mark II function points	1,060	106.00	1,000
8	FISMA function points	1,020	102.00	800
9	Unadjusted function points	890	89.00	750
10	Use case points	333	33.00	350
11	Story points	556	56.00	300
12	Feature points	1,000	100.00	200
13	Function points light	965	97.00	200
14	Pattern-match function points	1,000	100.00	200
15	RICE objects	4,714	471.00	150
16	Fast function points	970	97.00	100
17	Full function points	1,170	117.00	100
18	Legacy-mined function points	1,000	100.00	75
19	Engineered function points	1,015	102.00	50
20	SNAP nonfunctional metrics	155	24.00	40
	Total			118,515

Factor 42—Variations by application size: For reasons of cost and timing, most software benchmarks are for small projects below 1000 function points in size. Because function point analysis is slow and expensive, large systems above 10,000 function points are almost never included in benchmarks.

As it happens due to working for IBM, ITT, and for clients that built large applications, the author frequently studies applications between 10,000 and 100,000 function points in size. In fact, all of the lawsuits where the author has been an expert witness have been for large systems >10,000 function points. Table 1.37 illustrates typical results for three size plateaus: 1,000, 10,000, and 100,000 function points.

As can be seen, there are major shifts in effort as application size increases. Big systems have many more bugs and many more test cases. Big systems generate more paper documents than small applications. Big systems have greatly expanded team sizes and many more specialist occupations. Big systems also have more complicated development processes and carry out many different kinds of activities such as integration and configuration control that may not occur for small projects.

Factor 43—Mix of new development and modification projects: Currently, in 2017, new development projects are no longer the main kind of software project carried out by large companies. Presently, legacy enhancements and modifying commercial packages or enterprises resource planning (ERP) packages outnumber new development projects. Table 1.38 shows the distribution of 100 software projects in a Fortune 500 client covering the past 3 years.

Most benchmark studies focus on new development projects. But in today's world, ERP and Commercial Off-The-Shelf (COTS) modifications far outnumber new development projects.

Factor 44—Portfolio benchmarks: Most benchmark studies collect data on individual small projects usually below 1000 function points in size. From time to time the author and colleagues have been commissioned to carry out corporate portfolio benchmarks of all of the software owned by Fortune 500 companies. These large-scale benchmarks usually take more than a calendar year and may involve more than a dozen consultants who may have to visit 20 or more corporate software labs. Table 1.39 shows partial data on a Fortune 500 portfolio benchmark study. The full set of portfolio data is too large for this book.

Software is an important corporate asset, so every Fortune 500 company should commission portfolio benchmark studies at perhaps 5-year intervals.

Presently, in 2017, legacy renovation and repairs, and updates to COTS and ERP packages are the major activities of corporate portfolio management, while developing new applications is in decline.

Factor 45—Problem, code, and data complexity: There are more than 20 kinds of complexity that can impact software projects. The only two complexity metrics that are used fairly often for software are cyclomatic complexity and Halstead complexity.

Cyclomatic complexity was developed by Dr. Tom McCabe in 1976. It covers source code complexity and is based on a graph of control flow. Cyclomatic complexity is the sum of edges minus nodes plus two.

Table 1.37 Variations by Powers of 10 (100–10,000 Function Points)

Size in function points		100	1,000	10,000
Size in SNAP points		13	155	1,750
Examples		Medium update	Smartphone app	Local system
Team experience		Average	Average	Average
Methodology		Iterative	Iterative	Iterative
Sample size for this table		150	450	50
CMMI levels (0=CMMI not used)		0	1	1
Monthly burdened costs		$10,000	$10,000	$10,000
Major cost drivers (rank order)	1	Coding	Bug repairs	Bug repairs
	2	Bug repairs	Coding	Paperwork
	3	Management	Paperwork	Coding
	4	Meetings	Management	Creep
	5	Functional requirement	Functional requirement	Functional requirement
	6	Nonfunctional requirement	Nonfunctional requirement	Nonfunctional requirement
	7	Paperwork	Creep	Meetings

(Continued)

Table 1.37 (*Continued*) Variations by Powers of 10 (100–10,000 Function Points)

	8	Integration	Integration	Integration
	9	Creep	Meetings	Management
Programming language		Java	Java	Java
Source statements per function point		53	53	53
Size in logical code statements (SRM default for LOC)		5,300	53,000	530,000
Size in logical KLOC (SRM default for KLOC)		5.3	53	530
Size in physical LOC (not recommended)		19,345	193,450	1,934,500
Size in physical KLOC (not recommended)		19.35	193.45	1,934.50
Client planned schedule in calendar months		5.25	12.5	28
Actual schedule in calendar months		5.75	13.8	33.11
Plan/actual schedule difference		0.5	1.3	5.11
Schedule slip percentage (%)		9.61	10.43	18.26
Staff size (technical + management)		1.25	6.5	66.67
Effort in staff months		7.19	89.72	2,207.54

(*Continued*)

Table 1.37 (*Continued*) Variations by Powers of 10 (100–10,000 Function Points)

Work hours per month (U.S. value)		132	132	132
Unpaid overtime per month (software norms)		0	8	16
Effort in staff hours		949.48	11,843.70	291,395.39
IFPUG Function points per month		13.9	11.15	4.53
Work hours per function point		9.49	11.84	29.14
Logical LOC per month (Includes executable statements and data definitions)		736.83	590.69	240.09
Physical LOC per month (Includes blank lines, comments, headers, etc.)		2,689.42	2,156.03	876.31
Requirements creep (total percentage growth) (%)		1.00	6.00	15.00
Requirements creep (function points)		1	60	1,500
Probable deferred features to release 2		0	0	2,500
Client planned project cost		$65,625	$812,500	$18,667,600
Actual total project cost		$71,930	$897,250	$22,075,408
Plan/actual cost difference		$6,305	$84,750	$3,407,808

(*Continued*)

Table 1.37 (*Continued*) Variations by Powers of 10 (100–10,000 Function Points)

Plan/actual percentage difference		8.77%	9.45%	15.44%
Planned cost per function point		$656.25	$812.50	$1,866.76
Actual cost per function point		$719.30	$897.25	$2,207.54
Defect Potentials and Removal (%)				
Defect potentials		**Defects**	**Defects**	**Defects**
Requirements defects		5	445	6,750
Architecture defects		0	1	27
Design defects		25	995	14,700
Code defects		175	2,150	30,500
Document defects		11	160	1,650
Bad fix defects		15	336	3,900
Total defects		231	4,087	57,527
Defects per function point		2.31	4.09	5.75
DRE (%)		97.50	96.00	92.50
Delivered defects		6	163	4,313

(*Continued*)

Table 1.37 (*Continued*) Variations by Powers of 10 (100–10,000 Function Points)

High-severity defects		1	20	539
Security flaws		0	3	81
Delivered defects per function point		0.06	0.16	0.43
Delivered defects per KLOC		1.09	3.08	8.14
Test cases for selected tests		**Test Cases**	**Test Cases**	**Test Cases**
Unit test		101	1,026	10,461
Function test		112	1,137	11,592
Regression test		50	512	5,216
Component test		67	682	6,955
Performance test		33	341	3,477
System test		106	1,080	11,012
Acceptance test		23	237	2,413
Total		492	5,016	51,126
Test cases per function point		4.92	5.02	5.11
Probable test coverage (%)		95.00	92.00	87.00
Probable peak cyclomatic complexity		12	15	>25.00

(Continued)

Table 1.37 (*Continued*) Variations by Powers of 10 (100–10,000 Function Points)

Document sizing				
Document sizes		**Pages**	**Pages**	**Pages**
Requirements		40	275	2,126
Architecture		17	76	376
Initial design		45	325	2,625
Detail design		70	574	5,118
Test plans		23	145	1,158
Development plans		6	55	550
Cost estimates		17	76	376
User manuals		38	267	2,111
HELP text		19	191	1,964
Courses		15	145	1,450
Status reports		20	119	1,249
Change requests		18	191	2,067
Bug reports		97	1,048	11,467
Total		423	3,486	32,638

(*Continued*)

Table 1.37 (*Continued*) Variations by Powers of 10 (100–10,000 Function Points)

Document set completeness (%)		96.96	91.21	78.24
Document pages per function point		4.23	3.49	3.26
Project risks		**Risk (%)**	**Risk (%)**	**Risk (%)**
Cancellation		8.80	14.23	26.47
Negative ROI		11.15	18.02	33.53
Cost overrun		9.68	15.65	34.00
Schedule slip		10.74	18.97	38.00
Unhappy customers		7.04	11.38	34.00
Litigation		3.87	6.26	11.65
Technical debt/high COQ		5.00	16.00	26.21
Cyber-attacks		7.00	9.75	15.30
Financial risk		9.00	21.00	41.00
High warranty repairs/low maintainability		6.00	14.75	32.00
Risk average		7.83	14.60	29.22
Project staffing by Occupation Group		100	1,000	10,000
Programmers		1.91	6.23	43.53

(*Continued*)

Table 1.37 (*Continued*) Variations by Powers of 10 (100–10,000 Function Points)

Testers		1.85	5.66	38.58
Designers		0.51	2.13	18.00
Business analysts		0.00	2.13	9.00
Technical writers		0.44	1.05	7.00
Quality assurance		0.46	0.98	5.00
First-line managers		1.21	1.85	7.13
Database administration		0.00	0.00	3.68
Project office staff		0.00	0.00	3.19
Administrative support		0.00	0.00	3.68
Configuration control		0.00	0.00	2.08
Project librarians		0.00	0.00	1.72
Second-line managers		0.00	0.00	1.43
Estimating specialists		0.00	0.00	1.23
Architects		0.00	0.00	0.86
Security specialists		0.00	0.00	0.49
Performance specialists		0.00	0.00	0.49
Function point counters		0.00	0.07	0.49

(*Continued*)

Table 1.37 (*Continued*) Variations by Powers of 10 (100–10,000 Function Points)

Human factors specialists		0.00	0.00	0.49
Third-line managers		0.00	0.00	0.36
Average total staff		6.37	20.11	148.42
Project Activity Patterns				
		100	1000	10,000
		Function	Function	Function
Activities performed		**Points**	**Points**	**Points**
01 Requirements		X	X	X
02 Prototyping			X	X
03 Architecture				X
04 Project plans			X	X
05 Initial design			X	X
06 Detail design		X	X	X
07 Design reviews				X
08 Coding		X	X	X
09 Reuse acquisition		X	X	X

(Continued)

Table 1.37 (*Continued*) Variations by Powers of 10 (100–10,000 Function Points)

10 Package purchase				X
11 Code inspections				X
12 IIV&V				
13 Change control			X	X
14 Formal integration			X	X
15 User documentation		X	X	X
16 Unit testing		X	X	X
17 Function testing		X	X	X
18 Integration testing			X	X
19 System testing			X	X
20 Beta testing				X
21 Acceptance testing			X	X
22 Independent testing				
23 Quality assurance			X	X
24 Installation/training				X
25 Project management		X	X	X
Activities		8	17	23

Table 1.38 Distribution of 100 Projects in a Fortune 500 Company 2014–2017

	Type of Project	*Number*	*Function Points*	*Work Hours*	*Work Hours per FP*
1	Legacy enhancements	22	8,030	109,208	13.60
2	New development	7	7,350	84,525	11.50
3	ERP modifications	12	4,260	55,167	12.95
4	Legacy renovations	4	2,300	35,995	15.65
5	COTS modifications	16	2,160	31,644	14.65
6	Open-source modifications	3	285	2,950	10.35
7	Legacy repairs	30	270	2,025	7.50
8	Cyber-attack recovery	6	105	1,171	11.15
	Total	100	24,655	322,685	
	Average		247	3,227	13.09

Table 1.39 Sample Portfolio Benchmark Study

Overall Portfolio Size					
Initial internal applications in January	4,310				
Average language level	3.25				
Average size in FP	1,100				
Average size in LOC	108,308				
Initial size of portfolio in FP	4,015,000				
Initial size of portfolio in LOC	395,323,077				
Latent defects in portfolio	28,105				
Latent security flaws in portfolio	1,827				

(*Continued*)

Table 1.39 (*Continued*) Sample Portfolio Benchmark Study

Overall Portfolio Size					
Outsourced Activities					
Development (%)	10.00				
Enhancement (%)	10.00				
Maintenance (%)	75.00				
Customer support (%)	85.00				
Portfolio Size Distribution	**Number**	**Average Age (Years)**			
>10,000 function points	35	18.30			
5,000–10,000 FP	200	9.45			
1,000–5,000 FP	475	6.50			
500–1,000 FP	1,790	2.25			
<500 function points	1,810	2.00			
Total	4,310	7.70			
Portfolio Application Types	**Internal**	**COTS**	**Total**	**Percentage**	**Work Hours Per FP**
Information systems	1,460	96	1,556	28.35	8.60
Systems software	913	125	1,038	18.90	4.75
Embedded applications	548	125	673	12.25	5.20
Tools and support software	365	433	798	14.54	10.24
Manufacturing applications	365	20	385	7.02	9.78
Web applications	335	345	680	12.39	13.90
Cloud applications	125	34	159	2.90	16.40
End-user applications	*200*	0	200	3.64	22.40
Total	4,310	1,178	5,488	100.00	11.41

(*Continued*)

Table 1.39 (*Continued*) Sample Portfolio Benchmark Study

Overall Portfolio Size					
Portfolio Languages					
Information systems	COBOL, PL/I, SQL, QBE, ASP, Visual Basic, Mumps				
Systems software	Chill, Coral, Bliss, C, Objective C C++, Assembly, Ada, Jovial				
Embedded applications	Forth, Embedded Java, Assembly, Objective C, Java, J2ME				
Tools and support software	Objective C, C, C++, C#, Visual Basic. PERL, Ruby				
Manufacturing applications	Assembly, Objective C, C, C++, C#, Visual Basic				
COTS packages	Multiple				
Open source packages	Java, C, C++, Ruby				
End-user applications	Visual Basic, Java, Ruby, Perl				
Portfolio Platforms	**Internal**	**COTS**	**Total**	**Percent**	
Mainframe (IBM primarily)	2,190	343	2,533	43.77	
Client-Server	730	224	954	16.48	
Windows-based PC	335	236	571	9.87	
Custom/embedded	548	120	668	11.53	
SaaS	365	224	589	10.17	
Cloud	125	13	138	2.38	
Handheld/tablets	17	19	36	0.62	
Total	4,310	1,178	5,488	94.83	

Code that runs straight through with no branches has a cyclomatic complexity of 1. As branches increase, so does cyclomatic complexity. High levels of cyclomatic complexity above 25 correlate strongly with high numbers of coding defects and high numbers of "bad fixes" or new bugs accidentally included in bug repairs. There are a number of tools including several open-source free tools that measure cyclomatic complexity. The author recommends measuring cyclomatic complexity on all applications as the code is completed.

Halstead complexity was developed by Dr. Maurice Halstead in 1977. This metric counts operators and operands and generates several quantitative results such as program vocabulary, program length, program difficulty, and several others. Halstead metrics are not as widely used in 2017 as cyclomatic complexity, and don't have as many automated tools.

Unfortunately, both cyclomatic and Halstead complexities focus only on code and ignore all other forms of complexities such as combinatorial complexity, computational complexity, mnemonic complexity, and all of the others.

The author's SRM tool uses three forms of complexity that are evaluated on a scale that runs from 1 to 11 with a midpoint value of 6. We look at problem complexity, code complexity, and data complexity.

Problem complexity deals with the fundamental difficulty of the algorithms that need to be included in a software application.

Code complexity deals with branches and loops, somewhat similar to cyclomatic complexity.

Data complexity deals with the number of volume of data in the files used by the application.

Our complexity scoring method is shown in Table 1.40.

Although our complexity-scoring method uses subjective answers by clients, our method is intended to be used *before* projects start, so at that time there is no ability to actually count or measure anything since nothing exists when the inputs are used for estimates.

High complexity levels on these three topics are derived from observations of actual software projects. High complexity has these empirical impacts:

1. Defect potentials go up with high complexity
2. DRE comes down with high complexity
3. Test coverage comes down with high complexity
4. Numbers of test cases needed to top 90% coverage go up with high complexity
5. Bad-fix injections go up with high complexity
6. Software assignment scopes decrease with high complexity (more staff needed)
7. Software production rates decrease with high complexity (slower development)
8. Software risks increase with high complexity.

Sometimes high complexity is unavoidable, and especially for problem and data complexity. But developers and managers should know before starting that high complexity will cause problems and may cause cost and schedule overruns before the project is finished.

Factor 46—Deferred features: A very common problem is almost never estimated well, and often not measured well either. The problem is that of running late on the first release of a software application and making a management decision to defer some features until a later release.

Table 1.40 SRM Complexity Inputs

	Problem Complexity
1	Almost 100% extremely simple problems and algorithms
2	Simple problems, algorithms
3	About 75% simple but some average problems and algorithms
4	Mix of simple, and average problems, algorithms
5	Majority of simple or average algorithms with a few complex algorithms
6	Average problems and algorithms
7	Majority of average but a few complex algorithms
8	Mix of average and complex problems, algorithms
9	About 75% of algorithms and problems are complex
10	Highly complex problems and algorithms
11	Almost 100% extremely complex problems and algorithms
	Code Complexity
1	Most programming done with reuse, mashups, controls
2	Simple nonprocedural code plus reuse
3	Cyclomatic complexity <3
4	Mainly simple but some average code, control flows
5	Well-structured but some complex code, control flow
6	Well-structured, cyclomatic complexity <5
7	Fair structure with some large modules
8	Code and control flow is fairly complex
9	Complex code and control flow, large modules, cyclomatic complexity >10
10	Significant complex code, control flow, modules
11	Large modules, cyclomatic complexity >30
	Data Complexity
1	No permanent files or data kept by application
2	Only one simple file kept by application

(*Continued*)

Table 1.40 (*Continued*) SRM Complexity Inputs

	Problem Complexity
3	Two or three simple files
4	Three or four files, simple data interactions
5	Three to five files, mainly simple data interactions
6	Average files and data interactions, ≤5 logical files
7	Many files, some average, some complex
8	Multiple files (≥10) with many complex data interactions
9	Most files are complex, most data interactions are complex
10	Numerous, complex files and many complex data interactions
11	More than 20 files, majority of complex data interactions

Deferred features mean that estimates based on a specific size in function points or LOC have to be redone to match the actual size at delivery. If you had planned to deliver 10,000 function points but defer 1000 and only deliver 9000, then obviously values such as cost per function point need to be readjusted.

Table 1.41 shows normal volumes of deferred features based on planned size in function point metrics.

Deferred features are a common occurrence for large software projects but almost never occur below 1000 function points.

Table 1.41 shows that deferred features increase with overall application size in terms of function points. This is a common problem but one not covered well by the software literature. The author's SRM tool predicts probable size of deferred feature sets as a standard output from the tool, but users can either accept or reject the prediction.

Table 1.41 Deferred Features by Application Size

Size in Function Points	*Planned Size*	*Deferred Feature Size*	*Delivered Size*	*Deferred Percentage (%)*
10	10	—	10	0.00
100	100	—	100	0.00
1,000	1,000	50	950	5.00
10,000	10,000	950	9,050	9.50
100,000	100,000	18,000	82,000	18.00

Factor 47—Multinational development: Software is a global industry. Most major companies such as IBM, Apple, Microsoft, etc., have software development labs scattered around the world. Many companies do global outsourcing to lower cost countries such as Peru and China.

Estimating and measuring software projects developed in multiple countries is more complicated than single-country estimates because of large variations in local work hours and holidays. There are also travel costs to be considered, but work hour differences are the most complicated because they can throw off metrics such as function points per month.

Let's assume that an application of 1000 function points was developed in three countries: India, the United States, and the Netherlands. Each country handled 333.33 function points of the total application. Let's assume that the productivity rate in each country was 15.00 work hours per function point. That means that teams in each country all worked 4995 work hours. Assume that costs were $6000 per month in India, $10,000 per month in the U.S., and $12,000 per month in Belgium.

Table 1.42 shows the overall results of this project. Note that while the teams in all three countries worked on exactly the same size software package and worked exactly the same number of hours, there are very large differences in local costs and local costs per function point. Function points per month also have large differences. These differences are partly due to local work hours per month and partly due to local compensation costs.

As can easily be seen, estimating and measuring international software projects with multiple countries is a great deal more complicated than estimating and measuring projects done in one location and one country.

Factor 48—Special costs that are hard to estimate and measure: Most of the 50 factors discussed in this chapter can all be predicted before projects start, and can be measured fairly easily when projects are complete.

This discussion topic is more or less outside the scope of everyday estimates and benchmarks. There are quite a few software cost factors that are unpredictable before projects start, and also fairly hard to measure when they are complete.

Table 1.43 shows some 30 of these special costs that are outside the scope of normal project cost estimates, using either manual or parametric estimation.

Table 1.42 Impact of International Software Development

Country	*Monthly Work Hours*	*Total Work Hours*	*Work Months*	*Local Cost*	*$ per Function Point*
India	190	4,995	26.29	$157,737	$473.22
United States	132	4,995	37.84	$378,409	$1,135.24
Netherlands	115	4,995	43.43	$521,217	$1,563.67
Total	437	14,985	107.57	$1,057,363	$1,057.36

Table 1.43 Special Costs That Are Difficult to Estimate and Measure

1	Advertising costs for commercial software
2	Certification costs (FDA, FAA, etc.)
3	Cyber-attack insurance
4	Cyber-attack damages to equipment
5	Cyber-attack recovery costs
6	Cyber-attack reparations to clients
7	Employee/executive time spent on litigation
8	Equipment/supplies
9	Expert witness fees for litigation
10	External cyber security consultants
11	External estimating consultants
12	External function point counting consultants
13	External legal fees for noncompete litigation
14	External management consultants
15	External marketing consultants
16	External technical consultants
17	Governance costs (Sarbanes–Oxley)
18	Legal fees for litigation
19	Marketing costs for commercial software
20	Outsource contract legal fees
21	Patent filing fees
22	Patent licensing fees
23	Patent litigation costs (if any)
24	Personnel hiring agency fees
25	Personnel termination fees
26	Reusable component acquisition fees
27	Translation of documents into multiple languages
28	Travel costs to clients
29	Travel costs to development labs
30	Travel costs to subcontractors
	Total additional costs

Many of these 30 special costs don't occur at all, but when they do occur they can add quite a bit of costs to the total project. For example, an international telecom application was being built in eight countries in Europe plus two labs in the United States. This was a large system of about 20,000 function points with a technical staff of over 200 people around the world. This project had over 350 transatlantic flights by managers and technical personnel. The actual measured costs for air travel, hotels, and travel expenses for this system were larger than the cost of the source code!

The author and his colleagues use this check list during benchmark studies and ask clients to provide cost and effort data for any of the special costs that happened to occur for the projects being benchmarked.

If your goal is a complete understanding of software economics, then these special costs need to be included and studied since when they do occur they can add over $100 per function point to software development and over $150 per function point to total costs of ownership.

Factor 49—Software value and ROI: The value of a software application and the ROI for building it are very difficult to estimate. Part of the reason is that frequent schedule slips and cost overruns erode value and can turn ROI into a negative number.

Value is composed of a number of independent factors including revenue generation, operating cost reduction, and market expansion.

Software Revenue Generation

Direct revenues or the income derived from leading or selling software packages to companies or to individual users. Computer games and many utilities are good examples of software that generate direct revenues. For example, the author uses a commercial scientific calculator on an android smartphone that was purchased for $3.99.

Subscription revenues or the income derived from monthly or periodic fees to use software packages. Antivirus packages are a good example of subscription revenues, and so is the new Microsoft Office.

Software drag-along revenues, or sales of related packages that support a primary package. For example, if you buy an operating system, you will probably also buy an antivirus package.

Hardware drag-along revenues or sales of physical devices that support specific software packages. An example of hardware drag-along would be Oculus Rift, the 3-D viewer that works with android software and Samsung smartphone hardware, among others.

Consulting revenues or the income derived from consulting gigs in support of specific software applications. ERP packages such as SAP and Oracle usually need substantial consulting effort to get them installed and tuned up. The bigger the application, the more likely it is to need consulting support.

Training revenues or the income derived from teaching customers how to use specific software packages. Most complex software applications such as ERP packages, graphics design packages, and manufacturing support packages require fairly extensive training in order to use them effectively. Large applications and applications with important business impacts require training. For example, commercial human resource packages, commercial ERP packages, and Microsoft Office all have training available from both vendors themselves and third-party education companies.

Operating Cost Reductions

Direct savings or the result of software speeding up formerly manual processes. Examples of direct savings include automatic teller machines for dispensing cash, supermarket cash registers, automated car washes, computer-controlled machine tools, and computer-controlled manufacturing assembly lines.

Personnel savings or the savings associated with replacing live workers with some kind of automation. This is a continuing worry for clerical and administrative personnel because computers have greatly reduced staffing for labor-intensive activities such as insurance claims handling, inventory management, hotel reservations, and travel ticketing.

Market Expansion

Market share expansion or bringing in new customers based on unique software features. Apple's IPad is an example of a hardware/software product that created a new market. Microsoft's Visual Studio is another example.

New market creation: Google and Facebook are good examples of companies that have created brand new markets that did not exist before. This is actually the most exciting aspect of software engineering, and new markets in the virtual reality and artificial intelligence fields are both intellectually exciting and bring the promise of large revenue streams.

As of 2017, software value has been a mixed bag. Some software packages have generated many millions of dollars in revenue and made the developers hugely wealthy, as seen by Bill Gates, Jeff Bezos, Larry Ellison, Sergey Brin, Mark Zuckerberg, and quite a few others.

However, quite a few software packages barely make a profit and some eventually are withdrawn due to lack of sales revenues.

Value for internal software packages does not derive from external revenues, but rather from reducing operating costs and sometimes from opening up new markets, as seen by the telecommunications industry.

Factor 50—Summary overview of software measurements: The final table in this chapter shows the kinds of benchmark data collected by the author, but in summary form. Our full benchmarks are about five pages in size and some of the spreadsheets we provide to clients are too wide to fit the pages in this book.

Our benchmarks include our standard taxonomy, and then show software development, software quality, software documentation, software maintenance, software enhancements, software customer support, and TCO.

Table 1.44 summarizes the author's data collection for individual software projects. Usually we provide benchmarks for individual projects. The sample in this book shows two projects side by side, in order to illustrate how important a software taxonomy is for highlighting project differences. Some clients request three benchmarks for specific projects because they want to see best-case, average-case, and worst-case results. In that case we use a vertical format so that three projects can be squeezed into a normal page.

The project taxonomy for this telecom benchmark is as follows:

Table 1.44 Sample Software Benchmark 2017

• Project Name	Telecom Switching Best Case	
• Project Nature	New application	
• Project Scope	Central office switch	
• Project Class	External, bundled	
• Project Type	Telecommunications	
• Problem complexity	Above average	
• Code complexity	Below 10 cyclomatic complexity	
• Data complexity	Higher than average	
		Process used: TSP/PSP
	Project Cost-Driver Overview	**TSP/PSP Project Cost Drivers:**
		1. Requirements changes
		2. Programming or coding
		3. Producing paper documents

(Continued)

Table 1.44 (*Continued*) Sample Software Benchmark 2017

		4. Meetings and communications
		5. Finding and fixing bugs
		6. Project management
		6. Avoiding security flaws
	Benchmark Data Collection Cost Analysis	**Team/Personal Software Process (TSP/PSP)**
	Benchmark Client Effort	
	Client Benchmark Data Collection	
	Executive hours—data collection	1.00
	Management hours—data collection	2.00
	Team hours—data collection	8.00
	Total hours—data collection	11.00
	Client Benchmark Data Reporting	
	Executive hours—receiving benchmark data report	12.00
	Management hours—receiving benchmark data report/talk	4.00
	Team hours—receiving benchmark data report	16.00
	Total hours—data reporting	32.00
	Total client benchmark hours	43.00
	Benchmark Consulting Effort	
	Benchmark consulting project modeling with SRM	1.00

(*Continued*)

Table 1.44 (*Continued*) Sample Software Benchmark 2017

	Benchmark consultant hours—data collection	2.50
	Benchmark consultant hours—risk analysis	1.50
	Benchmark consultant hours—data analysis	6.00
	Benchmark consultant hours—preparing report	6.00
	Benchmark consultant hours—preparing presentation	3.00
	Benchmark consultant hours—presenting findings	4.00
	Total consulting hours	24.00
	Total benchmark effort	47.00
	Client benchmark collection costs	$1,250
	Client benchmark report costs	$3,636
	Consulting costs	$10,800
	Total benchmark costs	$15,687
	Benchmark costs per function point	$1.57
	Benchmark costs per KLOC	$36.76
	Additional fees not included	
	Legal fees for contract analysis	
	Travel costs if on-site work is needed	
	Special client requests not in standard benchmarks	
	Special factors (layoffs, management changes, etc.)	

(*Continued*)

Table 1.44 (*Continued*) Sample Software Benchmark 2017

	Translation of documents into multiple languages	
	Size and requirements creep	
	Size at end of requirements	
	New application size (function points)	9,037
	New application size (logical code)	385,609
	Size at delivery total requirements creep percentage	9.63%
	Size at delivery (function points)	10,000
	Size at delivery (logical code)	426,667
	Size at delivery (KLOC)	426.67
	Percentage reuse	26.02
	Major Software Project Risks	**Team/Personal Software Process (TSP/PSP)**
1	Cancellation of project	10.41%
2	Executive dissatisfaction with progress	12.80%
3	Negative ROI	11.65%
4	Cost overrun	15.75%
5	Schedule slip	16.21%
6	Unhappy customers	4.00%
7	Low team morale	4.50%
8	High warranty repair costs	6.74%
9	Litigation (for outsourced work)	5.30%
10	Competition releasing superior product	6.50%
	Average risks	9.39%

(*Continued*)

Table 1.44 (*Continued*) Sample Software Benchmark 2017

	Documentation Size Benchmarks (Pages)	TSP Pages
1	Requirements	2,230
2	Architecture	376
3	Initial design	2,753
4	Detail design	5,368
5	Test plans	1,158
6	Development plans	550
7	Cost estimates	376
8	User manuals	2,214
9	HELP text	1,964
10	Courses	1,450
11	Status reports	776
12	Change requests	1,867
13	Bug reports	14,084
	Total	35,166
	Total pages per function point	3.52
	Total pages per KLOC	82.42
	Total diagrams/graphics in documents	4,118
	Total words in documents	17,582,868
	Total words per function point	1,758.29
	Total words per line of code	41.21
	Work hours per page	1.05
	Paperwork months for all documents	271.50
	Paperwork hours for all documents	36,924.02

(*Continued*)

Table 1.44 (*Continued*) Sample Software Benchmark 2017

	Document work hours per function point	3.69
	Documentation costs	$4,196,046
	Documentation costs per function point	$419.60
	Document percentage of total costs	34.56%
	Project Meetings and Communications	**TSP/PSP Meetings**
	Conference calls	42
	Client meetings	26
	Architecture/design meetings	14
	Team status meetings	78
	Executive status meetings/phase reviews	18
	Problem analysis meetings	7
	Total meetings	184
		TSP/PSP Meeting Participants
	Conference calls	6
	Client meetings	5
	Architecture/design meetings	11
	Team status meetings	27
	Executive status meetings/phase reviews	9
	Problem analysis meetings	6
	Total participants	65
	Average participants	11

(*Continued*)

Table 1.44 (*Continued*) Sample Software Benchmark 2017

		TSP/PSP Meeting Clock Hours
	Conference calls	0.60
	Client meetings	2.70
	Architecture/design meetings	4.00
	Team status meetings	2.00
	Executive status meetings/phase reviews	4.00
	Problem analysis meetings	5.50
	Total duration hours	18.80
	Average duration hours	3.13
		TSP/PSP Meeting Total Hours
	Conference calls	152.78
	Client meetings	377.03
	Architecture/design meetings	631.42
	Team status meetings	4,249.92
	Executive status meetings/phase reviews	647.61
	Problem analysis meetings	213.71
	Total duration hours	6,272.47
		TSP/PSP Meeting Total Costs
	Conference calls	$17,362
	Client meetings	$42,846
	Architecture/design meetings	$71,754

(*Continued*)

Table 1.44 (*Continued*) Sample Software Benchmark 2017

	Team status meetings	$482,961
	Executive status meetings/phase reviews	$73,594
	Problem analysis meetings	$24,286
	Total meeting costs	$712,803
	Meeting percentage of total costs	5.87%
Quality and Defect Potentials (Req + Des + Code + Docs + Bad Fixes)		
	Per Function Point	**Best Case TSP**
1	Requirement defects	0.41
2	Design defects	0.60
3	Code defects	0.66
4	Document defects	0.07
5	Bad fixes (secondary defects)	0.02
	Defect potentials	1.76
		Below average for telecom
	Defect potentials (per KLOC-logical statements)	41.26
	Application Defect Potentials	**Best Case**
1	Requirement defects	4,086
2	Design defects	6,037
3	Code defects	6,629
4	Document defects	674
5	Bad fixes (secondary defects)	179
	Total defect potential of application	17,605

(*Continued*)

Table 1.44 (*Continued*) Sample Software Benchmark 2017

	DRE	Defects Remaining
1	JAD	27.00%
	Defects remaining	12,851
2	QFD	30.00%
	Defects remaining	8,996
3	Prototype	30.00%
	Defects remaining	6,297
4	Desk check	27.00%
	Defects remaining	4,597
5	Pair programming	0.00%
	Defects remaining	4,597
6	Static analysis	55.00%
	Defects remaining	2,069
7	Formal inspections	93.00%
	Defects remaining	145
8	Unit test	32.00%
	Defects remaining	98
9	Function test	35.00%
	Defects remaining	64
10	Regression test	14.00%
	Defects remaining	55
11	Component test	32.00%
	Defects remaining	37
12	Performance test	14.00%
	Defects remaining	32
13	System test	36.00%

(*Continued*)

Table 1.44 (*Continued*) Sample Software Benchmark 2017

	Defects remaining	21
14	Platform test	22.00%
	Defects remaining	16
15	Acceptance test	17.00%
	Defects remaining	13
	Delivered defects	13
	Defects per function point	0.0013
	Defects per KLOC	0.0313
	High severity defects	2
	Security flaws	0
	DRE	99.99%
	Defect removal work months	302.19
	Defect removal work hours	41,097
	Defect removal work hours per function point	4.11
	Defect removal costs	$4,670,314
	Defect removal costs per function point	$467.03
	Defect removal percentage of total costs	38.46%
	Test coverage percentage (approximate)	96.50%
	Test cases (approximate)	39,811
	Test case errors (approximate)	995
	Maximum cyclomatic Complexity	10
	Reliability (days to initial defect)	30.00
	Mean time between failures (MTBF) days	248.00

(*Continued*)

Table 1.44 (*Continued*) Sample Software Benchmark 2017

	Stabilization period (months) (months to approach zero defects)	0.72
	Error-prone modules (>5 indicates quality malpractice)	0
	Customer satisfaction	97.98%
	Development schedules, effort, productivity, costs (from requirements through delivery to clients)	**TSP**
	Average monthly cost	$15,000
	Average hourly cost	$113.64
	Overall Project	
	Development schedule (months)	22.19
	Staff (technical+management)	36.49
	Development effort (staff months)	809.51
	Development costs	$12,142,633
	Development Activity Work Months	**Work Months**
1	Requirements	36.98
2	Design	59.17
3	Coding	333.43
4	Testing	200.06
5	Documentation	26.90
6	Quality assurance	34.64
7	Management	118.34
	Totals	809.51

(*Continued*)

Table 1.44 (*Continued*) Sample Software Benchmark 2017

	Normalized Productivity Data	
	IFPUG function points per month	12.35
	IFPUG work hours per FP	11.01
	Logical LOC per month	527.07
	Development Activity Schedules	**Schedule Months**
1	Requirements	2.57
2	Design	3.11
3	Coding	9.64
4	Testing	6.54
5	Documentation	3.61
6	Quality assurance	5.71
7	Management	22.19
	Totals	31.17
	Overlap percentage	71.20%
	Overlapped net schedule	22.19
	Probable schedule without reuse	26.62
	Development Activity Staffing	**Development Staffing**
1	Requirements	14.42
2	Design	19.06
3	Coding	34.58
4	Testing	30.57
5	Documentation	7.46
6	Quality assurance	6.07
7	Management	6.83
	Totals	36.49

(*Continued*)

Table 1.44 (*Continued*) Sample Software Benchmark 2017

	Development Activity Costs	**Development Costs**
1	Requirements	$554,721
2	Design	$887,553
3	Coding	$5,001,405
4	Testing	$3,000,843
5	Documentation	$403,433
6	Quality assurance	$519,573
7	Management	$1,775,106
	Totals	$12,142,633
	Cost per function point	$1,214.26
	Cost per logical code statement	$28.46
8	Additional costs (patents, travel, consulting, etc.)	$1,575,000
	Project grand total	$13,718,876
	Grand total per function point	$1,371.89
	Grand total per LOC	$32.15
	Development Activity Productivity	**Work Hours per FP**
1	Requirements	0.50
2	Design	0.80
3	Coding	4.53
4	Testing	2.72
5	Documentation	0.37
6	Quality assurance	0.47
7	Management	1.61
	Total with reusable materials	11.01
	Totals without reusable materials	14.44

(*Continued*)

Table 1.44 (*Continued*) Sample Software Benchmark 2017

	Project Occupations and Specialists	**Normal Full-Time Staffing**
1	Programmers	28.7
2	Testers	25.5
3	Designers	12.4
4	Business analysts	12.4
5	Technical writers	5.6
6	Quality assurance	4.9
7	First-line managers	5.6
8	Database administration	2.7
9	Project office staff	2.4
10	Administrative support	2.7
11	Configuration control	1.6
12	Project librarians	1.3
13	Second-line managers	1.1
14	Estimating specialists	0.9
15	Architects	0.6
16	Security specialists	1.0
17	Performance specialists	1.0
18	Function point counters	1.0
19	Human factors specialists	1.0
20	Third-line managers	0.3

This benchmark report was one of a set of three telecommunications benchmarks provided to a client. This was the best-case benchmark example from the three. The other two showed average case and worst case. All sizes were the same at 10,000 function points for central office switching systems.

A problem with benchmarks in books is that in order to fit all of the data onto a page, some of it has to be converted from a horizontal landscape format to a vertical portrait format. The vertical format is a bit harder to read and understand, but is a technical necessity unless the pages were to be printed in a landscape format, which makes reading a book awkward.

Understanding software engineering economic and quality factors is about as complex as a full medical examination for a human patient. Unfortunately, the software industry is not as thorough as medical practice in learning important topics. Many software managers can't or don't want to absorb detailed data and demand short and simple explanations of topics that are actually quite complex.

One of the reasons this chapter shows so many factors individually is that they are easier to understand in isolation. But in reality, all of the factors in this chapter tend to interact and have synergistic impacts on software project results.

In a nutshell, good quality control tends to lead to optimal schedules and to minimum costs. Poor quality control stretches out schedules and raises costs. But good quality control needs more than just testing. It needs defect prevention, pretest defect removal such as static analysis, and formal testing using mathematical test-case design methods such as cause–effect graphs and design of experiments.

Summary and Conclusions on the 50 Software Economic Factors

For over 60 years the software industry has been running blind in terms of economic understanding. Both productivity and quality are poorly understood by the majority of companies and government groups due to metrics such as "LOC" and "cost per defect" that distort reality and conceal economic progress.

A total of 50 software economic factors probably seems like a lot, and perhaps too many, but if you want to understand what makes software productivity and quality good or bad they all need to be studied. Overall, software economic knowledge is sparse in 2017. Hopefully this book will lead to additional studies by other researchers on factors that are not well understood.

The essential lessons from studying software economics are that poor quality slows down productivity and increases both development and maintenance costs. Achieving good quality using defect prevention, pretest defect removal, and formal testing can shorten software schedules and reduce costs at the same time, and also lead to happier clients and more secure applications.

Custom development and manual coding are intrinsically expensive and error prone. The future of effective software engineering lies in moving away from custom software toward the construction of software from standard reusable materials.

Hopefully, some of the topics in this book can also reduce the troubling volume of canceled projects that are terminated and never finished, usually due to cost and schedule overruns so large that the ROI turns negative.

References

Abrain, A.; *Software Maintenance Management: Evolution and Continuous Improvement*; Wiley-IEEE Computer Society; Los Alamitos, CA; 2008.

Abrain, A.; *Software Metrics and Metrology*; Wiley-IEEE Computer Society; Los Alamitos, CA; 2010.

Abrain, A.; *Software Estimating Models*; Wiley-IEEE Computer Society; Los Alamitos, CA; 2015.

Albrecht, A.; *AD/M Productivity Measurement and Estimate Validation*; IBM Corporation, Purchase, NY; 1984.

Barrow, D., Nilson, S., and Timberlake, D.; *Software Estimation Technology Report*; Air Force Software Technology Support Center, Hill Air Force Base, UT; 1993.

Boehm, B.; *Software Engineering Economics*; Prentice Hall, Englewood Cliffs, NJ; 1981; 900 pages.

Brooks, F.; *The Mythical Man-Month*; Addison-Wesley, Reading, MA; 1995; 295 pages.

Brown, N. (editor); *The Program Manager's Guide to Software Acquisition Best Practices*; Version 1.0; U.S. Department of Defense, Washington, DC; 1995; 142 pages.

Bundschuh, M. and Dekkers, C.; *The IT Measurement Compendium*; Springer-Verlag, Berlin, Germany; 2008; 643 pages.

Chidamber, S.R. and Kemerer, C.F.; A metrics suite for object oriented design; *IEEE Transactions on Software Engineering*; 20; 1994; pp. 476–493.

Chidamber, S.R., Darcy, D.P., and Kemerer, C.F.; Managerial use of object oriented software metrics; Joseph M. Katz Graduate School of Business, University of Pittsburgh, Pittsburgh, PA; Working Paper # 750; November 1996; 26 pages.

Cohn, M.; *Agile Estimating and Planning*; Prentice Hall, Englewood Cliffs, NJ; 2005; ISBN 0131479415.

Conte, S.D., Dunsmore, H.E., and Shen, V.Y.; *Software Engineering Models and Metrics*; The Benjamin Cummings Publishing Company, Menlo Park, CA; 1986; ISBN 0-8053-2162-4; 396 pages.

DeMarco, T.; *Controlling Software Projects*; Yourdon Press, New York; 1982; ISBN 0-917072-32-4; 284 pages.

DeMarco, T.; *Why Does Software Cost So Much?* Dorset House Press, New York; 1995; ISBN 0-932633-34-X; 237 pages.

DeMarco, T.; *Deadline*; Dorset House Press, New York; 1997.

DeMarco, T. and Lister, T.; *Peopleware*; Dorset House Press, New York; 1987; ISBN 0-932633-05-6; 188 pages.

Department of the Air Force; *Guidelines for Successful Acquisition and Management of Software Intensive Systems*; Volumes 1 and 2; Software Technology Support Center, Hill Air Force Base, UT; 1994.

Dreger, B.; *Function Point Analysis*; Prentice Hall, Englewood Cliffs, NJ; 1989; ISBN 0-13-332321-8; 185 pages.

Gack, G.; *Managing the Black Hole—The Executive's Guide to Project Risk*; The Business Expert Publisher; Thomson, GA; 2010; ISBN10: 1-935602-01-2.

Galea, R.B.; *The Boeing Company: 3D Function Point Extensions, V2.0, Release 1.0*; Boeing Information Support Services, Seattle, WA; 1995.

Galorath, D.D. and Evans, M.W.; *Software Sizing, Estimation, and Risk Management*; Auerbach Publications, New York; 2006.

Garmus, D. and Herron, D.; *Measuring the Software Process: A Practical Guide to Functional Measurement*; Prentice Hall, Englewood Cliffs, NJ; 1995.

Garmus, D. and Herron, D.; *Function Point Analysis*; Addison-Wesley Longman, Boston, MA; 1996.

Grady, R.B.; *Practical Software Metrics for Project Management and Process Improvement*; Prentice Hall, Englewood Cliffs, NJ; 1992; ISBN 0-13-720384-5; 270 pages.

Grady, R. B. and Caswell, D.L.; *Software Metrics: Establishing a Company-Wide Program*; Prentice Hall, Englewood Cliffs, NJ; 1987; ISBN 0-13-821844-7; 288 pages.

Gulledge, T. R., Hutzler, W. P., and Lovelace, J.S. (editors); *Cost Estimating and Analysis-Balancing Technology with Declining Budgets*; Springer-Verlag; New York; 1992; ISBN 0-387-97838-0; 297 pages.

Harris, M. D.S., Herron, D., and Iwanacki, S.; *The Business Value of IT*; CRC Press, Auerbach Publications; Boca Raton, FL; 2009.

Hill, P.R. *Practical Software Project Estimation*; McGraw-Hill; New York; 2010.

Howard, A. (editor); *Software Metrics and Project Management Tools*; Applied Computer Research (ACR); Phoenix, AZ; 1997; 30 pages.

Humphrey, W. S.; *Managing the Software Process*; Addison-Wesley Longman, Reading, MA; 1989.

Humphrey, W.; *Personal Software Process*; Addison-Wesley Longman, Reading, MA; 1997.

Jones, C.; *Estimating Software Costs*; McGraw-Hill; New York; 2007.

Jones, C.; *Applied Software Measurement*; 3rd edition; McGraw-Hill; New York; 2008.

Jones, C.; *Software Engineering Best Practices*; McGraw-Hill; New York; 2010.

Jones, C.; *The Technical and Social History of Software Engineering*; Addison-Wesley; Boston, MA; 2012.

Jones, C.; *Comparing Sixty Software Development Methodologies*; CRC Press; Boca Raton, FL; 2017a.

Jones, C.; *Software Measurement Selection*; CRC Press; Boca Raton, FL; 2017b.

Jones, C. and Bonsignour, O.; *The Economics of Software Quality*, Addison-Wesley; Boston, MA; 2012.

Kan, S. H.; *Metrics and Models in Software Quality Engineering*; 2nd edition; Addison-Wesley Longman, Boston, MA; 2003; ISBN 0-201-72915-6; 528 pages.

Kemerer, C. F.; An empirical validation of software cost estimation models; *Communications of the ACM*; 30; 1987; pp. 416–429.

Kemerer, C.F.; Reliability of function point measurement: A field experiment; *Communications of the ACM*; 36; 1993; pp. 85–97.

Keys, J.; *Software Engineering Productivity Handbook*; McGraw-Hill, New York; 1993; ISBN 0-07-911366-4; 651 pages.

Laird, L. M. and Brennan, C. M.; *Software Measurement and Estimation: A Practical Approach*; John Wiley & Sons, Hoboken, NJ; 2006; ISBN 0-471-67622-5; 255 pages.

Love, T.; *Object Lessons*; SIGS Books, New York; 1993; ISBN 0-9627477 3-4; 266 pages.

Marciniak, J.J. (editor); *Encyclopedia of Software Engineering*; John Wiley & Sons, New York; 1994; ISBN 0-471-54002; in two volumes.

McConnell, S.; *Software Estimating: Demystifying the Black Art*; Microsoft Press, Redmond, WA; 2006.

Melton, A.; *Software Measurement*; International Thomson Press, London; 1995; ISBN 1-85032-7178-7.

Mertes, K.R.; Calibration of the CHECKPOINT model to the Space and Missile Systems Center (SMC) Software Database (SWDB); Thesis AFIT/GCA/LAS/96S-11, Air Force Institute of Technology (AFIT), Wright Patterson AFB, OH; September 1996; 119 pages.

Mills, H.; *Software Productivity*; Dorset House Press, New York; 1988; ISBN 0-932633-10-2; 288 pages.

Muller, M. and Abram, A. (editors); *Metrics in Software Evolution*; R. Oldenbourg Verlag GmbH, Munich; 1995; ISBN 3-486-23589-3.

Multiple authors; *Rethinking the Software Process*; (CD-ROM); Miller Freeman, Lawrence, KS; 1996. (This is a new CD ROM book collection jointly produced by the book publisher, Prentice Hall, and the journal publisher, Miller Freeman. This CD ROM disk contains the full text and illustrations of five Prentice Hall books: *Assessment and Control of Software Risks* by Capers Jones; *Controlling Software Projects* by Tom DeMarco; *Function Point Analysis* by Brian Dreger; *Measures for Excellence* by Larry Putnam and Ware Myers; and *Object-Oriented Software Metrics* by Mark Lorenz and Jeff Kidd.)

Park, R. E. et al; Software cost and schedule estimating: A process improvement initiative; Technical Report CMU/SEI 94-SR-03; Software Engineering Institute, Pittsburgh, PA; May 1994.

Park, R.E. et al; Checklists and criteria for evaluating the costs and schedule estimating capabilities of software organizations; Technical Report CMU/SEI 95-SR-005; Software Engineering Institute, Pittsburgh, PA; January 1995.

Paulk M. et al; *The Capability Maturity Model; Guidelines for Improving the Software Process*; Addison-Wesley, Reading, MA; 1995; ISBN 0-201-54664-7; 439 pages.

Perlis, A.J., Sayward, F.G., and Shaw, M. (editors); *Software Metrics*; The MIT Press, Cambridge, MA; 1981; ISBN 0-262-16083-8; 404 pages.

Perry, W.E.; *Data Processing Budgets: How to Develop and Use Budgets Effectively*; Prentice Hall, Englewood Cliffs, NJ; 1985; ISBN 0-13-196874-2; 224 pages.

Perry, W.E.; *Handbook of Diagnosing and Solving Computer Problems*; TAB Books, Inc.; Blue Ridge Summit, PA; 1989; ISBN 0-8306-9233-9; 255 pages.

Pressman, R.; *Software Engineering-A Practitioner's Approach*; McGraw-Hill, New York; 1982.

Putnam, L.H.; *Measures for Excellence: Reliable Software on Time, within Budget*; Yourdon Press-Prentice Hall, Englewood Cliffs, NJ; 1992; ISBN 0-13-567694-0; 336 pages.

Putnam, L.H and Myers, W.; *Industrial Strength Software: Effective Management Using Measurement*; IEEE Press, Los Alamitos, CA; 1997; ISBN 0-8186-7532-2; 320 pages.

Reifer, D. (editor); *Software Management*; 4th edition; IEEE Press, Los Alamitos, CA; 1993; ISBN 0 8186-3342-6; 664 pages.

Roetzheim, W.H. and Beasley, R.A.; *Best Practices in Software Cost and Schedule Estimation*; Prentice Hall, Upper Saddle River, NJ; 1998.

Royce, W.E.; *Software Project Management: A Unified Framework*; Addison-Wesley, Reading, MA; 1999.

Rubin, H.; *Software Benchmark Studies for 1997*; Howard Rubin Associates, Pound Ridge, NY; 1997.

Shepperd, M.; A critique of cyclomatic complexity as a software metric; *Software Engineering Journal*; 3; 1988; pp. 30–36.

Software Productivity Consortium; *The Software Measurement Guidebook*; International Thomson Computer Press; Boston, MA; 1995; ISBN 1-850-32195-7; 308 pages.

St-Pierre, D.; Maya, M.; Abran, A., and Desharnais, J.-M.; Full function points: Function point extensions for real-time software, concepts and definitions; University of Quebec, Quebec. Software Engineering Laboratory in Applied Metrics (SELAM); TR 1997-03; 1997; 18 pages.

Strassmann, P.; *The Squandered Computer*; The Information Economics Press, New Canaan, CT; 1997; ISBN 0-9620413-1-9; 426 pages.
Stukes, S., Deshoretz, J., Apgar, H., and Macias, I.; *Air Force Cost Analysis Agency Software Estimating Model Analysis*; TR-9545/008–2; Contract F04701-95-D-0003, Task 008; Management Consulting & Research, Inc., Thousand Oaks, CA; 91362; 1996.
Stutzke, R.D.; *Estimating Software Intensive Systems*; Addison-Wesley, Boston, MA; 2005.
Symons, C.R.; *Software Sizing and Estimating—Mk II FPA (Function Point Analysis)*; John Wiley & Sons, Chichester; 1991; ISBN 0 471-92985-9; 200 pages.
Thayer, R.H. (editor); *Software Engineering and Project Management*; IEEE Press, Los Alamitos, CA; 1988; ISBN 0 8186-075107; 512 pages.
Umbaugh, R.E. (editor); *Handbook of IS Management*; 4th edition; Auerbach Publications, Boston, MA; 1995; ISBN 0-7913-2159-2; 703 pages.
Whitmire, S.A.; 3-D function points: Scientific and real-time extensions to function points; *Proceedings of the 1992 Pacific Northwest Software Quality Conference*, June 1, 1992.
Yourdon, E.; *Death March: The Complete Software Developer's Guide to Surviving "Mission Impossible" Projects*; Prentice Hall, Upper Saddle River, NJ; 1997; ISBN 0-13-748310-4; 218 pages.
Zells, L.; *Managing Software Projects: Selecting and Using PC-Based Project Management Systems*; QED Information Sciences, Wellesley, MA; 1990; ISBN 0-89435-275-X; 487 pages.
Zuse, H.; *Software Complexity: Measures and Methods*; Walter de Gruyter, Berlin; 1990; ISBN 3-11-012226-X; 603 pages.
Zuse, H.; *A Framework of Software Measurement*; Walter de Gruyter, Berlin; 1997.

Chapter 2

The Origin and Evolution of Function Point Metrics

On a global basis, function points are the number one metric for software benchmarks. This is because some 35 out of 37 software benchmark organizations only support function point metrics. There are probably over 100,000 benchmarks using function point metrics circa 2017 if all the major forms of function points are included: the International Function Point Users Group (IFPUG), Finnish Software Metrics Association (FISMA), Netherland Software Metrics Association (NESMA), Common Software Measurement International Consortium (COSMIC), automated, and so on.

A number of national governments now mandate function point metrics for all government software contracts in both civilian and defense sectors. The countries that mandate function point metrics include Brazil, Italy, South Korea, Japan, and Malaysia. Several other countries may join this group in 2017. Eventually, I suspect all countries will do this.

The author happened to be working at IBM during the time when function point metrics were first developed. "Although I worked at IBM San Jose in California when function points were first developed at IBM White Plains in New York State, I happened to know a few members from the function point team." In later years, Al Albrecht, the principal function point developer, became a colleague and a personal friend at my first software company, Software Productivity Research (SPR).

The origin of function point metrics is interesting but not widely known. The first portion of this chapter discusses the reasons why IBM funded the development of function point metrics. The second portion discusses the evolution and many new uses of function point metrics. The final portion discusses the possibility of expanding function point logic into other fields that need accurate measures, such as perhaps creating a new "data point" metric based on function point logic and also a "value point" metric.

The Origins of Function Point Metrics at IBM

The author was working at IBM in the 1960s and 1970s, and was able to observe the origins of several IBM technologies such as inspections, parametric estimation tools, and function point metrics. This short section discusses the origins and evolution of function point metrics.

In the 1960s and 1970s, IBM was developing new programming languages such as APL, PL/I, and PL/S. IBM executives wanted to attract customers to these new languages by showing clients higher productivity rates.

As it happens, the compilers for various languages were identical in scope and also had the same feature sets. Some older compilers were coded in assembly language, whereas newer compilers were coded in PL/S, which was a new IBM language for systems software.

When we measured the productivity of assembly language compilers versus PL/S compilers using "lines of code" (LOC) per month, we found that even though PL/S took less effort, the LOC metric of "lines of code per month" favored assembly language. This was a surprise at the time, but it should not have been. The phenomenon was nothing more than a standard law of manufacturing economics dealing with the impact of fixed costs in the face of a reduction of units produced.

This problem is easiest to see when comparing products that are almost identical but merely coded in different languages. Compilers, of course, are very similar. Other products besides compilers that are close enough in feature sets to have their productivity negatively impacted by LOC metrics are Private Branch Exchange (PBX) switches, ATM banking controls, insurance claims handling, and sorts.

To show the value of higher level programming languages, the first IBM approach was to convert high-level language code statements into "equivalent assembly language." In other words, we measured productivity against a synthetic size based on assembly language instead of against true LOC size in the actual higher level languages. This method was used by IBM from around 1968 through 1972.

What enabled IBM to do this was an earlier quantification of the phrase "language level." Up until about 1970, the phrase "high-level language" was applied to many newer languages such as PL/I and APL, but it did not have an exact definition.

The author and some other researchers at IBM counted source code statements in various languages and developed a rule for assigning mathematical language levels. The rule was that the "level" of a language would be equal to the number of statements in basic assembly language needed to create one statement in the target language. For example, COBOL was deemed a "level 3" language because on average it took three assembly statements to create the functionality of one COBOL statement.

Later when function point metrics were developed, a second rule was created to show the average number of code statements per function point. For example, a level 3 programming language such as COBOL was equal to about 106.66 code statements per function point, and a level 6 programming language such as Java was equal to about 53.33 code statements per function point.

This rule is still widely used, and in fact, tables of conversion ratios between various programming languages and function points can be acquired from software benchmark groups. Some of these tables have over 1000 programming languages.

Going back in time to the early 1970s, the conversion of code statement counts into equivalent assembly counts worked mathematically and gave accurate results, but it was not elegant and seemed a bit primitive.

An IBM vice president, Ted Climis, said that IBM was investing a lot of money into new and better programming languages. Neither he nor the clients could understand why we had to use the old basic assembly language as the metric to show productivity gains for new languages. This was counterproductive to the IBM strategy of moving customers to better programming languages. He wanted a better metric that was language independent and could be used to show the value of all IBM high-level languages.

This led to the IBM investment in function point metrics and to the creation of a function point development team under Al Albrecht at IBM White Plains. The goal was to develop a metric that quantified the features or functionality of software and was totally independent of the programming languages used for development.

Function point metrics were developed by the team at IBM White Plains by around 1975 and used internally for a number of IBM software projects. The results were accurate and successful. In 1978, IBM placed function point metrics in the public domain and announced them via a technical paper given by Al Albrecht at a joint IBM/SHARE/GUIDE conference in Monterey, California.

Table 2.1 shows the underlying reason for the IBM function point invention based on the early comparison of assembly language and PL/S for IBM compilers. It also shows two compilers with one coded in assembly language and the other coded in the PL/S language.

The table shows productivity in the following four separate flavors:

1. Actual LOC in the true languages
2. Productivity based on equivalent assembly code
3. Productivity based on "function points per month"
4. Productivity based on "work hours per function point"

Note: Table 2.1 uses simple round numbers to clarify the issues noted with LOC metrics.

The three rows highlighted in boldface type show the crux of the issue. LOC metrics tend to penalize high-level languages and make low-level languages such as assembly look better than they really are.

Function points metrics, on the other hand, show tangible benefits from higher level programming languages, and this matches the actual expenditure of effort

Table 2.1 IBM Function Point Evolution circa 1968–1975 (Results for Two IBM Compilers)

	Assembly Language	*PL/S Language*
LOC	17,500	5,000
Months of effort	**30.00**	**12.50**
Hours of effort	3,960	1,650
LOC per month	**583.33**	**400.00**
Equivalent assembly	17,500	17,500
Equivalent assembly per month	583.33	1,400.00
Function points	100.00	100.00
Function points per month	**3.33**	**8.00**
Work hours per function point	39.60	16.50

and standard economic analysis. Productivity of course is defined as "goods or services produced per unit of labor or expense."

This definition has been used by all industries for over 200 years. It has been hard to apply to software because an LOC is not a very accurate choice for goods or services.

A basic law of manufacturing economics is as follows: "If a development process has a high percentage of fixed costs and there is a decline in the number of units produced, the cost per unit will go up."

If an LOC is selected as the manufacturing unit, and there is a change to a higher level language, then the number of units produced will decline. The noncoding work of requirements, design, etc. act like fixed costs and drive up the cost per LOC. This is a basic law of manufacturing economics.

The creation and evolution of function point metrics was based on a need to show IBM clients the value of IBM's emerging family of high-level programming languages such as PL/I and APL.

This is still a valuable use of function points because there are more than 3000 programming languages in 2017, and new languages are being created at a rate of more than 1 per month.

Another advantage of function point metrics vis-à-vis LOC metrics is that function points can measure the productivity of noncoding tasks such as creation of requirements and design documents. In fact, function points can measure all software activities, whereas LOC can only measure coding.

Up until the explosion of higher level programming languages in the late 1960s and 1970s, assembly language was the only language used for systems software.

(The author programmed in assembly language for several years when starting out as a young programmer.)

With only one programming language in common use, LOC metrics worked reasonably well. It was only when higher level programming languages appeared that the LOC problems became apparent. Even then the problems were not easy to identify. It just happened that IBM was studying compilers, which are functionally almost identical, and that made it easy to identify the problems caused by the LOC metric. If you were looking at unlike applications such as a smartphone application and a defense application of notably different types and sizes, the problem would be difficult to notice.

It was soon realized by the IBM research team that the essential problem with the LOC metric is really nothing more than a basic issue of manufacturing economics that had been understood by other industries for more than 200 years, as pointed out previously.

The software noncoding work of requirements, design, and documentation act like fixed costs. When there is a move from a low-level language such as assembly language to a higher level language such as PL/S or APL, the cost per unit will go up, assuming that LOC is the "unit" selected for measuring the product. This is because of the fixed costs of the noncoded work and the reduction of code units for higher level programming languages.

Function point metrics are not based on code at all, but are an abstract metric that defines the essence of the features that the software provides to users. This means that applications with the same feature sets will be the same size in terms of function points no matter what languages they are coded in. Productivity and quality can go up and down, of course, but they change in response to team skills, reusable materials, and methodologies used.

Once the function points were released to the public by IBM in 1978, other companies began to use them, and soon the IFPUG was formed in Canada.

In 2017, there are hundreds of thousands of function point users and quite a few thousands of benchmarks based on function points. There are also several other varieties of function points such as COSMIC, FISMA, NESMA, unadjusted, function points light, and backfired function points.

Overall, function points have proven to be a successful metric and are now widely used for productivity studies, quality studies, and economic analysis of software trends. Function point metrics are supported by parametric estimation tools and also by benchmark studies. There are several flavors of automatic function point tools and function point associations in most industrialized countries. There are also International Organization for Standardization (ISO) standards for functional size measurement.

(There was never an ISO standard for code counting, and counting methods vary widely from company to company and project to project. In a benchmark study performed for an LOC shop, we found four sets of counting rules for LOC in their software organization that varied by more than 500%.)

Table 2.2 shows the countries with increasing function point usage circa 2016, and it also shows the countries where function point metrics are now required for government software projects.

Several other countries will probably also mandate function points for government software contracts by 2017. Poland may be next because their government is discussing function points for contracts. Eventually, most countries will do this.

In retrospect, function point metrics have proven to be a powerful tool for software economic and quality analysis. As of 2017, function point metrics are the best and the most accurate metrics for software development yet created.

New and Old Function Point Business Models

Function point metrics are the most accurate and effective metrics yet developed for performing software economic studies, quality studies, and value analysis. But normal function point analysis is slow and expensive. Function point analysis performed by a certified function point consultant proceeds at a rate between 400 and 600 function points per day.

The cost per function point counted is around $6. Further, very small applications below about 15 function points in size cannot be counted with many function point methods. Function point analysis for applications larger than about 15,000 function points in size almost never occurs because the costs and schedule are larger than that most companies will fund.

In 2017 several forms of high-speed, low-cost function point analysis are either readily available or are under development. This report discusses the business value of high-speed, low-cost function point analysis. The goal of this function point analysis is to expand the usage of function points to 100% of software applications, corporate portfolios, and backlogs. In addition, function point analysis will also allow improved risk and value studies of both new applications and aging legacy applications.

Function point metrics were developed within IBM and put into the public domain in October 1978. The nonprofit IFPUG assumed responsibility for counting rules and function point definitions in 1984. In 2017, IFPUG has grown to more than 3000 members and has affiliates in 24 countries. Other function point groups such as COSMIC, FISMA, and NESMA also have expanding memberships.

In addition to standard IFPUG function points, no less than 24 function point variations have been developed, including backfired function points, COSMIC function points, Finnish function points, engineering function points, feature points, Netherlands function points (NESMA), unadjusted function points, function points light, and many others. There are some conversion rules between these variations and IFPUG function points, but full bidirectional conversion is still in the future. In the United States, IFPUG function points remain the primary

Table 2.2 Countries Expanding Use of Function Points 2017

1	Argentina	
2	Australia	
3	Belgium	
4	Brazil	Required for government contracts, 2008
5	Canada	
6	China	
7	Finland	
8	France	
9	Germany	
10	India	
11	Italy	Required for government contracts, 2012
12	Japan	Required for government contracts, 2014
13	Malaysia	Required for government contracts, 2015
14	Mexico	
15	Norway	
16	Peru	
17	Poland	
18	Singapore	
19	South Korea	Required for government contracts, 2014
20	Spain	
21	Switzerland	
22	Taiwan	
23	The Netherlands	
24	United Kingdom	
25	United States	

version, and about 85% of all projects are sized using the IFPUG model. This report is based on IFPUG function point metrics.

There are newer metrics such as story points and use case points that are expanding rapidly. These are "functional" in nature, but differ from function points in many respects. These newer metrics have no international standards from ISO or Object Management Group (OMG) and vary so widely from group to group that it is almost impossible to create accurate benchmarks.

The main uses of function point metrics include the following:

1. Baselines to measure rates of improvement in productivity, schedules, or costs
2. Benchmarks for software productivity
3. Benchmarks for software quality
4. Estimating new applications before development
5. Defining the terms of outsource agreements
6. Measuring the rates of requirements change

Function point metrics are far more useful than the older LOC metrics. For measuring quality, function points are also more useful than "cost per defect" metrics. Function points can measure noncoding activities such as project management and design. Also, function point metrics stay constant regardless of which programming language or languages are used. Since the industry has more than 3000 programming languages circa 2017, and almost every application uses multiple languages, the consistency of function point metrics allows economic studies that are not possible using any other metric.

The Costs and Limitations of Standard Function Point Metrics

Normal function point analysis is carried out by trained professionals who have passed a certification examination. There are about 750 function point counters in the United States, and this number increases at perhaps 50–75/year. Some of the other function point variations such as COSMIC and the Netherlands (NESMA) and the Finnish (FISMA) methods also have certification examinations.

The ISO has certified four function point metrics as being suitable for economic studies: IFPUG, COSMIC, NESMA, and the FiSMA function point metric. The measured accuracy of counts by certified function point counters is usually within about 5%.

The standard methods for counting function points count five major elements: inputs, outputs, interfaces, logical files, and inquiries. There are also adjustments and weighting factors for a few kinds of complexities. In practice, the weighting factors and adjustments cause a lower limit for function point analysis: applications smaller than about 15 function points cannot be counted.

There are also a number of small changes to software that do not affect function point totals. These are called "churn" as opposed to "creep." An example of churn would be shifting the location of a data item on a screen, but not adding or subtracting from the data itself. These cannot be counted using standard function point analysis but they obviously require effort.

To use standard function point analysis, it is necessary to have at least a fairly complete set of requirements for the software applications being counted. Additional materials such as functional specifications add value and rigor to function point analysis. This means that normal function point analysis cannot occur until somewhere between 1 month and 6 months after a software application is started. This starting point is too late for successful risk avoidance. It is also later than the usual need for an initial software cost and schedule plan.

Normal function point analysis by a trained counter proceeds at a rate of between 400 and 600 function points per day. At normal consulting fee rates, this means that the cost for function point analysis runs between $4 and $6 for every function point counted.

This is a rather significant cost that is so high that it has slowed down the use of function point analysis for applications larger than about 15,000 function points. For example, to count a really massive application such as an Enterprise Resource Planning (ERP) package at perhaps 300,000 function points in size, the number of days required for the count might total to 750 days of consulting time. The cost might total to $1,800,000. No company is willing to spend that much time and money for function point analysis.

The combined results of the lower limits of function point analysis and the high costs of function point analysis means that less than 10% of software applications have ever been counted, or are likely to be counted using normal manual counts.

The total effort corporations devote to small projects below 15 function points in size is close to 10% of their entire workload. This is because bug repairs and small changes and enhancements are almost always smaller than 15 function points in size.

Small changes are extremely common, and major corporations such as IBM, Microsoft, Electronic Data Systems (EDS) may carry out more than 30,000 of these each year. While individual changes may range from only about 1/50th of a function point up to 15 function points, the total volume of such changes can easily top 100,000 function points per calendar year.

For large applications at the upper end of the size spectrum, about 40% of a corporation's software workload is spent on applications that are 15,000 function points or larger in size. Although massive applications are few in number, they usually have development teams that run from 500 to more than 5000 personnel. These applications are also hazardous and prone to failure or enormous cost and schedule delays.

Large applications would actually be the top candidates for the power of function point analysis, because these applications are prone to failure, delays, and cost overruns. The use of function point metrics for early risk analysis and avoidance would be extremely valuable. Unfortunately, the high costs and high human effort of function point analysis have prevented function points from being used on the very applications that need them the most.

Mid-range applications between 15 and 15,000 function points in size comprise about 35% of a corporations' work load and technical staff. The average size of applications that are counted using standard function point analysis is about 1,500 function points. However, because of the high costs involved, only the more important applications typically use function point analysis.

If a corporation produces 100 applications per year in the size range between about 100 and 15,000 function points, probably only about 25% would actually be counted. Usually the counts would occur for a specific business purpose such as a formal benchmark study or a baseline study at the beginning of a process improvement program. Other than special studies, function point analysis is usually not performed because the costs and time required preclude day-to-day usage on ordinary projects.

Assume that a corporation is interested in comparing the results of a 10,000-function point applications against benchmark data published by the International Software Benchmark Standards Group (ISBSG). In order to do this, they commission a certified function point counter to perform function point analysis. If you assume a daily consulting fee of $3000 and a counting speed of 500 function points per day, the function point analysis will take 20 days, the consulting costs will be $60,000, and the cost per function point counted would be $6. Function point analysis is useful but was expensive in 2014.

In addition, all large organizations use commercial-off-the-shelf (COTS) applications such as Microsoft Vista, SAP, Oracle, Symantec antivirus, AutoCad, and hundreds of others. A study performed by the author and his colleagues found that a large manufacturing corporation owned more COTS packages than applications developed in-house.

It is suspected that the total size of these COTS packages in terms of function points may be much larger than the portfolio of applications developed by the companies themselves. (This is certainly true for small- and mid-sized corporations.)

These COTS packages are never counted using standard function point analysis because the vendors don't want them to be. Vendors provide no size data themselves, nor do they provide the raw materials for a normal function point analysis. About 15% of corporate software work is involved with maintenance and support of COTS packages.

Between small applications that cannot be counted, very large applications that are too expensive to count, mid-range applications that are not selected for function point analysis, and COTS packages that lack the available information for a count, less than 10% of the software owned by a major corporation or a large government organization is likely to have function point counts available. Given the power and usefulness of function point metrics for economic studies and quality analysis, this is an unfortunate situation.

To summarize:

1. Projects <15 function points cannot be counted using standard function points.
2. Projects >15,000 function points are seldom counted due to high costs.
3. Projects between 15 and 15,000 are counted less than 25% of the time.
4. COTS packages are never counted due to vendor reluctance.

Assume that a corporation owns a portfolio with a total size of 10,000,000 function points. Of these 5,000,000 function points represent 3,000 in-house applications and the other 5,000,000 represent some 2,500 COTS packages such as SAP, Oracle, AutoCAD, SAS, Microsoft applications, and other commercial packages.

As of 2017, it is likely that the corporation will have used function point analysis only on about 75 applications that total to perhaps 100,000 function points. This is a very small percentage of the total corporate portfolio. The total cost for performing standard function point analysis on the 75 applications would probably be about $600,000. This explains why the usage of function points is quite limited.

To perform standard function point analysis on the in-house portfolio of 5,000,000 function points would cost as much as $30,000,000. No company would spend such a large amount for ordinary function point analysis. To perform a function point analysis on the COTS portion of the portfolio was impossible in 2014 using normal function point counting. Thus, a major source of corporate expense is outside the current scope of standard function point analysis.

The bottom line is that standard function point analysis is too limited, too costly, and too slow to be used with 100% of the software developed and used by large organizations. As a result, the advantages of function point metrics for economic and quality studies were only partially available as of 2014.

Expanding the Role and Advancing the Start Time of Function Point Analysis

There is no question that function points provide the most effective metric for software applications that have ever been developed. There are no other metrics that perform as well as function points for economic studies, quality analysis, and value analysis. But to achieve optimum returns, it is necessary to modify function point practices in a number of important ways:

1. The starting point for function point analysis needs to be earlier by 6 months.
2. The cost of function point analysis needs to drop below 1¢ per function point.
3. The counting speed needs to be higher than 100,000 function points per day.
4. There should be no lower limit on application size, down to zero function points.
5. ERP packages above 250,000 function points should be counted.

6. COTS and open-source packages should be counted.
7. Smartphone, cloud, and tablet applications should be counted.
8. Defense and military software should be counted.
9. Canceled projects, if they occur, should be counted.
10. Changing requirements should be counted in real time as they occur.
11. Post-release feature growth should be counted for at least 5 years.
12. Deleted features should be counted in real time as they occur.
13. Deferred features should be counted as they occur.
14. Legacy applications should be counted, if not previously done so.
15. Updates to legacy applications should be counted before they are started.
16. Function point sizing should be possible as a service without on-site visits.
17. Function point size, effort, and quality need recalibration every fiscal year.
18. Both functional and nonfunctional requirements should be included.
19. Individual features as well as total packages should be counted.
20. Both reused code and custom code should be counted, and aggregated.

Fortunately, the technology for achieving these changes is starting to become available as of 2017, although some methods are still experimental and are not yet commercially available.

IFPUG developed a metric for counting nonfunctional requirements in 2012. This is called the SNAP metric and the results are SNAP points. Unfortunately, SNAP points are not equivalent and not additive to standard function points.

CAST Software has a high-speed function point tool that operates on legacy applications and produces function points via extraction and analysis of business rules from within the source code.

More recently, CAST has also developed an automated enhancement point (AEP) tool that solves chronic problems with measuring software enhancements. AEP might also be a surrogate for SNAP points for nonfunctional requirements.

TotalMetrics in Australia also has a high-speed function point method based in part on standard function points with some of the topics being streamlined via multiple-choice questions.

IntegraNova from Spain includes high-speed function point analysis as a standard feature of its application generator and requirements modeling tool. This tool shows the function point total for every application that it generates.

Relativity Technologies has announced a tool for sizing legacy applications called "Function Point Analyzer" or FPA. This tool operates by parsing source code in selected languages and extracting business rules that can then be analyzed using normal function point analysis. The assertion is that this new FPA tool provides counts with an accuracy equivalent to normal function point analysis, but is able to accomplish this in a matter of minutes rather than a matter of days or weeks. For legacy applications in languages that are supported, this tool provides an effective high-speed, low-cost approach for function point analysis.

SPR LLC has offered several methods for function point approximation as features of their commercial estimating tools, SPQR/20, Checkpoint and the more recent KnowledgePlan. (Unfortunately, SPR went out of business in 2015.)

An earlier tool developed by SPR generated function point counts automatically during the design process when using the Bachman Analyst design work bench. The closure of the Bachman company prevented widespread deployment of this tool, but it could be redeveloped for use with other methods such as use cases and the Rational Unified Process. Function points were automatically generated from the normal design process without requiring a separate function point analysis consultant.

William Roetzheim has developed a method called Fast Function points that is used in the CostXpert estimation tool. It speeds up IFPUG counting but generates results that are stated to be within a few percentage points of IFPUG counts.

Backfiring or mathematical conversion from logical code statements to function point metrics needs to be studied and analyzed for accuracy.

There is a "fast and easy" function point counting method used in Italy that simplifies counting rules. The comparison of results to IFPUG or COSMIC are stated to be close.

Namcook Analytics LLC and the author of this book have developed a method for high-speed function point predictions based on pattern matching. This method shifts the paradigm from traditional function point analysis to using historical data from applications already counted.

The author's method is embedded in an estimating tool called "Software Risk Master" (SRM) that is under development for a new release with expanded features as this report is being written. The sizing function can be tested on the website www.Namcook.com by requesting a password from within the site.

The author's sizing method takes only between 2 and 7 min to size applications of any size. Standard outputs include IFPUG function points and logical code statements. Optional outputs include use case points, story points, COSMIC function points, and a number of other alternate metrics. Currently, SRM produces size in a total of 23 metrics, but others such as data envelopment analysis (DEA) and the new CAST AEP might be included later in 2017.

As of 2017, more than 50,000 software applications have been counted using normal IFPUG function point analysis. Many of these application counts are commercially available from the ISBSG. Many others are available from software consulting groups such as SPR, Gartner Group, David Consulting Group, Quality and Process Management Group (QPMG), and few others.

As a result of the increasing volumes of measured projects, it is possible to go about function point analysis in a new way. If a proposed software project is mapped onto a standard taxonomy that includes the project's nature, class, type, scope, and several forms of complexity, it is possible to use the size of historical applications already counted as a surrogate for standard function point analysis. Applications that are similar in terms of taxonomy are also similar in terms of function point size. This method of applying pattern matching as used by the author rather than normal function point analysis has several advantages, as listed.

1. Pattern matching shifts the point in time at least 6 months earlier than function point analysis
2. It lowers the cost well below 1¢ per function point
3. It can be carried out at speeds above 100,000 function points per hour

Using algorithms and mathematical techniques, the pattern matching approach can also offer additional advantages compared to normal function point analysis.

4. It can be used for "micro function points" below the normal boundary of 15
5. It can be used to size specific features of applications such as security features
6. It can be used to size reusable components and object class libraries
7. It can be used for COTS packages as well as in-house development
8. It has no upper limit and can be used for applications >300,000 function points
9. It can be used as a front-end for commercial software estimation tools
10. It can be used for both risk and value analysis before applications are started
11. It can be done remotely as a service
12. It can deal with requirements changes during development and afterward

The availability of high-speed, low-cost function point sizing methods brings with it a need to examine the business model of function point analysis and develop a new business model using assumptions of much reduced costs and much greater speed than normal function point analysis.

The Current Business Model of Function Point Analysis in the United States

Since 1978 when IBM first released the original function point method through 2017, function point analysis has been a boutique industry. Of the approximate number of 1000 function point analysts in the United States, about 600 work for large corporations or government agencies and the other 400 are consultants. Many function point analysts who count function point metrics have been certified by IFPUG or one of the other certification groups such as COSMIC.

Assuming each counter completes 25 function point counts per year, about 25,000 projects per year are counted in the United States. However, the majority of these are for internal and proprietary uses. Only about 200 projects per year are added to public data sources such as the ISBSG database.

Assuming an average size of about 1,500 function points for the applications counted, about 3,750,000 function points per year are counted in the United

States. At a typical cost of $6per function point counted, the annual cost for U.S. function point analysis would be about $22,500,000. Most function point analysis is performed on the client sites by certified counting personnel.

Function points are normally used for benchmark and baseline studies. Assuming that about 500 of these studies are performed per year in the United States at an average cost of $50,000 per study, then about $25,000,000 per year is the revenue derived from benchmark and baseline studies.

Most of the benchmark data is proprietary and confidential, but some of the data is submitted to the ISBSG. As of 2012, ISBSG had perhaps 5,000 total projects from around the world, and is adding new data at a rate of perhaps 500 projects per year.

It is interesting to note that starting in 2008, the government of Brazil required that all software contracts be based on reliable metrics; this has led Brazil to become the major country for counting function points in the world. South Korea and Italy have also implemented as have Japan and Malaysia. No doubt other countries are watching to see how successful these governments will be. Eventually most or even all governments will mandate function points for government software contracts.

Function points are also used for other business purposes. For example, most commercial software cost-estimating tools use function points as a basic sizing method. Function points are also used in litigation, such as breach of contract lawsuits or taxes involving the costs and value of software applications.

Many commercial software cost-estimating tools use function points as a sizing method. Unfortunately, function points are usually not available until after the first and the most critical estimates need to be prepared. As a result, the software cost-estimation market is also a boutique market. The total number of copies of the major commercial software estimation tools (COCOMO, ExcelerPlan, KnowledgePlan, PriceS, SLIM, SRM and SEER) probably amounts to no more than 7,500 copies in 2017, with COCOMO having about a third because it is free.

There are also some specialized tools that support function point counts for individual applications and also portfolios. Here too, penetration of the market is small in 2017.

In 2012, IFPUG announced a new variation on function points for counting nonfunctional requirements. This new metric is called SNAP, which is an awkward acronym for "software nonfunctional assessment process."

The basic idea of SNAP is that software requirements have two flavors: (1) Functional requirements needed by users; (2) Nonfunctional requirements due to laws, mandates, or physical factors such as storage limits or performance criteria.

The SNAP committee view is that these nonfunctional requirements should be sized, estimated, and measured separately from function point metrics. Thus, SNAP and function point metrics are not additive, although they could have been.

Having two separate metrics for economic studies is awkward at best and inconsistent with other industries. For that matter, it seems inconsistent with standard economic analysis in every industry. Almost every industry has a single normalizing metric such as "cost per square foot" for home construction or "cost per gallon" for gasoline and diesel oil.

As of 2017, none of the parametric estimation tools have fully integrated SNAP and it is likely that they won't have it since the costs of adding SNAP are painfully expensive. As a rule of thumb, nonfunctional requirements are about equal to 15% of functional requirements, although the range is very wide.

The author's SRM tool predicts SNAP points, but assumes that nonfunctional requirements are a secondary topic and would not even exist without the required user functionality. Therefore, instead of showing SNAP costs separately, we add these costs to those of creating user functions.

If an application of 1,000 function points and 150 SNAP points spent $1,000,000 building user functions a $200,000 building nonfunctional features the total cost would be $1,200,000.

We say the total cost is $1,200 per function point on the grounds that the nonfunctional effort is a necessary and secondary attribute for building the application and the costs should be additive. This method allows us to compare recent data using SNAP to older projects sized and measured before SNAP was introduced in 2012.

When the total market for function point analysis and benchmark studies is summed, it is probably less than $50,000,000/year. That indicates a fairly small boutique industry with a number of small consulting groups plus individual certified function point counters. To summarize:

1. As of 2017, standard function point analysis is almost never used for applications >15,000 function points in size.
2. Standard IFPUG and SNAP function point analysis cannot be used for applications <15 function points in size due to limits in the adjustment factors, as of 2017. Even if function points are used for small applications, manual counting would be expensive if small changes exceeded 10,000/year, which is not uncommon.

   As of 2017, the high costs and low speed of standard function point analysis has been a barrier to widespread adoption.

The European business model is similar to the U.S. business model in overall size, with perhaps $50,000,000/year in revenues for function point analysis and benchmark studies. However, the European market is very fragmented and difficult to characterize because of the many function point variations in use in Europe.

Although IFPUG function points are number one in usage throughout most of Europe, there are a number of alternate function point metrics that are also utilized: the older British Mark II function point metric, the newer COSMIC function point metrics, and several national variations such as Finnish function points (FISMA) and NESMA. The use of function points in Russia is not yet as widespread as in Western Europe.

The Central and South American markets for function points are growing very rapidly. There are active function point users groups and hundreds of certified counters in Argentina, Brazil, Colombia, Costa Rica, Mexico, Paraguay, Uruguay, and other regional countries. Brazil has already hosted an annual conference for the IFPUG, and a past president for IFPUG is from Brazil.

The Asian business model is also similar, and currently seems to be slightly larger than either the U.S. or European markets when India, Japan, South Korea, China, and Singapore are aggregated into a single view.

Function point revenues are perhaps in the range of $75,000,000/year. The rate of growth of the Asian market is accelerating due to the very quick expansion of software companies and outsourcing business in both India and China.

Since local costs in both countries are below those of both Europe and the U.S., these two countries have favorable costs per function point. India also has very favorable quality data when measuring using "defects per function point" and "defect removal efficiency." However, inflation is very high in Asia and so cost differentials may change over time.

Other parts of Asia such as North Korea, Myanmar, Laos, and Cambodia no doubt use function points primarily for defense and weapons systems, although banks and telephone systems have computers also. There is no data on the function point business in these places.

The usage of function points in the Middle East is difficult to quantify. Israel has perhaps the largest number of certified function point counters, but it is hard to ascertain function point usage for that entire region.

At this point it is interesting to note what kind of a business model will emerge when the cost of function point analysis drops down to about 1¢ per function point counted, the starting point shifts 6 months earlier, and there are no longer either upper or lower size limits on function point analysis.

A New Business Model for Function Point Analysis

Once reliable high-speed, low-cost function point methods become widely available, the opportunities for function point consulting work should expand from the present day's small boutique industry into a mainstream industry with annual revenues that approach or exceed $2,600,000,000/year.

There are at least 50 business opportunities that will occur once function point costs drop down and speed goes up. It is also important that function points can be used before projects start rather than at the end of requirements. Many of these opportunities are new and do not exist today.

As the new business opportunities become visible, it can be expected that many of today's small function point companies will grow to substantial sizes. It can also be expected that major software companies such as Microsoft, Amazon, Apple,

EDS, Computer Associates, and IBM will move into the function point market in various ways, including acquisition of existing companies.

The expansion of IFPUG function point metrics shows promise of finally providing the software industry with solid, quantified data. Ambiguous and uncertain topics such as development schedules, development costs, maintenance costs, quality, reliability, process improvement results, process improvement costs, security, software value, and software renovation will become predictable and measurable. Entirely new topics such as the impact of software tools to knowledge workers will also be possible.

It is hypothesized by the author that every 50% reduction in the cost of counting function points will lead to a 100% expansion in function point usage. Currently, circa 2017, the cost of counting function points is about $6 per function point and the usage is only about 4,000 projects per year. For widespread or universal deployment of function point metrics, the costs will have to drop down to only a few cents per function point counted. Below 1 cent would be even better.

The goal of the new business model for function point metrics is to expand the use of function point metrics from perhaps 10% of mid-range software projects up to close to 100% of all software projects in the United States: new development, maintenance, enhancements, and creation and use of reusable components.

Even canceled projects and disasters can be measured once the cost of function point counting drops significantly. A related goal of the new business model is to allow function point analysis for software projects whose size ranges from less than 1 function point to more than 500,000 function points, as opposed to the current range of between about 15 function points and 15,000 function points.

Table 2.3 provides an overview of these new and evolving opportunities.

Following are short discussions of the 50 business opportunities highlighted in Table 2.3. These opportunities will begin to emerge in 2018 and should be fully available by 2027.

It can be expected that the number of companies involved with function point metrics will increase, and that new companies will probably be created as well. Large companies such as Accenture, KPMG, IBM, Microsoft, and others will also enter the expanding market for function point data.

As of 2017, there are perhaps 50 U.S. companies in the function point business arena. As function points come down in price, it can be expected that additional companies will enter the expanding market.

Not every opportunity will occur at the same time. Some opportunities may be much smaller than the predicted results in Table 2.6. But the bottom line is that once the emphasis of function point metrics switches from the mechanics of counting rules to the economic uses of function points, a major expansion will occur.

Opportunity 1. Early Risk/Value Analysis: Because both software risks and value are directly proportional to the sizes of applications measured in function points, the ability to perform early sizing prior to requirements will lead to a major new business opportunity. Every application larger than 1,500 function points should

Table 2.3 Business Opportunities from High-Speed, Low-Cost Function Point Analysis

	Activity	*Number per year*	*Days per Study*	*Cost per Study*	*Annual Revenues*
1	Early risk/value analyses	25,000	3	$15,000	$375,000,000
2	Real-time requirements	75,000	1	$4,000	$300,000,000
3	Competitive analyses	2,500	10	$50,000	$125,000,000
4	Quality analyses	10,000	3	$12,000	$120,000,000
5	Software usage studies	5,000	5	$20,000	$100,000,000
6	Activity analyses	5,000	5	$20,000	$100,000,000
7	Legacy renovations	7,500	3	$12,000	$90,000,000
8	Benchmark analyses	2,000	10	$40,000	$80,000,000
9	Baseline analyses	2,000	10	$40,000	$80,000,000
10	Feature/class analyses	5,000	3	$12,000	$60,000,000
11	Complete portfolios	750	20	$80,000	$60,000,000
12	Micro function points	3,000	5	$20,000	$60,000,000
13	Development methods	2,500	5	$20,000	$50,000,000
14	ITIL analyses	2,500	5	$20,000	$50,000,000
15	Development tool studies	2,500	5	$20,000	$50,000,000
16	Maintenance tool studies	2,500	5	$20,000	$50,000,000
17	Mass update analyses	5,000	2	$10,000	$50,000,000
18	Backlog analyses	3,000	4	$16,000	$48,000,000
19	Normal function points	3,500	4	$12,000	$42,000,000
20	COTS risk analyses	5,000	2	$8,000	$40,000,000
21	ERP deployment analyses	1,000	10	$40,000	$40,000,000
22	Reengineering analyses	5,000	2	$8,000	$40,000,000
23	Customer support	5,000	2	$8,000	$40,000,000
24	SOA/reuse analyses	2,000	5	$20,000	$40,000,000

(*Continued*)

Table 2.3 (*Continued*) Business Opportunities from High-Speed, Low-Cost Function Point Analysis

	Activity	*Number per year*	*Days per Study*	*Cost per Study*	*Annual Revenues*
25	Canceled projects	1,500	5	$20,000	$30,000,000
26	Agile analyses	2,500	3	$12,000	$30,000,000
27	CMMI analyses	1,500	5	$20,000	$30,000,000
28	Security risk analyses	2,000	3	$15,000	$30,000,000
29	Outsource contracts	2,500	3	$12,000	$30,000,000
30	Litigation (contracts)	250	35	$100,000	$25,000,000
31	Mergers/acquisitions	500	10	$50,000	$25,000,000
32	Occupation groups	2,000	3	$12,000	$24,000,000
33	Test tool analyses	2,000	3	$12,000	$24,000,000
34	Embedded analyses	2,000	3	$12,000	$24,000,000
35	Website analyses	3,000	2	$8,000	$24,000,000
36	Balanced scorecards	2,000	3	$12,000	$24,000,000
37	Metrics conversion	5,000	1	$4,000	$20,000,000
38	Defense analyses	1,000	5	$20,000	$20,000,000
39	Supply chain analyses	1,000	5	$20,000	$20,000,000
40	International analyses	100	30	$150,000	$15,000,000
41	Marketing analyses	300	10	$50,000	$15,000,000
42	Business processes	300	10	$50,000	$15,000,000
43	Litigation (tax cases)	100	35	$140,000	$14,000,000
44	Venture capital analyses	500	5	$25,000	$12,500,000
45	Programming languages	1,000	3	$12,000	$12,000,000
46	Earned value analyses	1,000	3	$12,000	$12,000,000
47	Infrastructure analyses	1,000	3	$12,000	$12,000,000
48	Educational uses	2,500	1	$4,000	$10,000,000
49	Delivery channels	200	10	$50,000	$10,000,000
50	National studies	50	40	$200,000	$10,000,000
	Total	218,050			$2,607,500,000

have a formal risk/value analysis before any funding occurs. The current consulting companies that perform function point analysis should evolve into more sophisticated risk and value consultancies.

Risk and value analysis consulting is needed because there is a widespread lack of understanding and training for both the risk and value topics. If the software industry knew how to deal with risks, there would not be so many outright failures and so many overruns.

As of 2017, almost one-third of the software personnel in the United States are working on projects that will be canceled due to excessive cost and schedule overruns. The major risks analyzed will be outright cancellations, schedule delays, cost overruns, security vulnerabilities, poor customer satisfaction, poor quality, poor test coverage, and the odds of litigation occurring.

The major value topics analyzed will be both financial and intangible values such as customer satisfaction and corporate prestige. Applications >10,000 function points in size fail more often than they succeed, so better risk and value analysis are urgently needed by the software industry. The cost for a basic risk/value analysis would be about $15,000 or 3 days of time by an experienced risk/value consultant. The average cost for a canceled project of 10,000 function points is over $37,000,000. Spending $15,000 up front in order to eliminate a $37,000,000 failure is a very good return on investment. This business opportunity could involve 50 consulting companies, 1,000 trained risk consultants, and generate annual revenues of perhaps $375,000,000/year. Although some risk analysis can be done as a service or with tools, the U.S. needs more live consultants and project managers who actually know how to navigate large projects >10,000 function points to a successful conclusion.

Opportunity 2. Real-Time Requirements Estimates: A major source of litigation and also a major source of schedule delays and cost overruns has been that of rapidly changing requirements that are not properly integrated into project cost and schedule estimates. Now that function point analysis can be performed in minutes rather than days, every requirements change can be analyzed immediately and its impact integrated into cost and schedule estimates.

Both commercial software estimating tools and consulting groups need to be ready to deal with the fact that requirements changes occur at rates of between 1% and 3% per calendar month. Requirements changes on a nominal 10,000-function point application can add another 2,000 function points during development. If these additions are not properly planned for, failure or major overruns are likely to occur. This opportunity can be done remotely as a service, or on-site using quick sizing tools.

Opportunity 3. Competitive Analyses: This is a new business opportunity that is not part of current function point analysis. Using the quick sizing method based on pattern matching, it will be possible to analyze the software portfolios of direct competitors in the same industry. Thus, it will be possible to quantify the software portfolios of every major company in industries such as insurance, banking, pharmaceuticals, oil and energy, and manufacturing.

The companies with the largest and most sophisticated portfolios are likely to have a competitive edge. These studies can also be stepped up to measure the defense capabilities of national governments and even the specific capabilities of aircraft, missiles, and weapons systems. The most probable form of competitive analysis will be studies commissioned by a single company of major competitors in the same industry.

Opportunity 4. Quality analyses: The weakest technology in all of software is that of quality control. Function point metrics coupled with measures of defect removal efficiency levels provide the most effective quality control methods yet developed. Every software project and every requirements change and enhancement should use early function point sizing to predict defect potentials. Since the total volumes of defects range between about 4 and 7 defects per function point and only about 85% are removed, this information can be used to plan successful defect prevention and defect removal strategies.

The U.S. average for defect removal efficiency in 2017 is only 92.5%, which is a professional embarrassment. For every $1 spent on software development, more than $0.40 goes to fixing bugs that should not even have been there. During the maintenance period, perhaps $0.35 goes to bug repairs for every $1 spent. Some large applications never stabilize and have bug reports their entire lives of 20 years or more.

Better quality control is a major business opportunity that will benefit the software industry and the United States economy as a whole. Improved quality control will lower the odds of cancellations, cost and schedule overruns, and also breach of contract litigation. Improved quality will also expand market penetration and improve market shares. It would be very beneficial if the major quality associations such as the American Society of Quality endorsed function points. It would also be useful for testing companies, Six Sigma consultants, inspection instructors, and quality educators to begin to express quality results in terms of function points for defect potentials and also to measure defect removal efficiency levels.

Unfortunately, the economics of quality has been blinded for many years by two of the worst business metrics in history: LOC and cost per defect. The LOC metric is useless for measuring defects in requirements, design, and noncode sources that are often greater in volume than coding defects. It also penalizes high-level and object-oriented programming languages. The cost per defect metric penalizes quality and achieves its lowest values for the buggiest applications. Both these metrics are bad enough to be viewed as professional malpractices. Examples of the failure modes of LOC and cost per defect are given later in this report.

Opportunity 5. Software Usage Studies: This is a new business opportunity that is not part of current function point analysis. Function point metrics can be used to study the use of software as well as its development. For example, a well-equipped project manager will need about 5000 function points of estimating and scheduling tools. A well-equipped test engineer will need about 15,000 function points of test library control and test coverage tools. There are significant correlations between efficiency, effectiveness, and the volume of tools available in every form of knowledge work.

Now that tools and COTS packages can be sized in terms of function points, it will be possible to carry out studies that relate tool availability to performance. Not only software personnel but also other knowledge workers use software. Lawyers have Nexis/Lexis available and a number of specialized tools amounting to perhaps 50,000 function points. Physicians on staff in well-equipped hospitals with magnetic resonance imaging (MRI) and other diagnostic instruments are now supported by more than 5,000,000 function points. A modern combat pilot has more than 25,000 function points of onboard navigation and weapons control applications. New kinds of economic studies can now be performed that have never been done before. This is a new opportunity not only for the function point community, but for the entire software industry.

Opportunity 6. Activity Analyses: To date there is little reliable data available on the performance and schedules of individual activities that are part of software development. How long does it take to gather requirements? How much does it cost to do design? How long will coding take? How much does it cost to translate a user's guide into another language? What are the costs and effectiveness of software quality assurance activities? What are the costs and effectiveness of database administrators? Questions such as these can be explored once function point analysis becomes a cheap and easy technique.

Every software activity can be studied using function point assignment scopes and function point production rates. Studies at the activity level were not possible using older metrics such as LOC. Although such studies are possible using standard function point analysis, they were seldom performed. The ISBSG has started to perform some phase-level studies, but with more than 25 development activities, 30 maintenance activities, and hundreds of tasks, much more detailed studies are needed.

Opportunity 7. Legacy Renovation Analyses: Software does not age gracefully. Software life expectancies are directly proportional to application size measured in function point metrics. Some large applications are now more than 25 years old. As a result, there are dozens of large and decaying applications in the >10,000-func tion point size range. Most of these do not have function point counts, so it is hard to do studies that quantify defects, rates of change, or cost of ownership. Now that legacy applications can be sized quickly and cheaply, it is possible to carry out economic studies of various forms of geriatric care and renovation. Among these are data mining to extract business rules, surgical removal of error-prone modules, restructuring, refactoring, clarifying comments, and eventual replacement. The costs and value of these geriatric approaches can now be measured using function points, which was not feasible using manual function point analysis. Not only were the legacy applications large, but many of them had obsolete or missing specifications so there was no documentation available for standard function point analysis.

Opportunity 8. Benchmark Analyses: Benchmarks, or comparisons of software applications to similar applications produced elsewhere, have been one of the main uses of function point metrics since 1978. However, due in part to the high costs of function point analysis, only about 200 projects each year get submitted to various benchmarking organizations such as ISBSG, SPR, and others. Now that function point analysis can be done quickly and cheaply, it is anticipated that the number of benchmarked projects submitted will rise from 200 to perhaps 2000/year. Indeed, as many as 20,000 benchmark projects per year might occur as the cost of function point analysis goes down. However, gaps and leakages from resource-tracking systems remain a problem that needs to be solved to ensure that the benchmarks are accurate enough to be depended upon.

New kinds of benchmarks such as individual features can also occur. At the high end of the scale, portfolio benchmarks of all software owned by major corporations will also be possible. Special benchmarks, such as tool usage and specific methodologies, will also occur in significant numbers. Benchmark data can be collected either remotely as done by ISBSG or via on-site interview as done by SPR. Remote collection is less expensive, but on-site collection is more thorough because errors can be corrected during the team interviews.

Opportunity 9. Baseline Analyses: Before starting a process improvement program, it is valuable to measure at least a sample of applications using function point analysis. Then as the process improvement program unfolds, the productivity and quality of the original baseline can be used to measure rates of progress. Now that high-speed, low-cost function points are available, the number of existing applications used for baseline purposes can be increased. Not only mid-range applications can be measured, but also very large projects > 100,000 function points can be studied. In addition, the thousands of small updates <15 function points in size can now be evaluated using micro function points. Thus, baselines can increase in number and sophistication at the same time that they are dropping down in cost. Establishing current productivity and quality levels by means of function point metrics should be a standard initial step in all software process improvement programs.

Opportunity 10. Feature/Class Analyses: The current taxonomy that allows rapid function point sizing deals with complete applications. If this taxonomy is extended down another level, then it will be possible to perform sizing of the individual features that comprise software applications. Feature analysis is an important prerequisite for increasing the volumes of reused materials in software applications. It is also valuable to know the sizes of the contents of object-oriented class libraries. Feature size analysis is also an important topic for massive applications such as ERP and Service-Oriented Architecture (SOA) packages. For example, it is important to know the size of customer resource management components or the size of accounts payable components. Feature analysis will improve the accuracy of estimating, and will also facilitate planning for software reuse at the macro level, such as SOA systems. This is a new business opportunity that has not been part of the older function point business model.

Opportunity 11. Portfolio Analyses: Because software is a taxable asset, there is a business value associated with knowing how much software a corporation owns. Portfolio analysis has not been possible with standard function points because the costs and timing of normal function point analysis are too high. Also, at least half of corporate portfolios are in the form of COTS packages acquired from external vendors. Now that high-speed, low-cost function point counts are possible, it can be expected that most Fortune 500 companies will want to know exactly how much software they own, its rate of growth, the retirement rate of aging applications, and the split between in-house development, outsourced development, and COTS packages. Portfolio analysis is a major new business opportunity for function point consultants.

It was not possible to perform a full portfolio analysis using standard function point counting methods because the cost for a portfolio of 10,000,000 function points might exceed $60,000,000. In order to be cost-effective, the consulting fees for a corporate portfolio analysis need to be less than $100,000 or roughly 1¢ per function point. Now that function point costs are low, portfolio analysis and long-range portfolio planning are major new business opportunities. As of 2017, there are probably fewer than 10 companies in the world who actually know how much software they own and use. There are probably no government organizations at any level (national, state, county, or city) who have any knowledge of the software they own and use. Considering that software is one of the most expensive and valuable assets in history, the lack of economic quantification is a serious economic deficiency.

Opportunity 12. Micro function points: A surprising amount of work takes place in the form of small updates that range between a fraction of a function point up to perhaps 15 function points in size. Almost all bug repairs and many maintenance changes are found in the size range of <15 function points, and therefore cannot be measured using standard function point analysis. Now that micro function points are available, a number of important new studies can be carried out on the economics of small changes. Topics that need definitive answers are those of productivity rates, bad-fix injection rates, and total volumes of small changes on an annual basis. Since the total volume of small updates in large corporations can top 100,000 function points per year, the lack of any ability to study update economics is a gap that needs to be filled. This is a new business opportunity for micro function point metrics and has not been part of standard function point analysis due to the lower boundary of the counting rules.

Opportunity 13. Development Methods: One of the important values of function point metrics is the ability to study the productivity and quality results from various development methods such as the Rational Unified Process (RUP), Watts Humphrey's Team Software Process (TSP), the Agile methods, object-oriented development, and others. Now that high-speed, low-cost function points are available, the numbers of methods studied will increase exponentially. In addition, the number of samples measured for each methodology will increase. To judge the effectiveness of a methodology at least 50 projects are needed, and their sizes should range from a low of perhaps 100 function points up to a high of perhaps 100,000 function points. Evaluating development is a significant new opportunity for function point analysis. There are more than 100 methods of software development in existence circa 2017, so this topic will of necessity be complex.

The author of this book has just published another book with CRC press that quantifies 60 software development methodologies. The title is *Quantifying Software Methodologies*, and is a few months from publication as this text is written.

Opportunity 14. ITIL Analyses: The Information Technology Infrastructure Library (ITIL) is among the fastest growing software technologies. ITIL deals with post-release maintenance, service, and change control. The ITIL concept itself is still undergoing change and evolution. A major gap in the current ITIL literature is any reference to the sizes of applications expressed in terms of function points. Since both defect volumes and rates of change are directly proportional to application size measured with function points, this is a gap that needs to be filled as quickly as possible.

Opportunity 15. Development Tool Analyses: As of 2017, there are hundreds of software development tools available. These include design tools, compilers for programming languages, test tools, change control tools, configuration control tools, Web development tools, and many others. Some of these tools have been sized in terms of function point metrics. Few have been studied under controlled conditions. Now that high-speed, low-cost function point metrics are available, it will be possible to measure the size of every tool using function point metrics. There are already some rough correlations that show that tool availability and tool usage benefits quality and productivity, but these studies need to be done in greater numbers and with greater precision. From partial data, it looks as though a fully equipped software engineer for mainframe development will have about 25,000 function points of tools; Web developers will have about 50,000 function points of tools.

Opportunity 16. Maintenance Tool Studies: Although maintenance in the dual forms of bug repairs and small enhancements is now the dominant activity of the software world, it is among the least covered in terms of quality and productivity studies. Part of the reason for this lack of empirical data is the fact that standard function point analysis does not work below a size of about 15 function points. Most bug repairs are between 1/50th of a function point and 2 function points in size. Most small enhancements are between 1 and 10 function points in size. Although each of these is small individually, corporations and government agencies can top 30,000 such changes per calendar year with a total volume of changes in excess of 100,000 function points. More than 50% of the total staff of many corporations now work in the areas of maintenance and enhancement.

In order to study the effectiveness of maintenance tools, such as code-restructuring tools or renovation work benches, it is necessary to have a micro function point that can deal with small updates to legacy applications. It is also necessary to measure the size of maintenance tools using standard function point metrics.

Preliminary data indicates that a typical maintenance programmer is supported by about 15,000 function points of maintenance tools. Monthly bug repairs and small enhancements proceed at a rate of perhaps 25 function points per staff month. This is a major new opportunity for function point analysis and was not part of the older function point business model because small updates could not be measured.

Opportunity 17. Mass-Update Analyses: The phrase "mass update" refers to problems that cause simultaneous changes to thousands of applications throughout an industry or throughout the world. The Y2K problem was the most widely studied mass update, but it is only one of dozens of such problems. Other mass updates included the roll out of the Euro in 1999 and the Sarbanes–Oxley legislation in 2004. In 2038, the Unix internal calendar will expire triggering a new problem similar to that of Y2K.

Almost every year government mandates or changes in tax laws trigger mass updates of financial applications. In the future, we can expect significant mass updates such as the need to add digits to telephone numbers or the need to expand social security numbers.

Function point metrics are the best tool for estimating the costs and schedules for such updates. The caveat is that while estimates for thousands of individual updates are needed, it is also necessary to produce overall estimates for each company and industry. Standard function point analysis is too slow and time-con suming, so newer high-speed methods will be needed to deal with mass update economics.

Opportunity 18. Backlog Analyses: Large corporations and government agencies have "backlogs" of applications that are awaiting development at some future time, but where available resources are not in the current year's budget. The arrival of high-speed, low-cost function point analysis now makes it possible to size the entire backlog. The size data can be used to prioritize the applications awaiting development. Backlog analysis was not feasible using standard function point analysis because the requirements for many backlogged applications are not fully defined. The total size of the backlog in a large company can approach 100,000 function points, which is too large for conventional function point analysis. Backlog sizing and prioritization are significant new business opportunities for function point analysis. Every corporate and government backlog should be sized and have preliminary risk and value data available.

Opportunity 19. Normal Function Point Analyses: Surprisingly, the advent of high-speed, low-cost function point methods will not eliminate standard function point analysis. This is because the high-speed methods are not as accurate as normal function point analysis. Although some of the high-speed methods can be used 6 months before requirements are complete, when the requirements finally are complete, it will be necessary to validate the original function point predictions by means of normal function point analysis. Rather than curtailing normal function point analysis, the high-speed methods are likely to increase normal function point counts by possibly 100%, or by doubling the current business.

Opportunity 20. COTS Risk Analyses: If a company buys software from Microsoft, Oracle, SAP, Symantec, or essentially any other vendor, it will be delivered with latent bugs. The vendors will not provide data on how many bugs are likely to be there. However, by using high-speed, low-cost function point analysis of COTS applications and looking at industry data for defect removal efficiency levels, it is possible to predict the number of latent bugs in COTS packages. Because COTS packages make up at least 50% of all the software owned by major corporations, there is a strong business need to predict the bugs in these applications and the amount of internal effort required to install and support the applications. This is a new business opportunity for function point analysis.

Opportunity 21. ERP Deployment Analyses: ERP packages are the largest commercial applications ever developed. The main products from SAP and Oracle top 300,000 function points in size. These massive packages are hard to debug, so each release is delivered with thousands of latent bugs. As a result, installation and successful deployment of an ERP package often fails, and is never cheap or easy. The vendors do not provide such information. However, now that ERP packages can be sized using function point metrics, it is possible to predict the number of latent bugs at delivery and the probable amount of time to install and adjust the ERP packages, which may be more than one calendar year. Every ERP deployment should be supported by a formal risk and value assessment coupled with a function point analysis of potential defects. This is a major new opportunity for function point analysis.

Note that the author's SRM estimating tool has an ERP deployment estimating feature.

Opportunity 22. Reengineering Analyses: Many legacy applications contain error-prone modules, sections of dangerously high complexity, and lack suitable comments. Some are written in obscure or dead languages where only few programmers are available and even working compilers or interpreters may be hard to find. If these legacy applications still have business value, then it will be necessary to reengineer them and plan for replacements. Because the cost and schedules of replacements correlate directly with application size, it will be necessary to perform a function point analysis as part of the reengineering process. In particular, the key features and algorithms will need to have their sizes known. Also, it may be possible to find replacements from libraries of certified reusable components, or possibly to replace some aging applications with SOA versions. The legacy applications should be mined for business rules and useful algorithms at the same time.

Opportunity 23. Customer Support Analyses: The area of customer support is a weak link in the software industry. With the exception of Apple Computers and a few other high-technology companies such as Advanced Bionics, customer support is hard to access (telephone wait time often exceeds 10 min). Once customer support is reached, only a small percentage of problems can be fixed in a single call. Because customer support is often outsourced to countries with low labor costs, it is sometimes difficult for hard-of-hearing customers to understand the accents of the support personnel. In fact, customer support for the deaf is extremely limited. Customer support staffing and skill requirements are partly based on applications size in function points, and partly on numbers of users or customers.

As a rule of thumb, one customer support staff member can support an application of about 5000 function points in size, with perhaps 150 customers or users. As size and/or usage increase, additional customer support personnel will be needed or support will degrade to unsatisfactory levels.

Customer support planning is a new opportunity for high-speed, low-cost function point metrics. Standard function points were of little use in the past because most heavy duty commercial packages are larger than 10,000 function points and hence did not use standard function point analysis. The new high-speed versions can size applications in excess of 100,000 function points so that problem is no longer an issue.

Opportunity 24. SOA and Reuse Analyses: The new SOA applications are composed of large segments that range between 1,000 and 10,000 function points in size. By contrast, normal reusable modules or objects run between 10 and 100 function points in size. Function point metrics can be used to predict latent defects in both SOA components and smaller reusable modules. They can be used to predict development costs and construction costs as well. Function point metrics can also be used as an aid in determining the taxable value of reusable components of any size. Related topics such as integration and system testing of SOA and reusable material will also benefit from high-speed, low-cost function point analysis.

Opportunity 25. Canceled Project Analyses: On the basis of depositions and discovery materials during litigation, an "average" canceled project is larger than 10,000 function points in size, almost a year late at the point of cancellation, and around 50% over budget when terminated. An average cost for a canceled project in the 10,000-function point range is about $37,000,000.

Not many people know that canceled projects actually cost more than successful projects of the same size. In-house canceled projects that don't go to court are almost never studied and analyzed. Now that low-cost, high-speed function points are available, it would be a sound business practice to perform a formal postmortem on every canceled project. This is possible due to high-speed, low-cost function points. Corporations would not pay the fees for normal function point analysis on projects that were canceled due to excessive cost and schedule overruns.

Opportunity 26. Agile Development Analyses: To date the Agile approaches have been expanding rapidly in terms of usage and numbers of projects completed. A major gap in the Agile method has been lack of solid empirical data regarding development schedules, costs, and quality. Because Agile is so new, there is an almost total lack of data on Agile maintenance. While some Agile projects do measure, they tend to use specialized metrics such as story points and velocity that cannot be used for benchmarks because there is little or no data available. Further, it is not possible to do a side-by-side study if one project used function points and the other used story points.

The availability of high-speed, low-cost function point metrics should allow benchmarks of thousands of Agile projects within a few years. Even retrospective studies can now be performed by using function point metrics on existing Agile projects that were developed over the past 5 years.

One of the more valuable features of high-speed function point analysis is the ability to size each "sprint" in an Agile environment. Traditional function point analysis is far too sluggish to be a good match to the Agile approach, but being able to size features and sprints in less than 1 min should make function point metrics a standard part of Agile development in the future.

Opportunity 27. Capability Maturity Model Integrated (CMMI) Analyses: The Software Engineering Institute (SEI) was a late adopter of function point metrics. The defense community also has very sparse usage of function points. As a result, there is little solid data on the value of ascending the CMMI scale. When function point metrics are used to analyze the various levels of the CMMI, the results are very convincing that the higher CMMI levels have lower defect potentials, higher levels of defect removal efficiency, shorter schedules, and higher productivity. Function point metrics are the best method for such studies. In fact, the older LOC metric behaves so badly that its usage should be classified as professional malpractice. Not only development but maintenance improves at the higher CMMI levels. It is now possible to show the results of each CMMI level for an entire sequence of both development and multiyear maintenance period culminating in total cost of ownership (TCO). The TCO of an application developed at CMMI level 5 is only about one-fourth of the TCO of the same application developed at CMMI level 1. The long-range economic pictures should be useful in expanding the usage of the CMMI approach.

Opportunity 28. Security risk Analyses: By interesting coincidence, the number of security vulnerabilities applications correlates with application size measured with function points. Security vulnerabilities also correlate, with less certainty, with the defect potentials of applications also measured in terms of function points. Because of the high costs of conventional function point analysis, there have been few studies of the relationships between bugs and security flaws in applications >10,000 function points. Now that high-speed, low-cost function point analysis is available, security studies should soon increase in number.

Opportunity 29. Outsource Contract Analyses: A significant percentage of outsource contracts utilize function point metrics for size, productivity, and quality requirements. It is not uncommon for the sizes of outsourced applications to be defined in terms of function points. As the costs of function point analysis decline, this practice may become universal. It would also be valuable to include clauses in outsource agreements that deal with defect potentials and delivered defects measured in terms of function point metrics. Requirements changes can also be mea sured in terms of function point metrics. Defect removal efficiency should also be included in outsource agreements. Function points are used today in perhaps 10% of outsource contracts in the United States. In the future, the use of function point metrics will probably approach 100% of U.S. software outsource agreements. Early sizing before full requirements will be useful for outsource contract analysis.

Opportunity 30. Litigation (Breach of Contract): About 5% of outsource contracts end up in court for breach of contract litigation. In all these cases, it is important to know the size of the applications expressed in function point form. Although standard function point analysis is sometimes used during lawsuits, it is too slow and costly to be used for lawsuits involving big applications that are >100,000 function points in size. The time required for the function point analysis would exceed the time period assigned by the courts for discovery. However, low-cost, high-speed function point counts will probably become widely used in software breach of contract litigation.

Opportunity 31. Merger and Acquisition Analyses: This is a new business opportunity that has not been part of the current function point business model. When two companies merge and one or both possess significant quantities of software, it is appropriate to use function point metrics as part of the due diligence process. Standard function point analysis is too slow and costly to be used with companies that own more than 1,000,000 function points of software. However, the new high-speed, low-cost methods of function point counting allow entire portfolios of software applications to be considered even if those portfolios top 10,000,000 function points. Function point metrics can quantify the value of portfolios and also predict future maintenance costs, warranty costs, defect levels, and other factors that should be studied as part of a due diligence process.

Opportunity 32. Occupation Group Analyses: The software industry employs more than 90 different occupations and specialty groups. Among these can be found architects, software engineers, database administrators, quality assurance personnel, testers, technical writers, function point analysts, project librarians, Web master, SCRUM masters, cost estimators, and specialized engineers such as telecommunications engineers. There is a shortage of empirical data as to how many of these various specialists are needed. There is an even greater shortage of empirical data on their productivity rates, value to projects, and other business topics. For an application of 10,000 function points, how many architects, software engineers, quality assurance personnel, testers, and technical writers are going to be needed? Questions such as this can be explored using function point metrics for assignment scopes and production rates. Another aspect of occupation group analysis could also be carried out, with perhaps alarming results. One of the first uses of computers was to displace clerical workers in tasks that dealt primarily with paper handling. Thus, starting in the 1970s, it was noted that about 10,000 function points could be substituted for one insurance claims clerk. In robotic manufacturing, it takes about 150,000 function points of automated machine tools to displace one factory worker. As software becomes more sophisticated, it is of some importance to be able to quantify the numbers of workers of various kinds that might be displaced.

Opportunity 33. Test Tool Analyses: There are important questions about test tools that need better answers than those available today. How effective are automated test tools vs. manual testing for unit test, stress test, or security testing? What is the code coverage of various forms of testing? How big are test tools measured in terms of function point metrics? How many test cases are needed to fully test an application of 10,000 function points in size? Preliminary data suggests that about 15,000 function points of test tools are required to fully support a professional software tester. A somewhat depressing topic is that the measure defect removal efficiency level of most forms of testing is only about 35%, in that the tests only find one bug out of three that are present. An important question that can be studied using function point metrics is that of how many test cases are needed to raise the removal efficiency levels up to 950%? Also, how many test cases are needed to raise test coverage to 90%.

Opportunity 34. Embedded Software Analyses: The embedded software community has lagged in adopting function point metrics because of a mistaken idea that "function points only work for IT projects." In fact, function point metrics are the best available metric for demonstrating facts about embedded software, such as the fact that embedded software's defect potentials and defect removal efficiency are better than that of IT projects. Now that function point costs and speed have reached useful levels, there is no reason why function point metrics cannot be used routinely on every kind of embedded application: medical equipment, aerospace equipment, automotive equipment, military equipment, machine tools, and even computer gaming equipment, MP3 players, and wrist watches.

Opportunity 35. Website Analyses: Web applications and websites have been growing explosively for the past 10 years, and show no sign of slowing down their growth over the next 10 years. Web applications have lagged in using function point metrics, or any other kind of metrics for that matter. When Web applications are studied, they have higher-than-average productivity but lower-than-average quality. It is now possible to use function points for ascertaining the size of very large and complex websites such as Amazon and eBay in the 20,000-function point size range. This will allow useful economic studies of development costs, maintenance costs, rates of growth, and TCO.

Opportunity 36. Balanced Scorecard Analyses: Dr. Robert Kaplan and Dr. David Norton of the Harvard Business School are the originators of the balanced scorecard measurement approach. The balanced scorecard includes four measurement topics: (1) the learning and growth perspective; (2) the business process perspective; (3) the customer perspective; and (4) the financial perspective. Function point metrics are useful in all four perspectives. The most intriguing possibility is to use the new capability of using function points to measure software usage, and combine that view with software development, also using function points. For example, a tool like Microsoft Project is about 5,000 function points in size. If it is used by 100 managers in a corporation and saves each manager 150 h in developing project plans, then the economic value can be quantified. One of the steps is to consider the total usage, which in this example would be 100 people each using 5,000 function points or a 500,000-function point daily usage value.

Opportunity 37. Metrics Conversion: As of 2017, there are about 25 variations in counting function point metrics and 5 variations in counting LOC metrics. Some of the function point variations include IFPUG, COSMIC, NESMA, Finnish, Australian, the older Mark II method, backfired function points, feature points, Web-object points, Gartner function points, unadjusted function points, and many more. The major variations in code counting include counting physical lines (with or without comments and blanks), counting logical statements, and counts of code that either include or exclude reused sections, macro calls, and many other topics. One of the advantages of the high-speed, low-cost function point methods is that it is technically straightforward to include conversion logic from one metric to another. However, the effort to include accurate conversion logic is not trivial, so conversion will only be added if there is a legitimate business need. Obviously, there is a need to convert old data expressed in various flavors of LOC into function point form. There may also be a business need to convert COSMIC function points into IFPUG function points, or vice versa. Since the bulk of all reliable benchmark data is now expressed in terms of IFPUG function points, the most solid business case is for a need to convert other metrics into the IFPUG format.

Opportunity 38. Defense Analyses: The U.S. Department of Defense (DoD) owns more software than any other organization on the planet. In spite of constant attempts to improve and upgrade performance, the DoD continues to be troubled by a large number of software failures and by an even larger number of massive cost overruns and schedule slippages. It is not coincidental that the DoD has also lagged in use of function point metrics, in studies of defect potentials, in studies of defect removal efficiency levels, studies of maintenance performance, and studies of the costs and reliability of COTS packages. The older LOC metric cannot do any of these studies. In fact, LOC results are bad enough to be viewed as professional malpractice. It can be stated that the DoD will gain more value from the widespread utilization of function point metrics than any other industry segment, because of the huge volumes of software already deployed and planned for the future. The DoD urgently needs accurate measurements of quality, reliability, schedules, costs, and staffing levels. These are not possible using LOC metrics, so migration to function points is important. However, with a portfolio of existing software that approaches 100,000,000 function points, the DoD cannot possibly afford the high costs of normal function point analysis.

Opportunity 39. Supply Chain Analyses: As a manufactured product moves from the raw material stage to the finished product stage, there may be more than a dozen companies involved. Each of these companies will use software to add value to the finished product. It is an interesting but unanswered question as to whether products that utilize very large amounts of software are more efficient or have better or worse quality levels than products using traditional manual methods. This is a new business opportunity because the total quantity of software involved in manufacturing a complex product such as an automobile or an aircraft could top 1,000,000 function points and as many as 50 suppliers.

Opportunity 40. International Analyses: A topic of great importance to the software industries and national governments of many countries is how software productivity and quality compare to those of other countries. Function point metrics combined with process assessments provide the best approaches for measuring this important topic. Additional factors need to be recorded, however. Basic topics such as the number of work days per year, number of work hours per day, and the amount of unpaid overtime have a very significant effect on national performance. For example, software engineers in Japan put in more unpaid overtime than almost any country. Software workers in France, Canada, or Germany, by contrast, put in very few hours of unpaid overtime. Within any given country there are also regional and industrial differences in work patterns, and these need to be recorded too. Although the well-known ISBSG database of software benchmarks does record countries that provide data, the country information has not yet been released due to the fear that it might be used to gain undue competitive advantages.

Opportunity 41. Marketing Analyses: It obviously costs more to develop an application of 10,000 function points than it does to develop an application of 1,000 function points. But is it harder or easier to market large applications? Up until now marketing analysts have never had reliable data on the sizes of software applications. Now that it is known that applications such as Windows 7 are about 200,000 function points in size, whereas Linux is only about 20,000 function points in size, it will be possible to use this information to study interesting market trends. It will also be possible to carry out side-by-side feature analyses of similar applications, such as comparing Microsoft Office to Open Office to the Google office suite. In order to do this, the new ability size individual features of large applications will be used. This is a new opportunity for function point metrics that was not part of the older business model. Further, now that COTS applications can be sized it will be possible to analyze business topics such as retail costs per function point, which have never been possible before. This is likely to benefit competition and perhaps lower some of the excessive pricing structures that are now part of the software business.

Opportunity 42. Business Process Analyses: One of the reasons why business process reengineering does not usually accomplish very much is lack of quantification of the relative proportions of manual effort and software usage to accomplish basic business tasks. For example, when a customer places an order by phone with a company such as Dell, how many people and software are involved until the computer that was ordered actually leaves the dock for the customer? There are hundreds of business processes that need careful analysis of the relative proportions of manual effort vs. software measured in terms of function points. Some additional examples include basic activities such as accounts payable, accounts receivable, purchasing, government activities such as renewing drivers' licenses, or assessing real estate value. By including software size in terms of function points, it will be possible to do more sophisticated forms of process analysis. Another aspect of business process analysis that has historically been weak is the failure to consider the effects of bugs or defects in the software itself. Thus, quality control has been a missing ingredient. As mentioned previously, one of the first uses of computers was to displace clerical workers in tasks that dealt primarily with paper handling. Thus, starting in the 1970s it was noted that about 10,000 function points could be substituted for one insurance claims clerk. In robotic manufacturing, it takes about 150,000 function points of automated machine tools to displace one factory worker. As software becomes more sophisticated, it is of some importance to be able to quantify the numbers of workers of various kinds that might be displaced.

Opportunity 43. Litigation (tax cases): Function point metrics are often used in tax litigation where the value of software assets has been challenged by the Internal Revenue Service (IRS). One of the largest software tax cases in history involved the valuation of the software assets owned by EDS when the company was acquired by General Motors. Function point metrics are the best choice for dealing with software creation expenses and software value analysis. Since the IRS already uses function points for software tax issues, this is a current business opportunity that should expand in the future.

Opportunity 44. Venture Capital Analyses: When a venture capital company is about to invest money in a software company, what are the risks involved? Given the high failure rate and frequent overruns of software projects, what are the odds of a successful software application being developed and released before the money runs out? Currently, about 90% of venture-funded companies fail within 2 years. By using high-speed, low-cost function points to predict the development speed, costs, and quality of the software that is being venture-funded, it would be possible to weed out weak candidates. If so, venture failure rates might drop below 50% and hence permit larger investments in companies with good chances of business success. The new high-speed function point method combined with quality and risk analyses should give a much better understanding of both the risks and potential values of the situation. Both the venture capitalists and the entrepreneurs are likely to commission studies of schedules, costs, quality, and value using function point metrics. This is a new business opportunity and not a significant part of the older function point business model, in part because the software under consideration may not have complete requirements at the time the venture company makes a first round investment.

Opportunity 45. Programming Languages: The topic of "backfiring" or mathematical relationships between function point metrics and LOC metrics has been studied since about 1975. As of 2017, there are more than 3,000 programming languages in existence. As of 2017, over 50% of all software applications utilize at least two separate programming languages, and some applications contain as many as 15 different languages. There is a continuing need to explore topics such as the relationship between programming languages and function point metrics. Now that low-cost, high-speed function point sizing is possible, the costs of such studies will become so low that backfiring ratio analyses could be carried out by university students as research projects.

Opportunity 46. Earned-Value Analyses: The earned-value approach is widely used in defense contracts, and is starting to move into the area of civilian contracts as well. To date, function point metrics have not been part of the earned-value approach, but they should soon become widely used for early estimates before development begins and then used again to certify that the components meet contractual agreements as they are delivered.

Opportunity 47. Infrastructure and substrate or technical requirements analyses: Many years ago (before function points were invented) IBM did a study of accounting software and found that only about 25% of the code in accounting packages had anything to do with accounting at all. The other 75% was the code needed to make accounting work on a computer. As time passed, improvements in operating systems and utilities have reduced the overhead needed to put business problems on a computer, but even in 2017, between 15% and 50% of the effort involved with software development is concerned with technical platform issues rather than the business issues of the applications themselves.

As of 2017, Web development has the lowest overhead, and embedded software has the highest overhead. This is one of the reasons why embedded software has a higher cost per function point than Web applications. It is useful to know the kinds of activities and costs associated with putting business features onto various platforms. As Web-enabled applications, SOA, software as a service (SaaS), and certified reusable components become more widespread, the overhead will probably decline still further. However, it is significant to quantify the overhead items. This is a new opportunity that has only recently started to be studied with care. An interesting question is whether the substrate software underneath the business application itself can be measured with function point metrics, or needs a separate metric such as the new SNAP metric for nonfunctional requirements.

Function point metrics can be used if the user view is expanded from end users to the designers of operating systems and hardware platforms. An analogy is that infrastructure costs are similar to building codes in home construction. For example, in many coastal states such as Rhode Island, homes within a mile of the ocean now are required to use hurricane-proof windows, at considerable expense. Home construction can still be estimated using cost per square foot, but for homes near the ocean, the new windows will add to those costs.

Opportunity 48. Educational uses of function points: Knowledge of software economics is distressingly bad in the United States. Neither software engineering schools nor MBA programs have effective training in software economics, in part because of the paucity of reliable data on costs, schedules, quality, and maintenance expenses. Now that high-speed function point metrics are available and are part of high-speed estimating tools, it is possible to make real advances in software economic education. Some of the topics that can now be illustrated in only a few minutes are the economic values of the CMM approach, how Agile projects compare to older waterfall projects, what happens when formal inspections are added to a defect removal series, how software project requirements grow, and dozens of topics that should be understood but are not. Postmortems of canceled projects can be displayed side-by-side against successful projects of the same size and type. Very few project managers in the United States know that canceled projects cost more than successful projects. Very few software engineers or project managers know that adding inspections to the defect removal steps will shorten total development schedules and lower development costs. Although the educational uses of high-speed function points will not be a major source of income, in the long run the educational uses of function points will be the most valuable for the software industry and possibly for the U.S. economy as a whole.

Opportunity 49. Delivery Channel Analyses: Up until the past few years, software was always delivered on physical media such as tapes or disks. In recent years, more and more software is delivered via downloads from the Web. In fact, software may not even be delivered at all to a client's computer, but can be run remotely from the vendor's website or host computer. Because delivery by physical media is expensive, vendors prefer electronic delivery. Software is almost the only expensive product that can be delivered electronically at high speeds. But electronic delivery demands high-speed connections to be cost-effective for users. As software becomes a global commodity and usage expands in developing countries, it is interesting to perform studies on the most cost-effective and secure delivery channels. For example, a floppy disk can only hold about 5000 function points, but a CD or a DVD disk can hold millions of function points. Downloading via a modem at 56K baud only allows about 500 function points per minute to be downloaded, while a high-speed connection allows about 50,000 function points per minute. This is a new form of analysis that was not part of the older function point business model. This question is of more than casual importance. If website and broadband deliveries increase as fast as they have been to date, by about 2020 a substantial portion of the available bandwidth may be tied up in distributing and using function points. The business model for SaaS assumes almost infinite bandwidth. If the bandwidth is finite and the speed is slow, then SaaS cannot reach its projected business potentials. Applications residing and being used in the cloud also will benefit from high-speed, low-cost function point analysis.

Opportunity 50. National Studies: As of 2017, the United States produces and consumes more software than any other country. The U.S. probably produces in the range of 100,000,000 function points per calendar year. When consumption is considered, every U.S. citizen probably uses more than 100,000 function points per day without even knowing it. Everyone who has a cell phone, a digital watch, a digital camera, modern kitchen appliances, and a fairly new automobile is surrounded by dozens of small computers and large quantities of software. The production and sale of software either as software itself or as embedded in manufactured products is a major component of the U.S. economy, and an increasingly important component of countries such as China, India, and Russia.

Function point metrics can be used to measure the economic importance of software to both the developed countries and the developing countries. In order to use function points for this purpose, it is necessary to know the sizes of every form of software that a country produces and uses. This was not possible using manual function point analysis because many applications were too large or too specialized for function point counts. Now that function points can be applied quickly to any form of software, an important new form of large-scale economic analysis is now possible.

Examples of topics that can now be studied using function point metrics include: (1) The relative manufacturing efficiencies of Ford vs. Toyota; (2) Correlations between insurance and medical function point usage and health care costs; (3) The relative performance and efficiencies of state government operations; (4) The relative capabilities of U.S. combat troops, aircraft, and Naval vessels vs. those of other countries; (5) The economic value of software exports and imports; (6) The relative volumes of open-source software as part of overall national software development and consumption. This last topic is of some importance. Due to Microsoft, IBM, Symantec, and a few other companies, the U.S. exports more commercial fee-based software than all other countries put together, and the value of these exports is significant to the overall U.S. economy. If open-source software should begin to supplant fee-based software, it will have a perceptible impact on both Microsoft and the national economy.

As of 2017, open-source software seems to be less than 2% of overall domestic software usage. If that number should increase to perhaps 20%, it would cause a significant economic perturbation throughout the world. National studies of software using function point metrics are a brand-new kind of economic analysis that has only just begun to be considered. National usage of software measured with function points and transnational distribution, sales, and purchasing of software quantified by function points are an interesting indicator of relative economic performance.

The 50 business opportunities discussed here are not an exhaustive list. They are only examples of some of the kinds of business opportunities that need low-cost and high-speed function point counts to become commercially viable. For example, yet another kind of study would examine and quantify the quality levels of open-source applications such as Linux and Firefox compared to fee-based applications such as Vista and Internet Explorer.

Another very important kind of study will be a form of "Consumer Reports" analysis that examines the cost per function point of various commercial software packages and reports to customers on products that may be priced artificially high for the services and features they provide. For example, Microsoft Office Professional is perhaps 95,000 function points in size and costs about $500 for a cost of $0.052 per function point. Google Office is only about 50,000 function points and the Open Office Suite is about 75,000 function points, but both are available for no cost. Word Perfect's office suite is perhaps 80,000 function points and available for about half the cost of Microsoft Office. These are only hypothetical examples, but they show an interesting new kind of analysis that has not been possible before. Once this kind of information becomes available, it is likely to benefit competition and perhaps to lower the retail prices of applications that have been priced artificially higher than their competitors.

Not only can the price per function point be compared, but also the features of various applications in the same market space. Historically, Microsoft applications tend to have the largest number of features of commercial applications in any general area. But since not all features are necessary for all users, a similar application with fewer features and a lower cost may be a more attractive product for many customers.

Although pricing studies of COTS software is likely to become an important use of function point metrics, it is not listed as a major revenue opportunity because of the assumption that this kind of study would be carried out by either nonprofit organizations or by universities. However, vendors in competitive markets may well commission function point feature analyses of other tools and software packages to demonstrate that they either have a lower cost per function point, provide more features, or have better quality levels. If this occurs, then the revenues from COTS pricing studies could top $50,000,000/year.

In addition, there will probably be dozens of new or updated software tools such as function point sizing tools, quality-estimating tools, cost-estimating tools, maintenance-estimating tools, ERP deployment-estimating tools, and many more.

The Hazards and Errors of LOC Metrics

The reason that function point metrics will expand rapidly once costs go down is because the older LOC metric has been found to violate standard economic assumptions when analyzed carefully. Few people in the industry realize that LOC metrics penalize high-level and object-oriented languages. Very few people have actually studied the problems of LOC metrics.

One of the oldest and the most widely used metrics for software has been that of lines of code, which is usually abbreviated to LOC. Unfortunately, this metric is one of the most ambiguous and hazardous metrics in the history of business. LOC metrics are ambiguous because they can be counted using either physical lines or logical statements. LOC metrics are hazardous because they penalize high-level programming languages, and can't be used to measure noncoding activities.

The development of Visual Basic and its many competitors have changed the way many modern programs are developed. Although these visual languages do have a procedural source code portion, quite a bit of the more complex kinds of "programming" are done using button controls, pull-down menus, visual worksheets, and reusable components. In other words, programming is being done without anything that can be identified as an LOC for measurement or estimation purposes. In 2008, perhaps 60% of new software applications were developed using either object-oriented languages or visual languages (or both). Indeed, sometimes as many as 12–15 different languages are used in the same applications.

Table 2.4 Rank Order of Large System Software Cost Elements

1. Defect removal (inspections, testing, finding, and fixing bugs)
2. Producing paper documents (plans, specifications, user manuals)
3. Meetings and communication (clients, team members, managers)
4. Programming
5. Project management

For large systems, programming itself is only the fourth most expensive activity. The three higher cost activities cannot be measured or estimated effectively using the LOC metric. Also, the fifth major cost element, project management, cannot easily be estimated or measured using the LOC metric either. Table 2.4 shows the rank order of software cost elements for large applications in descending order.

The usefulness of a metric such as LOC, which can only measure and estimate one out of the five major software cost elements of software projects, is a significant barrier to economic understanding.

Following is an excerpt from the third edition of the author's book *Applied Software Measurement* (McGraw Hill, 2008) that illustrates the economic fallacy of KLOC metrics:

"The reason that LOC metrics give erroneous results with high-level languages is because of a classic and well known business problem: the impact of fixed costs. Coding itself is only a small fraction of the total effort that goes into software. Paperwork in the form of plans, specifications, and user documents often cost much more. Paperwork tends to act like a fixed cost, and that brings up a well-known rule of manufacturing: "When a manufacturing process includes a high percentage of fixed costs and there is a reduction in the number of units manufactured, the cost per unit will go up."

Here are two simple examples, showing both the LOC results and the function point results for doing the same application in two languages: basic assembly and C++. In Case 1, we will assume that an application is written in assembly. In Case 2, we will assume that the same application is written in C++.

Case 1: Application Written in the Assembly Language

Assume that the assembly language program required 10,000 LOC, and the various paper documents (specifications, user documents, etc.) totaled to 100 pages. Suppose that coding and testing required 10 months of effort, and writing the paper documents took 5 months of effort. The entire project totaled 15 months of effort, and thus has a productivity rate of 666 LOC per month. At a cost of $10,000 per staff month, the application cost $150,000. Expressed in terms of cost per source line, the costs are $15 per line of source code.

Case 2: The Same Application Written in the C++ Language

Assume that C++ version of the same application required only 1,000 LOC. The design documents probably were smaller as a result of using an O-O language, but the user documents were the same size as the previous case: assuming that a total of 75 pages were produced. Suppose that coding and testing required 1 month and document production took 4 months. Now we have a project where the total effort was only 5 months, but productivity expressed using LOC has dropped to only 200 LOC per month. At a cost of $10,000 per staff month, the application cost $50,000 or only one-third as much as the assembly language version. The C++ version is a full $100,000 cheaper than the assembly version, so clearly the C++ version has much better economics. But the cost per source line for this version has jumped to $50.

Even if we measure only coding, we still can't see the value of high-level languages by means of the LOC metric: The coding rates for both the assembly lan guage and the C++ versions were both identical at 1,000 LOC per month, even though the C++ version took only 1 month as opposed to 10 months for the assembly version.

Since both the assembly and C++ versions were identical in terms of features and functions, let us assume that both versions were 50 function points in size. When we express productivity in terms of function points per staff month, the assembly version had a productivity rate of 3.33 function points per staff month. The C++ version had a productivity rate of 10 function points per staff month. When we turn to costs, the assembly version had a cost of $3000 per function point, while the C++ version had a cost of $1000 per function point. Thus, function point metrics clearly match the assumptions of standard economics, which define productivity as goods or services produced per unit of labor or expense.

LOC metrics, on the other hand, do not match the assumptions of standard economics, and in fact show a reversal. LOC metrics distort the true economic case by so much that their use for economic studies involving more than one programming language might be classified as professional malpractice.

The only situation where LOC metrics behave reasonably well is when two projects utilize the same programming language. In that case, their relative productivity can be measured with LOC metrics. But if two or more different languages are used, the LOC results will be economically invalid.

A Short History of LOC Metrics

It is interesting to consider the history of LOC metrics and some of the problems with LOC metrics that led IBM to develop function point metrics. Following is a brief history from 1960 through today, with projections to 2010.

Circa 1960: When the LOC metric was first introduced, there was only one programming language and that was basic assembly language. Programs were small and coding effort comprised about 90% of the total work. Physical lines and logical statements were the same thing for basic assembly. In this early environment, LOC metrics were useful for both economic and quality analysis. Unfortunately, as the software industry changed, the LOC metric did not change and so became less and less useful, until by about 1980 it had become extremely harmful without very many people realizing it.

Circa 1970: By 1970, basic assembly had been supplanted by macro assembly. The first generation of higher level programming languages such as COBOL, FORTRAN, and PL/I were starting to be used. The first known problem with LOC metrics was in 1970 when many IBM publication groups exceeded their budgets for that year. It was discovered (by the author) that technical publication group budgets had been based on 10% of the budget for programming. The publication projects based on assembly language did not overrun their budgets, but manuals for the projects coded in PL/S (a derivative of PL/I) had major overruns. This was because PL/S reduced coding effort by half, but the technical manuals were as big as ever. The initial solution was to give a formal mathematical definition to language levels. The level was the number of statements in basic assembly language needed to equal the functionality of 1 statement in a higher level language. Thus, COBOL was a level 3 language because it took 3 basic assembly statements to equal 1 COBOL statement. Using the same rule, SMALLTALK is a level-18 language. The documentation problem was one of the reasons IBM assigned Allan Albrecht and his colleagues to develop function point metrics. Also, macro assembly language had introduced reuse, and had also begun the troublesome distinction between physical LOC and logical statements. The percentage of project effort devoted to coding was dropping from 90% down to about 50%, and LOC metrics were no longer effective for economic or quality studies. After function point metrics were developed circa 1975, the definition of language level was expanded to include the number of logical code statements equivalent to 1 function point. COBOL, for example, requires about 105 statements per function point in the procedure and data divisions. This expansion is the mathematical basis for backfiring or direct conversion from source code to function points. Of course, individual programming styles make backfiring a method with poor accuracy.

Circa 1980: By about 1980, the number of programming languages had topped 50, and object-oriented languages were rapidly evolving. As a result, software reusability was increasing rapidly. Another issue circa 1980 was the fact that many applications were starting to use more than one programming language, such as COBOL and SQL. In the middle of this decade, the first commercial software cost-estimating tool based on function points had reached the market, SPQR/20. By the end of this decade, coding effort was below 35% of the total efforts, and LOC was no longer valid for either economic or quality studies. LOC metrics could not quantify requirements and design defects, which now outnumbered coding defects. LOC metrics could not be used to measure any of the noncoding activities such as requirements, design, documentation, or project management. The response of the LOC users to these problems was unfortunate: they merely stopped measuring anything but code production and coding defects. The bulk of all published reports based on LOC metrics cover less than 35% of development effort and less than 25% of defects, with almost no data being published on requirements and design defects, rates of requirements creep, design costs, and other modern problems.

Circa 1990: By about 1990, not only were there more than 500 programming languages in use, but some applications were written in 12–15 different languages. There were no international standards for counting code, and many variations were used sometimes without being defined. A survey of software journals in 1993 found that about one-third of published articles used physical lines, one-third used logical statements, and the remaining third used LOC metrics without even bothering to say how they were counted. Since there is about a 500% variance between physical LOC and logical statements for many languages, this was not a good situation. Even worse, the arrival of Visual Basic introduced a class of programming language where counting LOC was not possible. This is because a lot of Visual Basic programming was not done with procedural code but rather with buttons and pull-down menus. In the middle of this decade, a controlled study was done that used both LOC metrics and function points for 10 versions of the same application written in 10 different programming languages including 4 object-oriented languages. This study was published in American Programmer in 1994. This study found that LOC metrics violated the basic concepts of economic productivity and penalized high-level and Object Oriented (OO) languages due to the fixed costs of requirements, design, and other noncoding activities. This was the first published study to state that LOC metrics constituted professional malpractice if used for economic studies where more than one programming language was involved. By the 1990s, most consulting studies that collected benchmark and baseline data used function points. There are no large-scale benchmarks based on LOC metrics. The ISBSG was formed in 1997 and only publishes data in function point form. By the end of the decade, some projects were spending less than 20% of the total effort on coding, so LOC metrics could not be used for the 80% of efforts outside the coding domain. The LOC users remained blindly indifferent to these problems, and continued to measure only coding, while ignoring the overall economics of complete development cycles that include requirements, analysis, design, user documentation, project management, and many other noncoding tasks. By the end of the decade, noncoding defects in requirements and design outnumbered coding defects almost 2 to 1. But since noncode defects could not be measured with LOC metrics, the LOC literature simply ignores them.

Circa 2000: By the end of the century the number of programming languages had topped 700 and continues to grow at more than 1 new programming language per month. Web applications are mushrooming, and all these are based on very high-level programming languages and substantial reuse. The Agile methods are also mushrooming, and also tend to use high-level programming languages. Software reuse in some applications now tops 80%. LOC metrics cannot be used for most Web applications and are certainly not useful for measuring Scrum sessions and other noncoding activities that are part of Agile projects. Function point metrics have become the dominant metric for serious economic and quality studies. But two new problems have appeared that have kept function point metrics from actually becoming the industry standard for both economic and quality studies. The first problem is the fact that some software applications are now so large (>300,000 function points) that normal function point analysis is too slow and too expensive to be used. The second problem is that the success of function points has triggered an explosion of function point "clones." In 2008, there were at least 24 function point variations. This makes benchmark and baseline studies difficult, because there are very few conversion rules from one variation to another. Although LOC metrics continue to be used, they continue to have such major errors that they constitute professional malpractice for economic and quality studies where more than one language is involved, or where noncoding issues are significant.

Circa 2010: It would be nice to predict an optimistic future, but if current trends continue, within a few more years the software industry will have more than 3,500 programming languages, of which about 3,250 will be obsolete or be becoming dead languages, more than 20 variations for counting LOC, more than 50 variations for counting function points, and probably another 20 unreliable metrics such as cost per defect or percentages of unstable numbers.

Future generations of sociologists will no doubt be interested in why the software industry spends so much energy on creating variations of things, and so little energy on fundamental issues. No doubt large projects will still be canceled, litigation for failures will still be common, software quality will still be bad, software productivity will remain low, security flaws will be alarming, and the software literature will continue to offer unsupported claims without actually presenting quantified data. What the software industry needs is actually fairly straightforward: (1) measures of defect potentials from all sources expressed in terms of function points; (2) measures of defect removal efficiency levels for all forms of inspection and testing; (3) activity-based productivity benchmarks from requirements through delivery and then for maintenance and customer support from delivery to retirement using function points; (4) certified sources of reusable material near the zero-defect level; (5) much improved security methods to guard against viruses, spyware, and hacking; and (6) licenses and board certification for software engineering specialties. But until measurement becomes both accurate and cost-effective, none of these are likely to occur. An occupation that will not measure its own progress with accuracy is not a true profession.

Circa 2020: The increasing number of national governments that now mandated function points for government software contracts is a good sign that software metrics are starting to be taken seriously. Congratulations to the governments of Brazil, Italy, South Korea, Japan, and Malaysia for mandating function point metrics.

In the best case, LOC metrics will be on the way out in 2020 and function point metrics will be on a steep acceleration and perhaps topping 50% of all U.S. projects and perhaps 20% of global projects.

In the worst case, LOC will continue to blindfold the software industry and cost per defect will continue to distort software quality economics. In other words, the software industry will be almost as bad in measurements and quality as they are in 2017 and have been for 60 years.

The Hazards and Errors of the Cost per Defect Metric

The well-known and widely cited "cost per defect measure" also violates the canons of standard economics. Although this metric is often used to make quality claims, its main failing is that it penalizes quality and achieves the best results for the buggiest applications! Furthermore, when zero-defect applications are reached, there are still substantial appraisal and testing activities that need to be accounted for. Obviously, the cost per defect metric is useless for zero-defect applications.

Because of the way cost per defect is normally measured, as quality improves, cost per defect steadily increases until zero-defect software is achieved, at which point the metric cannot be used at all.

As with KLOC metrics, the main source of error is that of ignoring fixed costs. Three examples will illustrate how cost per defect behaves as quality improves.

In all three cases, A, B, and C, we can assume that test personnel work 40 h per week and are compensated at the rate of $2,500 per week or $62.50/h. Assume that all three software features that are being tested are 100 function points in size.

Case A: Poor Quality

Assume that a tester spent 15 h writing test cases, 10 h running them, and 15 h fixing 10 bugs. The total hours spent was 40 and the total cost was $2,500. Since 10 bugs were found, the cost per defect was $250. The cost per function point for the week of testing would be $25.

Case B: Good Quality

In this second case, assume that a tester spent 15 h writing test cases, 10 h running them, and 5 h fixing one bug, which was the only bug discovered. However, since no other assignments were waiting and the tester worked a full week, 40 h were charged to the project. The total cost for the week was still $2,500, so the cost per defect has jumped to $2,500. If the 10 h of slack time are backed out, leaving 30 h for actual testing and bug repairs, the cost per defect would be $1,875. As quality improves, cost per defect rises sharply. Let us now consider cost per function point.

With the slack removed, the cost per function point would be $18.75. As can easily be seen, cost per defect goes up as quality improves, thus violating the assumptions of standard economic measures. However, as can also be seen, testing cost per function point declines as quality improves. This matches the assumptions of standard economics. The 10 h of slack time illustrate another issue: When quality improves, defects can decline faster than personnel can be reassigned.

Case C: Zero Defects

In this third case, assume that a tester spent 15 h writing test cases and 10 h running them. No bugs or defects were discovered. Because no defects were found, the cost per defect metric cannot be used at all. But 25 h of actual effort were expended writing and running test cases. If the tester had no other assignments, he or she would still have worked a 40-h week and the costs would have been $2,500. If the 15 h of slack time are backed out, leaving 25 h for actual testing, the costs would have been $1,562. With slack time removed, the cost per function point would be $15.63. As can be seen again, testing cost per function point declines as quality improves. Here too, the decline in cost per function point matches the assumptions of standard economics.

Time and motion studies of defect repairs do not support the aphorism that "it costs 100 times as much to fix a bug after release as before." Bugs typically require between 15 min and 4 h to repair. There are some bugs that are expensive and these are called "abeyant defects" by IBM. Abeyant defects are customer-reported defects that cannot be recreated by the repair center, due to some special combination of hardware and software at the client site. Abeyant defects comprise less than 5% of customer-reported defects.

Because of the fixed or inelastic costs associated with defect removal operations, cost per defect always increases as numbers of defects decline. Because more defects are found at the beginning of a testing cycle than after release, this explains why cost per defect always goes up later in the cycle. It is because the costs of writing test cases, running them, and having maintenance personnel available act as fixed costs. In any manufacturing cycle with a high percentage of fixed costs, the cost per unit will go up as the number of units goes down. This basic fact of manufacturing economics is why both cost per defect and LOC are hazardous and invalid for economic analysis of software applications.

The Hazards of Multiple Metrics without Conversion Rules

There are many sciences and engineering disciplines that have multiple metrics for the same values. For example, we have nautical miles, statute miles, and kilometers for measuring speed and distance. We have Fahrenheit and Celsius for measuring temperature. We have three methods for measuring the octane ratings of gasoline. However, other engineering disciplines have conversion rules from one metric to another.

The software industry is unique in having more metric variants than any other engineering discipline in history, combined with an almost total lack of conversion rules from one metric to another. As a result, producing accurate benchmarks of software productivity and quality is much harder than for any other engineering field.

The author has identified five distinct variations in methods for counting LOC, and 25 distinct variations in counting function point metrics. New variations are occurring almost faster than they can be counted. There are no standard conversion rules between any of these variants.

For sociological reasons, the software industry uses more different size and productivity metrics than any industry in human history. This plethora of overlapping and competing metrics slows technical progress because actual economic analysis is impossible when faced with no less than 30 inconsistent size and productivity metrics. This is professionally embarrassing.

Here is an example of why this situation is harmful to the industry. Suppose you are a consultant who has been commissioned by a client to find data on the costs and schedules of producing a certain kind of software, such as a PBX switching system.

You scan the literature and benchmark databases and discover that data exists on 90 similar projects. You would like to perform a statistical analysis of the results for presentation to the client. But now the problems begin when trying to do statistical analysis of the 90 PBX samples:

1. Three were measured using LOC and counted only physical LOC.
2. Three were measured using LOC and included blanks and comments.
3. Three were measured using LOC and counted logical statements.
4. Three were measured using LOC and did not state the counting method.
5. Three were constructed from reusable objects and only counted custom code.
6. Three were constructed from reusable objects and counted both reuse and custom.
7. Three were measured using IFPUG function point metrics.
8. Three were measured using IFPUG and SNAP metrics.
9. Three were measured using CAST automated function points.
10. Three were measured using Relativity Technology automated function points.
11. Three were measured using COSMIC function point metrics.
12. Three were measured using full function points.
13. Three were measured using Mark II function point metrics.
14. Three were measured using FESMA function points.
15. Three were measured using NESMA function points.
16. Three were measured using unadjusted function points.
17. Three were measured using engineering function points.
18. Three were measuring using weighted micro function points.
19. Three were measured using Web-object points.

20. Three were measured using Function points light.
21. Three were measured using backfire function point metrics.
22. Three were measured using Feature points.
23. Three were measured using Story points.
24. Three were measured using Agile velocity metrics.
25. Three were measured using Use Case points.
26. Three were measured using MOOSE object-oriented metrics.
27. Three were measured using goal-question metrics.
28. Three were measured using TSP/PSP task hours.
29. Three were measured using RTF metrics.
30. Three were measured using DEA.

As of 2017, there are no effective conversion rules among most of these metrics. There is no effective way of performing a statistical analysis of results of 90 projects measured with 30 different metrics. Why the software industry has developed so many competing variants of software metrics is an unanswered sociological question.

Developers of new versions of function point metrics almost always fail to provide conversion rules between their new version and older standard metrics such as IFPUG function points. In the author's view, it is the responsibility of the developers of new metrics to provide conversion rules to older metrics. It is not the responsibility of organizations such as IFPUG to provide conversion rules to scores of minor variations in counting practices.

The existence of five separate methods for counting source code and at least 25 variations in counting function points with almost no conversion rules from one metric to another is a professional embarrassment to the software industry. As of 2017, the plethora of ambiguous metrics is slowing progress toward a true economic understanding of the software industry.

Function point metrics are the most effective metrics yet developed for economic studies, quality studies, maintenance studies, and comparative studies of industries and even countries. However, the very slow speed of about 400 function points per day, and the high cost of about $6 per function point have limited the role of function point metrics to a few specialized benchmark and baseline studies. Less than 200 projects per year are being added to the available public benchmark data collections.

For function point metrics to achieve their full promise as the most effective software metrics in history, the speed of function point analysis needs to top 10,000 function points per day, and the costs need to drop down to about 1¢ per function point counted.

As of 2017, high-speed, low-cost function point methods are starting to become available. This discussion of a possible new business model for function point metrics considers what kinds of opportunities will become available when function point metrics can be applied to 100% of software development projects and 100% of maintenance and enhancement projects.

Extending Function Point Logic into New Domains

Function point metrics are the most accurate and effective metrics yet developed for performing software economic studies, quality studies, and value analysis. The success of function point metrics for software applications leads to the conclusion that the logic of function point metrics should be applied to a linked suite of similar metrics that can size other business and technical topics.

This portion of Chapter 2 suggests that an integrated suite of 14 functional metrics be created that would encompass not only software via function points but also data points, risk points, value points, service points, website points, security points, hardware function points, nonfunctional size (SNAP points), and software usage points.

The reason for this suggestion is to enable large-scale economic analysis of complex systems that involve software, data, hardware, websites, and other business topics that need concurrent sizing, planning, estimating, and economic analysis.

From their first publication in 1978, function point metrics have proven their value for software application sizing, cost estimation, quality predictions, benchmarks, and overall economic studies.

The IFPUG has become the largest software measurement association in the world. There are also other function point variants that are growing rapidly too, including COSMIC function points, NESMA function points, FISMA function points, and a number of others.

Yet software does not exist in a vacuum. There are many related business topics that lack effective size metrics. One critical example is that of the data used by software applications.

Most large companies own more data than they do software. The costs of acquiring data and maintaining it are at least as high as software development and maintenance costs. Data migration from legacy applications to new applications can take more than three calendar years. Data quality is suspected to be worse than software quality, but no one really knows because there is no effective size metric for quantifying database volumes or measuring data quality.

It would seem to be useful to apply the logic of function point metrics to other critical business topics, and create an integrated suite of functional metrics that could encompass not only software, but the related areas of data, websites, hardware devices, and also risk and value.

Software and online data are among the most widely utilized commodities in human history. If you consider the total usage of various commodities, the approximate global rank in terms of overall usage would be:

1. Water
2. Salt
3. Rice
4. Wheat
5. Bread

6. Corn
7. Fish
8. Plastics
9. Wood
10. Clothing
11. Shoes
12. Cell phones
13. Alcoholic beverages
14. Software
15. Online Web data
16. Electricity
17. Gasoline and oil
18. Aluminum
19. Narcotics
20. Automobiles

(The sources of data for this table include a number of websites and government tables. The importance is not actual rankings, but the fact that software and online data in 2017 are used so widely that they can be included in the list.)

The expansion of software (and online data) to join the world's most widely used commodities means that there is an urgent need for better metrics and better economic analysis.

Because of the widespread deployment of software and the millions of software applications already developed or to be developed in the future, software economic studies are among the most critical of any form of business analysis. Unfortunately, lack of an integrated suite of metrics makes software economic analysis extremely difficult.

This portion of Chapter 2 proposes a suite of related metrics that are based on the logic of function points, but expanding that logic to other business and technical areas. The metrics are hypothetical and additional research would be needed to actually develop such a metrics suite.

Potential Expansion of Function Points to Other Business Topics

In spite of the considerable success of function point metrics in improving software quality and economic research, there are a number of important topics that still cannot be measured well or even measured at all in some cases. Here are some areas where there is a need for related metrics within a broad family of functional metrics:

1. Application function point metrics
2. Component feature point metrics

3. Hardware function point metrics
4. COTS application point metrics
5. Micro function point metrics
6. Data point metrics
7. Website point metrics
8. Software usage point metrics
9. Service point metrics
10. Risk point metrics
11. Value point metrics
12. Security point metrics
13. Configuration point metrics—developed by IBM
14. Nonfunction size (SNAP points)—developed by IFPUG

This combination of a related family of functional metrics would expand the ability to perform economic studies of modern businesses and government operations that use software, websites, data, and other business artifacts at the same time for the same ultimate goals. Let us now consider each of these metrics in turn.

Topic 1: The Need for Application Function Point Metrics

From their first external publication outside of IBM in 1978, function point metrics have become the de facto standard for quantifying software applications. As of 2017, the usage of function points encompasses international benchmark studies, outsource agreements, economic analysis, quality analysis, and many other important business topics. In 2017, the governments of Brazil, Japan, Malaysia, Mexico, and South Korea now require function point metrics for software contracts. The major topics found within function points as originally defined by Allan Albrecht include:

- *Function Points*
  Inputs
  Outputs
  Inquiries
  Logical files
  Interfaces
  Complexity adjustments

There are several tools available for counting function points, but human judgment is also needed. Both IFPUG and the other major function point user groups provide training and also examinations that lead to the position of "certified function point analysts."

Function points are now the most widely used metric for quantifying software application size, for quantifying productivity and quality, and for quantifying application development costs. There is only sparse data on application maintenance costs, but that situation is improving. The ISBSG now includes software maintenance data. Several companies such as the Software Improvement Group (SIG), Relativity Technologies, CAST Software, Optimyth, and Computer Aid measure and evaluate maintainability.

It should be noted that software is treated as a taxable asset by the IRS in the United States and by most other international tax organizations. Function point metrics are now widely used in determining the taxable value of software when companies are bought or sold.

Note 1: In 2012, the IFPUG released information on a new metric for nonfunctional requirements called SNAP. As of early 2017, there is still sparse empirical data on the volume of SNAP points relative to normal function points in the same application. As data becomes available, it will be added to software-estimating tools such as the author's SRM tool. As of 2017, SRM predicts SNAP point size.

Also, function point metrics have a tendency to be troublesome for maintenance and multi-tier software where quite a bit of work involves dealing with surrounding software packages.

Note 2: This chapter is based on function points as defined by IFPUG. There are a number of alternative function point metrics including but not limited to:

COSMIC function points
Engineering function points
FISMA function points
Mark II function points
NESMA function points
Unadjusted function points
Story points
Use case points

These function point variations all produce different results from IFPUG function points. Most produce larger results than IFPUG for unknown reasons.

Topic 2: The Need for Component Feature Point Metrics

While function points are the dominant metric for software applications, currently, in 2017, applications are often created from libraries of reusable components, objects, and other existing software segments. While some of these may have been counted via normal function point analysis, most are of unknown size.

There is a need to extend normal function point analysis down at least one level to be able to size reusable modules, objects, and the contents of class libraries. To avoid confusion with the term function points, which normally apply to entire applications, it might be better to use a different term such as "component feature points."

- *Component Feature Points*
  Inputs
  Outputs
  Inquiries
  Logical files
  Interfaces
  Complexity adjustments

Examples of the kinds of specific features that might be sized using component feature points would include, but not be limited to:

1. Input validation	(25–50 component feature points)
2. Output formatting	(10–30 component feature points)
3. Query processing	(3–15 component feature points)
4. Currency exchange rate calculation	(5–15 component feature points)
5. Inflation rate calculation	(5–10 component feature points)
6. Compound interest calculation	(5–25 component feature points)
7. Sensor-based input monitoring	(10–35 component feature points)
8. Earned-value calculations	(30–75 component feature points)
9. Internal rate of return (IRR)	(5–15 component feature points)
10. Accounting rate of return (ARR)	(5–15 component feature points)
11. Password processing	(5–15 component feature points)

The basic idea is to assemble a taxonomy of standard components that are likely to be acquired from reusable sources rather than being custom developed. In other words, component feature points shift the logic of functional analysis from the external applications themselves to the inner structure and anatomy of applications.

As of 2017, the total number of possible reusable components is unknown, but probably is in the range of about 500 to 2500. There is also a lack of a standard taxonomy for identifying the specific features of software components. These are problems that need additional research.

The best way to develop an effective taxonomy of application features would probably be a forensic analysis of a sample of current software applications, with the intent of establishing a solid taxonomy of specific features including those inserted from reusable materials.

Component feature points would adhere to the same general counting rules as standard function points, but would be aimed at individual modules and features that are intended to be reused in multiple applications. Because some of the smaller components may be below the boundary line for normal function point analyses, see the section on micro function points later in this chapter. Component feature points would be additive to ordinary IFPUG function points; i.e., for an application with 25 features, the sum of the feature sizes would be equal to total function point size.

Topic 3: The Need for Hardware Function Point Metrics

The U.S. Navy, the U.S. Air Force, the U.S. Army, and the other military services have a significant number of complex projects that involve hardware, software, and microcode. Several years ago, the Navy posed an interesting question: "Is it possible to develop metric-like function points for hardware projects, so that we can do integrated cost analysis across the hardware/software barrier?"

In addition to military equipment, there are thousands of products that feature embedded software: medical devices, smartphones, GPS units; cochlear implants, hearing aids, pacemakers, MRI devices, automobile antilock brakes; aircraft control systems; and countless others. All these hybrid devices require sizing and estimating both the software and hardware components at the same time.

The ability to perform integrated sizing, cost, and quality studies that could deal with software, hardware, databases, and human service and support activities would be a notable advance indeed. A hypothetical engineering point metric might include the following factors:

- *Hardware function points*
  Inputs
  Outputs
  Constraints
  Innovations
  Algorithms
  Subcomponents

Integrated cost estimates across the hardware/software boundary would be very welcome in many manufacturing and military domains. These hardware function points would be utilized for embedded applications such as medical devices, digital cameras, and smart appliances. They would also be used for weapons systems and avionics packages. They would also be used for all complex devices such as automobile engines that use software and hardware concurrently. Hardware function points would be a useful addition to an overall metrics suite.

Topic 4: The Need for COTS Function Point Metrics

Many small corporations and some large ones buy or acquire more software than they build. The generic name for packaged applications is "commercial off-the-shelf software," which is usually abbreviated to COTS.

COTS packages could be sized using conventional function point analysis if vendors wished to do this, but most do not. As of 2017, it is technically possible to size COTS packages using pattern matching. For example, the SRM sizing tool of Namcook Analytics LLC can size COTS software such as Windows 10, Quicken, the Android operating system, and all others. The same is true for sizing open-source applications. The open-source business sector is growing rapidly, and many open-source applications are now included in corporate portfolios.

The concept of pattern matching uses a formal taxonomy of applications types that includes the class of the application (internal or external), the type (embedded software, information technology, systems or middleware, etc.), and several other parameters. An application to be sized is placed on the taxonomy. Applications that have the same "pattern" on the taxonomy are usually of almost the same size in function points. The pattern matching approach uses a combination of a standard taxonomy and mathematical algorithms to provide a synthetic function point total, based on historical applications whose sizes already exist. While normal function points are in the public domain, the pattern matching approach is covered by a patent application. Some of the other metrics in this chapter may also include patentable algorithms.

The pattern matching approach substitutes historical data for manual counting, and to be effective, the patterns must be based on a formal taxonomy. Pattern matching applies some of the principles of biological classification to software classification.

A study performed by the author of the corporate portfolio of a major Fortune 500 corporation noted that the company owned software in the following volumes:

Application Types	*Ownership*
Information systems	1360
COTS packages	1190
Systems software	850
Embedded applications	510
Tools (software development)	340
Manufacturing and robotics	310
End-user developed	200
Open-source	115
SaaS applications	5
TOTAL	4880

As can be seen, COTS packages ranked number two in the corporation's overall portfolio and comprised 24.4% of the total portfolio. This is far too important a topic to be excluded from sizing and economic analysis. For one thing, effective "make or buy" analysis or determining whether to build software or acquire software packages needs the sizes of both the COTS packages and the internal packages to ensure that features sets are comparable. In fact, both function points and component feature points would be valuable for COTS analysis.

Note that the pattern matching method can also size SaaS applications such as Google Docs. Essentially, any software application can be sized using this method so long as it can be placed on the basic taxonomy of application types. Of course, the complexity questions will have to be approximated by the person using the sizing method, but most can be assumed to center on "average" values.

Right now, COTS packages and SaaS packages (and most open-source applications) are outside the boundaries of normal function point metrics primarily because the essential inputs for function point analysis are not provided by the vendors.

It would be useful to include COTS packages in economic studies if vendors published the function point sizes of commercial software applications. This is unlikely to happen in the near future. A COTS, SaaS, and open-source pattern matching metric based on pattern matching might include the following factors:

- *COTS, SaaS, and Open-Source application points*
  Taxonomy
  Scope
  Class
  Type
  Problem complexity
  Code complexity
  Data complexity

The inclusion of COTS points is desirable for dealing with make or buy decisions in which possible in-house development of software is contrasted with possible acquisition of a commercial package.

In the present-day world, many large and important applications are combinations of custom code, COTS packages, open-source packages, reusable components, and objects. There is a strong business need to be able to size these hybrid applications.

There is also a strong business need to be able to size 100% of the contents of corporate portfolios, and almost 50% of the contents of portfolios are in the form of COTS packages, open-source packages, SaaS services, and other kinds of applications whose developers have not commissioned normal function point analysis.

Topic 5: The Need for Micro Function Point Metrics

A surprising amount of software work takes place in the form of very small enhancements and bug repairs that are below about 10 function points in size. In fact, almost 20% of the total effort devoted to software enhancements and about 90% of the effort devoted to software bug repairs deal with small segments below 10 function points in size and quite a few are below 1 function point in size.

Individually, these small features are cheap and are built rapidly. Readers might wonder why size is even needed for them. But large companies in the Amazon, IBM, and General Motors class may carry out over 25,000 small updates per calendar year. Government agencies such as the Department of Defense may carry out over 75,000 small updates per year. Individually none are very expensive, but thousands of them can have expenses of quite a few million dollars per year.

The original IFPUG function point metric had mathematical limits associated with the complexity adjustment factors that made small applications difficult to size. Also, the large volume of small enhancements and the even larger volume of software defect repairs would be time-consuming and expensive for normal function point analysis.

Some of the other function point metrics such as COSMIC function points can size smaller sizes, but these are not widely used in the United States.

The same method of pattern matching can easily be applied to small updates and bug repairs, and this form of sizing takes only a few minutes.

There are three possibilities for micro function points: (1) normal function point analysis with changes to eliminate the lower boundaries of adjustment factors; (2) pattern matching; (3) backfiring or mathematical conversion from counts of logical code statements.

- *Micro function points using normal counts*
  Inputs
  Outputs
  Inquiries
  Logical files
  Interfaces
  Revised complexity adjustments
- *Micro function points using pattern matching*
  Taxonomy
  Scope
  Class
  Type
  Problem complexity
  Code complexity
  Data complexity

There is another way of sizing small features that has existed since 1975, because it was first developed by Al Albrecht and the original IBM function point team. This method is called backfiring. The term backfire means mathematical conversion from function points to logical code statements, or vice versa. Data is available on ratios of function points to logical code statements for over 1,000 programming languages in 2017. Table 2.5 illustrates a small sample for common languages.

Table 2.5 Sizing Small Features by Backfiring

Language Level	*Sample Languages*	*Source Code per Function Point*
1	Basic assembly	320
2	C	160
3	COBOL	107
4	PL/I	80
5	Ada95	64
6	Java	53
7	Ruby	46
8	Oracle	40
9	Pearl	36
10	C++	32
11	Delphi	29
12	Visual Basic	27
13	ASP NET	25
14	Eiffel	23
15	Smalltalk	21
16	IBM ADF	20
17	MUMPS	19
18	Forte	18
19	APS	17
20	TELON	16
10	AVERAGE	58

Backfiring or mathematical conversion from logical code statements is as old as function point analysis. The first backfire results were published by Allan Albrecht in the 1970s based on simultaneous measurements of logical code statements and function points within IBM.

Surprisingly, none of the function point organizations have ever analyzed backfire data. Backfiring is not as accurate as normal function point analysis due to variations in programming styles but it remains a popular method due to the high speed and low cost of backfiring compared to normal function point analysis.

There are published tables of ratios between logical code statements and function points available for about 1,000 programming languages. In fact, the number of companies and projects that use backfiring circa 2017 is probably larger than the number of companies that use normal function point analysis.

As an example of why micro function points are needed, a typical software bug report when examined in situ in the software itself is usually between about 0.1 and 4 function points in size: much too small for normal function point analysis.

Individually, each of these bugs might be ignored, but large systems such as Windows 7 or SAP can receive more than 50,000 bug reports per year. Thus, the total volume of these tiny objects can top 100,000 function points and the costs associated with processing them can top $50,000,000/year. There is a definite need for a rapid and inexpensive method for including thousands of small changes into overall software cost and economic analyses.

Since normal function point analysis tends to operate at a rate of about 400 function points per day or 50 function points per hour, counting a typical small enhancement of 10 function points would require perhaps 12 min.

The pattern matching method operates more or less at a fixed speed of about 1.5 min per size calculation, regardless of whether an ERP package of 300,000 function points or an enhancement of 10 function points is being sized. Therefore, pattern matching would take about 1.5 min.

What would probably be a suitable solution would be to size a statistically valid sample of several hundred small bug repairs and small enhancements, and then simply use those values for sizing purposes. For example, if an analysis of 1000 bugs finds the mean average size to be 0.75 function points, that value might be used for including small repairs in overall economic studies.

It might be noted that the author's Software Risk Master tool can size applications over a range that spans from less than 1 function point to more than 300,000 function points. Further, the time required to size the application is independent of the actual size and averages about 1 min and 30 s per application.

Topic 6: The Need for Data Point Metrics

In addition to software, companies own huge and growing volumes of data and information. In fact, "big data" is one of the hottest topics in software engineering as of 2017.

As topics such as repositories, data warehouses, data quality, data mining, and online analytical processing (OLAP) become more common, it is obvious that there are no good metrics for sizing the volumes of information that companies own. Neither are there good metrics for exploring data quality, the costs of creating data, migrating data, or eventually retiring aging legacy data.

A metric similar to function points in structure but aimed at data and information rather than software would be a valuable addition to the software domain. A hypothetical data point metric might include the following factors:

- *Data points*
  Logical files
  Entities
  Relationships
  Attributes
  Inquiries
  Interfaces

Surprisingly, database and data warehouse vendors have performed no research on data metrics. Each year more and more data is collected and stored, but there are no economic studies of data costs, data quality, data life expectancy, and other important business topics involving data.

If you look at the entire portfolio of a major corporation such as a large bank, they probably own about 3,000 software applications with an aggregate size of perhaps 7,500,000 function points. But the volume of data owned by the same bank would probably be 50,000,000 data points, if there were an effective data point metric in existence.

It is a known fact that the average number of software defects released to customers in 2017 is about 0.45 per function point. No one knows the average number of data errors, but from analysis of data problems within several large companies, it is probable that data errors in currently active databases approach 2.5 defects per "data point" or almost four times as many errors as software itself.

There is a very strong economic need to include data acquisition costs, data repair costs, and data quality in corporate financial analyses. The data point metric would be probably as useful and as widely utilized as the function point metric itself. Lack of quantification of data size, data acquisition costs, data migration costs, and data quality are critical gaps in corporate asset economic analysis. A data point is important enough so that it might well be protected by a patent.

Data is already a marketable product and hundreds of companies sell data in the form of mailing lists, financial data, tax information, and the like. If data is treated as a taxable asset by the IRS, then the need for a data point metric will be critical for tax calculations, and for use in determining the asset value of data when companies are bought or sold.

Since the theft of valuable data is now one of the most common crimes in the world, an effective data point metric could also be used in ascertaining the value of lost or stolen data.

Topic 7: The Need for Website Point Metrics

In today's business world of 2017, every significant company has a website, and an ever-growing amount of business is transacted using these websites.

While function points can handle the software that lies behind the surface of websites, function points do not deal with website content in the forms of graphical images, animation, and other surface features. There is a strong business need to develop "website points" that would be able to show website development costs, maintenance costs, and website quality.

Some of the topics that would be included in website points would be:

- *Website points*
  Transactions
  Inquiries
  Images
  Text
  Audio
  Animation

An examination of any of today's large and complex websites, such as Amazon, Google, state governments, and even small companies immediately demonstrates that sizing and quantification are needed for many more topics than just the software that controls these websites.

From a rudimentary analysis of website economics, it appears that the cost of the content of websites exceeds the cost of the software controlling the website by somewhere between 10 to 1 and 100 to 1. Massive websites such as Amazon are at the high end of this spectrum. But the essential point is that websites need formal sizing methods and reliable economic methods.

The software that controls the Amazon website is probably about 18,000 function points in size. But the total Web content displayed on the Amazon site would probably top 25,000,000 website points if such a metric existed.

Topic 8: The Need for Software Usage Point Metrics

Function point metrics in all of their various flavors have been used primarily to measure software development. But these same metrics can also be used to measure software usage and consumption.

In order to come to grips with software usage patterns, some additional information is needed:

- Is the software used by knowledge workers such as physicians and lawyers?
- Is the software used for business transactions such as sales?
- Is the software used to control physical devices such as navigational instruments?
- Is the software used to control military weapons systems?

Table 2.6 illustrates the approximate usage patterns noted for 30 different occupation groups in 2016.

Software usage points are identical to normal function points, except that they are aimed at consumption of software rather than production of software. Software usage patterns play a major role in quantifying the value of many software applications. Software usage can be calculated using either normal function point analysis or pattern matching.

- *Usage points using normal function point counts*
  Inputs
  Outputs
  Inquiries
  Logical files
  Interfaces
  Revised complexity adjustments
  Knowledge usage
  Operational usage
  Transactional usage
  Indirect usage (in embedded devices)
- *Usage points using pattern matching*
  Taxonomy
  Scope
  Class
  Type
  Problem complexity
  Code complexity
  Data complexity
  Knowledge usage
  Operational usage
  Transactional usage
  Indirect usage (in embedded devices)

Usage points are not really a brand-new metric but rather are function points augmented by additional information and aimed in a different direction.

Table 2.6 Daily Software Usage by 30 Occupation Groups (Size Expressed in Terms of IFPUG Function Points, Version 4.3)

	Occupation Groups	*Size in Function Points*	*Number of Packages*	*Hours used per Day*	*Value to Users*
1	NSA analysts	7,500,000	60	24	10
2	Military planners	5,000,000	50	7.50	9
3	Astronaut (space shuttle)	3,750,000	50	24	10
4	Physicians	3,500,000	25	3	9
5	Ship captains (naval)	2,500,000	60	24	8
6	Aircraft pilots (military)	2,000,000	50	24	10
7	FBI Agents	1,250,000	15	3	7
8	Ship captains (civilian)	1,000,000	35	24	7
9	Biotech researchers	1,000,000	20	4.50	6
10	Airline pilots (civilian)	750,000	25	12	7
11	Movie special effects engineer	750,000	15	6	9
12	Air traffic controllers	550,000	5	24	9
13	Attorneys	325,000	12	2.50	5
14	Combat officers	250,000	12	10	6
15	Accountants	175,000	10	3	4
16	Pharmacists	150,000	6	3.50	4
17	U.S. congress staff	125,000	15	6	4
18	Electrical engineers	100,000	25	2.50	5
19	Combat troops	75,000	7	18.00	6
20	Software engineers	50,000	20	6.50	8
21	Police officers	50,000	6	8	4
22	Corporate officers	50,000	10	1.50	3
23	Stock brokers	50,000	15	10	5
24	Project managers	35,000	15	2	5

(Continued)

Table 2.6 (*Continued*) Daily Software Usage by 30 Occupation Groups (Size Expressed in Terms of IFPUG Function Points, Version 4.3)

	Occupation Groups	*Size in Function Points*	*Number of Packages*	*Hours used per Day*	*Value to Users*
25	IRS tax agents	35,000	12	8	6
26	Civil engineers	25,000	10	2	6
27	Airline travel reservations	20,000	3	12	9
28	Railroad routing and control	15,000	3	24	9
29	Customer support (software)	10,000	3	8	4
30	Supermarket clerks	3,000	2	7	4
	Averages	18,000	5.50	10.17	6.33

Incidentally, examining software usage patterns led to placing software as number 10 on the list of widely used commodities at the beginning of this chapter.

Topic 9: The Need for Service Point Metrics

The utility of function points for software studies has raised the question as to whether or not something similar can be done for service groups such as customer support, human resources, sales personnel, and even health and legal professionals.

Once software is deployed, a substantial amount of effort is devoted to responding to customer request for support. This service effort consists of answering basic questions, dealing with reported bugs, and making new information available to clients as it is created.

The cost drivers of software service are based on five primary factors:

1. The size of the application in function points
2. The number of latent bugs in the application at release
3. The number of clients using the application
4. The number of translations into other national languages
5. The planned response interval for customer support contacts

What would be useful would be a metric similar in structure to function points, only aimed at service functions within large corporations. Right now, there is no easy way to explore the lifetime costs of systems that include extensive human service components as well as software components. A hypothetical service point metric might include the following factors:

- *Service points*
  Customers (entities)
  Countries where the application is used
  Latent defects at deployment
  Desired response time for customer contacts
  Inquiries
  Reference sources
  Rules and Regulations (constraints)

Experiments with variations on the function point metric have been carried out for software customer support groups. The results have been encouraging, but are not yet at a point for formal publication.

The U.S. is now largely a service-oriented economy. Software has a significant amount of TCO tied up in service-related activities.

Topic 10: The Need for Value Point Metrics

One of the major weaknesses of the software industry has been in the areas of value analysis and the quantification of value. All too often what passes for "value" is essentially nothing more than cost reductions or perhaps revenue increases. While these are certainly important topics, there are a host of other aspects of value that also need to be examined and measured: customer satisfaction, employee morale, national security, safety, medical value, and a host of other topics. A hypothetical value point metric might include the following factors:

- *Value points*
  Safety improvement
  Military and defense value
  National security improvement
  Health and medical improvement
  Patents and intellectual property
  Risk reduction
  Synergy (compound values)
  Cost reduction
  Revenue increases
  Market share increases
  Schedule improvement
  Competitive advantages
  Customer satisfaction increase
  Staff morale increase
  Mandates or statutes

Note that although cost reduction and revenue increases are both tangible value factors, a host of other less tangible factors also need to be examined, weighted, and included in a value point metric.

Intangible value is the current major lack of today's methods of value analysis. There is no good way to quantify topics such as medical value, security value, or military value. For example, medical devices such as cochlear implants improve quality of life. Military devices such as a shipboard gun control system improve crew and vessel defenses. Better encryption algorithms in software improve security of classified and all kinds of private information.

A value point metric would assign points for: (1) Direct revenues; (2) Indirect revenues; (3) Transaction rate improvements; (4) Operational cost reduction; (5) Secondary cost reduction; (6) Patents and intellectual property; (7) Enterprise prestige; (8) Market share improvements, (9) Customer satisfaction improvement; (10) Employee morale improvements; (11) Medical condition benefits; (12) National security benefits. In other words, both financial and nonfinancial values would be assigned value points. The sum total of value points would include both financial and nonfinancial values such as medical and military values.

Topic 11: The Need for Risk Point Metrics

Software projects are nothing if not risky. Indeed, the observed failure rate of software projects is higher than almost any other manufactured product. While software risk analysis is a maturing discipline, there are still no metrics that can indicate the magnitude of risks. Ideally, both risks and value could be analyzed together. A hypothetical value risk point metric might include the following factors:

- *Risk points*
  Risks of death or injury
  Risks to national security
  Risks of property destruction
  Risks of theft or pilferage
  Risks of litigation
  Risks of business interruption
  Risks of business slowdown
  Risks of market share loss
  Risks of schedule delays
  Risks of cost overruns
  Risks of competitive actions
  Risks of customer dissatisfaction
  Risks of staff dissatisfaction

Large software projects fail almost as often as they succeed, which is a distressing observation that has been independently confirmed.

It is interesting that project management failures in the form of optimistic estimates and poor quality control tend to be the dominant reasons for software project failures.

The bottom line is that risk analysis supported by some form of risk-point quantification might reduce the excessive number of software project failures that are endemic to the production of large software applications.

As it happens, there is extensive data available on software risks. A number of risks correlate strongly to application size measured in function points. The larger the application, the greater the number of risks will occur and the more urgent the need for risk abatement solutions.

Risk points could be combined with value points, function points, and data points for determining whether or not to fund large and complex software projects that might not succeed. While function points are useful in funding decisions, the costs of data migration and data acquisition need to be considered too, as do risk factors.

Topic 12: The Need for Security Points

Software and the data processed by software now control most of the major assets of the industrialized world. All citizens now have proprietary information stored in dozens of databases: birth dates, social security number, bank account numbers; mortgages, debts, credit ratings, and dozens of other confidential topics are stored in numerous government and commercial databases.

Hacking, worms, denial-of-service attacks, and identity thefts are daily occurrences, and there is no sign that they will be reduced in numbers in the future.

These facts indicate a strong need for a "security point" metric that will provide quantification of the probable risks of both planned new applications and also legacy applications that process vital information.

- *Security points*
  Value of the information processed
  Volume of valuable information (using data points)
  Consequences of information theft or loss
  Consequences of disruption or denial of service
  Security flaw prevention methods
  Security attack monitoring methods
  Immediate responses for security attacks

Security as of 2017 is not as thorough as it should be. Hopefully, the development of a security point metric will encourage software developers, executives, and clients to be more proactive in avoiding security risks, and more effective in dealing with security attacks.

The purpose of security points is twofold: one is to identify in a formal manner all the security risk topics; the second is to identify in a formal manner all the known security solutions. It is obvious that security cannot be fully effective by using only firewalls and external software to intercept viruses, worms, and other malware. Software needs to have a stronger immune system that can fight off invading malware due to better internal controls and eliminating the current practice of transferring control and exposing confidential information.

Topic 13: The Need for Configuration Points

This 13th metric was not developed by the author but was provided by George Stark of the IBM Global Technology Center in Austin, TX. IBM has been a pioneer in metrics research since the original function point metrics were developed at IBM White Plains in the middle 1970s.

The configuration point metric is used to predict the work effort for deploying complex suites of software and hardware that need to operate together. Unlike some of the prior metrics in this report, configuration points have existed since 2006 and have been used on a number of actual installations and seem to generate useful information.

- *Configuration Points*
  Cabling
  Software assets and configurations
  Computing assets
  Communication assets
  External interfaces
- *Value-added adjustments*
  Security
  Installation ease
  Common components
  Environment complexity
  Customizations
  External services
  Staff experience

When used for deploying large and complex combinations of software and devices, the ranges of component points to date have been between about 30,000 and 70,000. When comparing component points to standard function points, it can be seen that this metric is clearly aimed at the problems of deploying fairly massive combinations of features.

Topic 14: The Need for Nonfunctional Size (SNAP Points)

Software is primarily constructed to satisfy user requirements. But in addition to user requirements, there may be government mandates, laws, or technical topics that also need to be included. These are called "nonfunctional" requirements.

In 2012, the IFPUG started work on a metric to quantify nonfunctional requirements called SNAP for software nonfunctional assessment process.

Unfortunately, the SNAP definition committee decided not to make SNAP points additive to function points. If an application is 1,000 function points in size and 150 SNAP points in size, you can't add the two numbers together and get a single size.

This was an unfortunate choice because before SNAP ordinary function points could be used to quantify the effort for a large number of work activities such as requirements, design, architecture, coding, testing, quality assurance, management, etc.

The work of building nonfunctional requirements should be additive to all the other kinds of work, and SNAP points should have been additive to function points. In fact, in this section on 14 expanded metrics, SNAP is the ONLY metric that is not additive.

As a result of the difficulty of including SNAP, most parametric estimation tools may predict size for SNAP separately, but the effort and costs are added to function point costs, which makes good economic sense.

When you are building a house, you will have costs for the foundation, the walls, roofing, plumbing, electricals, septic system, windows, etc. But all these disparate costs can be subsumed into a metric such as cost per square foot or cost per square meter.

Software needs a single unified cost metric, so the author adds SNAP costs and functional costs together, on the grounds that nonfunctional costs would not even exist unless user functions are constructed, so it is right and proper to subsume nonfunctional costs into cost per function point.

If functional costs for 1,000 function points are $1,000 per function point or $1,000,000, and nonfunctional SNAP points total to 200 at $1,000 per SNAP point or $200,000, then the total cost for the application will be $1,200,000. We state that the overall cost is $1,200 per function point with a "tax" included for nonfunctional work.

Example of Multi-Metric Software Economic Analysis

Because this proposed suite of metrics is hypothetical and does not actually exist as of 2017, it might be of interest to show how some of these metrics might be used. (In this small example, some of the metrics aimed at large applications such as configuration points are not shown.) Let us consider an example of a small embedded device such as a smartphone or a handheld GPS that utilizes a combination of hardware, software, and data in order to operate (Table 2.7).

Table 2.7 Example of Multi-Metric Economic Analysis

Development Metrics	*Number*	*Cost*	*Total*
Function points	1,000	$1,000	$1,000,000
SNAP points ** (Not additive)	150	$1,100	$165,000
Development Metrics	*Number*	*Cost*	*Total*
Function points	1,000	$1,000	$1,000,000
Data points	1,500	$500	$750,000
Hardware function points	750	$2,500	$1,875,000
Subtotal	3,250	$4,000	$3,625,000
Annual Maintenance Metrics			
Enhancements (micro function points)	150	$750	$112,500
Defects (micro function points)	750	$500	$375,000
Service points	5,000	$125	$625,000
Data maintenance	125	$250	$31,250
Hardware maintenance	200	$750	$150,000
Annual Subtotal	6,235	$2,375	$1,293,750
TCO			
(Development+5 years of usage)			
Development	3,250	$1,115	$3,625,000
Maintenance, enhancement, service	29,500	$189	$5,562,500
Data maintenance	625	$250	$156,250
Hardware maintenance	1,000	$750	$750,000
Application total TCO	34,375	$2,304	$10258,750
Risk and Value Metrics			
Risk points	2,000	$1,250	$2,500,000
Security points	1,000	$2,000	$2,000,000
Subtotal	3,000	$3,250	$4,500,000
Value points (revenues, savings)	45,000	$2,000	$90,000,000
Net Value	10,625	$7,521	$79,906,250
Return on investment			$8.92

**Mean that IFPUG does not want SNAP added to ordinary function points.

As can be seen, normal function points are used for the software portion of this product. But since it also has a hardware component and uses data, hardware points, data points, and service points are also part of the cost of the application.

While smartphones are security risks, GPS devices are not usually subject to hacking in a civilian context. Therefore, the risk and security totals are not high.

Value points would be based on a combination of direct revenues, indirect revenues for training and peripherals. There might also be drag-along revenues for additional services such as applications.

Note that software development itself is less than one-tenth of the TCO. Note also that economic value should be based on TCO for the entire product, and not just the software component.

The Probable Effort and Skill Sets to Create Additional Metrics

Allan Albrecht, John Gaffney, and other IBM colleagues worked on the development of function point metrics for several years before reaching a final version that achieved consistently good results.

Each of the proposed metrics in this chapter would probably require a team that includes both function point experts and domain experts in topics such as data structures, hardware engineering, accounting, and other relevant topics. A single inventor might be able to derive some of these metrics, but probably a multidisciplinary team would have more success.

Because function points already exist, creating a family of metrics that utilize similar logic would not be trivial, but would probably not be quite as difficult as the original development of function points in IBM in the 1970s. Following are the probable team sizes, skill sets, and schedules for creating a family of functional metrics.

Metric and Skills	*Team Size*	*Schedule Months*
1. Application function point metrics[a] Software engineering Accounting and finance Statistical analysis	6	24
2. Component feature point metrics[b] Function point analysis Software engineering Taxonomy construction	4	12

Metric and Skills	*Team Size*	*Schedule Months*
3. Hardware function point metrics Function points Electrical engineering Mechanical engineering Aeronautical engineering Accounting and finance	6	18
4. COTS application point metrics[a] Function point analysis Taxonomy construction Software engineering	1	6
5. Micro function points[b] Function point analysis Maintenance of software	3	3
6. Data point metrics Function point analysis Data structure analysis Data normalization methods Accounting and finance	6	18
7. Website point metrics Function point analysis Website design Web content sources Graphical design Accounting and finance	6	18
8. Software usage point metrics[a] Function point analysis Accounting and finance	1	3
9. Service point metrics Function point analysis Info. Tech. Infrastructure. Library	4	9
10. Risk point metrics Function point analysis Software risks Software risk abatement Accounting and finance	4	6

Metric and Skills	*Team Size*	*Schedule Months*
11. Value point metrics Function point analysis Accounting and finance Software engineering Economic modeling Multivariate analysis	6	9
12. Security point metrics Software security principles Costs of security breaches Function point analysis	6	6
13. Configuration points[a] (Developed by IBM)	NA	NA
14. SNAP points[a] (Developed by IFPUG)	NA	NA
Total	53	132

[a] Metric currently exists.
[b] Metric exists in prototype form.

As can be seen, the set of possible functional metrics discussed in this chapter requires substantial research. This kind of research would normally be performed either by a university or by the research division of a major company such as IBM, Microsoft, Google, Oracle, and the like. Indeed, configuration points are a recent metric developed by IBM.

For example, as a database company, Oracle should certainly be interested in data point metrics and should already have data about migration costs, data quality, and the like. But as of 2017, database and ERP installation routinely cost more than expected, while data migration efforts routinely run late and encounter data quality problems. Data economics remains a critical unknown in corporate economic studies.

The purpose of this chapter is to illustrate that while function points are valuable metrics for software economic analysis, software does not exist in a vacuum and many other business and technical topics would benefit from the logic of functional metrics.

The most critical gaps in metrics as of 2017 are the lack of effective metrics for dealing with data size and data quality, and the lack of effective metrics that can integrate tangible and intangible value.

It goes without saying that the suite of metrics cannot be developed in isolation. They need to be considered as a set, and they also need to be commensurate with standard function points so that the various functional metrics can be dealt with mathematically and be used for statistical analysis as a combined set of related metrics.

The 13 proposed metrics discussed in this chapter are not necessarily the only additional metrics that might be useful. The fundamental point is that the function point community should expand their vision from software alone and begin to address other critical business problems that lack effective metrics and measurement techniques.

Size and Cost Growth over Multiple-Year Periods

A topic that is not covered well in the metrics and function point literature is that of continuous growth and change in size. Software requirements tend to grow at rates of about 1% to 2% per calendar month during development. After release, software applications continue to grow at about 8% to 12% per calendar year. Every few years commercial software will add mid-life kickers or big increases in functionality, which of course adds to function point totals.

SRM uses a proprietary sizing engine that predicts and accumulates size from the start of requirements through up to 10 years of post-release maintenance and enhancements.

Although this chapter concentrates on quality and the initial release of a software application, the SRM sizing algorithms actually create 15 size predictions. The initial prediction is for the nominal size at the end of requirements.

SRM also predicts requirements creep and deferred functions for the initial release. After the first release, SRM predicts application growth for a 10-year period.

To illustrate the full set of SRM size predictions, the following table shows a sample application with a nominal starting size of 10,000 function points. All the values are in round numbers to make the patterns of growth clear (Table 2.8).

As can be seen from the table, software applications do not have a single fixed size, but continue to grow and change for as long as they are being used by customers or clients. Namcook Analytics and SRM renormalize productivity and quality data on an annual basis due to changes in application size over time.

Because the table is based on commercial software, two mid-life kickers with extra functionality were added during years 4 and 8. Commercial software vendors tend to do this for competitive reasons.

Table 2.8 SRMs Multiyear Sizing

	Nominal Application Size in IFPUG Function Points and SNAP Points	*IFPUG 10,000*	*SNAP 150*	
		Function Points	*SNAP Points*	
1	Size at end of requirements	10,000	1,500	
2	Size of requirement creep	2,000	300	
3	Size of planned delivery	12,000	1,800	
4	Size of deferred functions	–4,800	–720	
5	Size of actual delivery	7,200	1,080	
6	Year 1	12,000	1,800	
7	Year 2	13,000	1,950	
8	Year 3	14,000	2,100	
9	Year 4	17,000	2,550	Kicker 1
10	Year 5	18,000	2,700	
11	Year 6	19,000	2,850	
12	Year 7	20,000	3,000	
13	Year 8	23,000	3,450	Kicker 2
14	Year 9	24,000	3,600	
15	Year 10	25,000	3,750	

Summary and Conclusions on Function Points and Expanded Functional Metrics

The value of function point metrics for economic analysis of software applications is good enough to suggest that the same logic might usefully be applied to other business topics that are difficult to measure.

The two most difficult measurement topics other than software are data and value, as of 2017. Data lack any metrics whatsoever, and there is no reliable information on data costs or data quality. Value has metrics for revenues and cost reduction, but no effective metrics for handling nonfinancial value such as medical value, military value, and many others.

A suite of 14 integrated metrics should open up new kinds of economic analysis for hybrid hardware/software products, and also for hybrid software/big data applications.

Readings and References on Metrics and Function Point Analysis

Boehm, B. Dr.; *Software Engineering Economics*; Prentice Hall, Englewood Cliffs, NJ; 1981; 900 pages.

Brooks, F.; *The Mythical Man-Month*; Addison-Wesley, Reading, MA; 1974, rev. 1995.

DeMarco, T.; *Why Does Software Cost So Much?* Dorset House, New York; 1995; ISBN 0-9932633-34-X; 237 pages.

Fleming, Q.W. and Koppelman, J.M.; *Earned Value Project Management*; 2nd edition; Project Management Institute, New York; 2000; ISBN 10 1880410273; 212 pages.

Galorath, D.D. and Evans, M.W.; *Software Sizing, Estimation, and Risk Management: When Performance Is Measured Performance Improves*; Auerbach, Philadelphia, PA; 2006; ISBN 10-0849335930; 576 pages.

Garmus, D. and Herron, D.; *Function Point Analysis*; Addison Wesley, Boston, MA; 2001; ISBN 0-201069944-3; 363 pages.

Garmus, D. and Herron, D.; *Measuring the Software Process: A Practical Guide to Functional Measurement*; Prentice Hall, Englewood Cliffs, NJ; 1995.

Jones, C.; Program quality and programmer productivity; IBM Technical Report TR 02.764, IBM San Jose, CA; January 1977.

Jones, C.; Sizing up software; *Scientific American*; New York; Dec. 1998, Vol. 279 No. 6; December 1998; pp. 104–109.

Jones, C.; *Software Assessments, Benchmarks, and Best Practices*; Addison Wesley Longman, Boston, MA; 2000; 659 pages.

Jones, C.; Conflict and litigation between software clients and developers; Version 6; Software Productivity Research, Burlington, MA; June 2006; 54 pages.

Jones, C.; *Estimating Software Costs*; McGraw Hill, New York; 2nd edition, 2007; ISBN13: 978-0-07-148300-1; 644 pages.

Jones, C.; *Applied Software Measurement*; McGraw Hill, 3rd edition; 2008; ISBN 978-0-07-150244-3; 575 pages.

Jones, C.; *Software Engineering Best Practices*; McGraw Hill; New York; 2010.

Jones, C. and Bonsignour, O.; *The Economics of Software Quality*; Addison Wesley; Boston, MA; 2011.

Jones, C.; *The Technical and Social History of Software Engineering*; Addison Wesley; Boston, MA; 2014.

Kan, S.H.; *Metrics and Models in Software Quality Engineering*, 2nd edition; Addison Wesley Longman, Boston, MA; 2003; ISBN 0-201-72915-6; 528 pages.

Kaplan, R.S. and Norton, D.B.; *The Balanced Scorecard*; Harvard University Press, Boston, MA; 2004; ISBN 1591391342.

Love, T.; *Object Lessons: Lessons Learned in Object-Oriented Development Projects*; SIG Books Inc., New York; 1993; ISBN 0-9627477-3-4; 266 pages.

McConnell, S.; *Software Estimation: Demystifying the Black Art*; Microsoft Press, Redmond, WA; 2006; ISBN 10: 0-7356-0535-1.

Parthasarathy, M.A.; *Practical Software Estimation: Function Point Methods for Insourced and Outsourced Projects*; Addison Wesley, Boston, MA; 2007; ISBN 0-321-43910-4; 388 pages.

Putnam, L.H.; *Measures for Excellence: Reliable Software On-Time Within Budget*; Yourdon Press, Prentice Hall, Englewood Cliffs, NJ; 1992; ISBN 0-13-567694-0; 336 pages.

Putnam, L. and Myers, W.; *Industrial Strength Software: Effective Management Using Measurement*; IEEE Press, Los Alamitos, CA; 1997; ISBN 0-8186-7532-2; 320 pages.

Strassmann, P.; *The Squandered Computer*; Information Economics Press, Stamford, CT; 1997.

Stutzke, R.D.; *Estimating Software-Intensive Systems:Projects, Products, and Processes*; Addison Wesley, Boston, MA; 2005; ISBN 0-301-70312-2; 917 pages.

Yourdon, Ed.; *Outsource: Competing in the Global Productivity Race*; Prentice Hall, Upper Saddle River, NJ; 2004; ISBN 0-13-147571-1; 251 pages.

Yourdon, Ed.; *Death March: The Complete Software Developer's Guide to Surviving "Mission Impossible" Projects*; Prentice Hall, Upper Saddle River, NJ, 1997; ISBN 0-13-748310-4.

Chapter 3

Software Information Needed by Corporate Executives

Software is a critical factor in many aspects of corporate operations. Unlike most aspects of corporate operations, software has been difficult to bring under full executive control. Many chief executive officers (CEOs) and other C-level executives have only a limited knowledge about both computers and software.

Many CEOs don't regard their software teams as being as professional as other operating units. A survey found that many CEOs don't fully trust their software teams due to high cancellation rates and frequent cost and schedule overruns.

This chapter discusses 60 key questions that C-level executives should ask about, to ensure that the software their companies depend upon will be an asset and not a liability to the corporations they control. It also discusses 25 targets for improvement that are technically possible. Although specific answers within specific companies are needed, general answers to the questions are given to illustrate the ranges of possibilities.

The term "C-level executive" refers to a variety of top corporate executives whose titles begin with the word "chief": chief executive officer (CEO), chief operating officer, chief financial officer (CFO), chief information officer (CIO), chief technical officer (CTO), chief legal officer (CLO), chief security officer, and quite a few more similar titles.

Software has been a difficult technology to bring under executive control. In spite of the importance of software to corporate operations, it has remained an intractable technology, which causes almost as much trouble as it brings benefits.

CEOs and other C-level officers of modern corporations are responsible for more software and more software employees than any other occupation groups in human history. Due to government laws and mandates such as Sarbanes–Oxley in 2004, some top executives may be at personal risk of criminal charges if software governance is poor, security breaches occur, or software is misused for fraudulent purposes.

Many CEOs and other C-level officers are 40–70 years of age, and their background and training did not include much information about software and how to deal with it. This is also true of many vice presidents of operating units such as manufacturing, sales, marketing, and human resources. It is also true of other C-level executives such as CFOs and CLOs.

Today, every major corporation is largely automated for financial reporting, billing, marketing, and often manufacturing as well. Software has become a critical factor for the time to market of many new products. For a surprising number of high-technology products, software is actually embedded in the product itself and therefore may be subject to warranties.

This chapter discusses 60 of the major software topics that C-level officers should know about, to ensure that the software their companies depend upon will be an asset and not a liability to the corporations they control.

There are a number of questions that CEOs should ask their CIOs, CTOs, chief risk officers (CROs), and other key software executives. The questions are divided into six sections:

1. Current "hot" topics regarding software
2. Security, quality, and governance topics
3. Software usage, user satisfaction, and value with the corporation
4. Employee satisfaction and demographic questions
5. The economic impact of software on the corporation
6. Competitive analysis of how other corporations deal with software

Not only should the CEO know the answer to these questions, but also should the CIO, CTO, vice president of software engineering, vice president of engineering, and any other senior executive with software staffs and responsibilities know.

The answers to the questions would normally come from a CIO, CTO, vice president of information resources, or vice president of software engineering, or from an equivalent executive position.

However, senior executives themselves would be able to get answers to many of the questions available only if the corporation has a fairly sophisticated software measurement system and uses function point metrics. The answers also assume that software organizations exist that are chartered to analyze and improve software performance, i.e., one or more of these kinds of organizations are present in the company:

- Software quality assurance
- Software process improvement
- Software measurement and metrics

These three kinds of organizations are usually associated with leading-edge corporations. If such organizations are not present and if the questions are unanswerable within a company, it is a sign that software is not fully under executive control.

Leading-edge companies probably would have the answers to all these software questions readily at hand, due to annual surveys of software-related topics that leading companies tend to carry out.

Lagging companies or companies where the CEO has no interest in, or knowledge of, the significance of software face an uncomfortable future when dealing with more knowledgeable competitors.

For the purposes of this chapter, the questions are given first, and then the general answers to the questions are discussed. Readers are urged to put themselves in the position of the CEO of a major corporation and see how many of these important questions they can answer for their own corporations.

Question Set 1: Current "Hot Topic" Questions Involving Software

1. How many service-oriented architecture (SOA) and cloud applications do we have?
2. What are we doing with Information Technology Infrastructure Library (ITIL), service-level agreement (SLA), and customer support?
3. What percentage of our key business is supported by enterprise resource planning (ERP) packages?
4. What is our current distribution between development and maintenance?
5. How many of our applications are more than 10 years old and are becoming obsolete?
6. What percentage of our portfolio comes from commercial off-the-shelf (COTS) packages?
7. What percentage of our portfolio comes from open-source applications?
8. What percentage of our portfolio is being handled by outsourcers?
9. What are we doing with big data and cloud-based software?
10. What are we doing with Agile and other forms of software development?

Question Set 2: Security, Quality, and Governance Questions

1. How do we prevent security vulnerabilities and security attacks?
2. How much do we spend per year on security prevention and control?
3. How many times per year do hackers, viruses, or spyware cause damage?
4. How safe are our customer and financial data from hacking or spyware?
5. How safe are our customer and financial data from in-house theft?
6. What are our defect potentials and defect removal efficiency (DRE) in 2017?
7. What is our released defect density in terms of defects per function point?
8. How good is our software governance?
9. How many audit problems are reported against our software?
10. What is our cost of quality (COQ) per year fixing bugs or defects?

Question Set 3: Software Usage, Value, and User Satisfaction Questions

1. How many of our employees and managers use computers and software today?
2. What do our people think about the value of the software in doing their jobs?
3. What are the top complaints our people make about software?
4. Are our customers satisfied or dissatisfied with the software we deliver?
5. Are we suing any software outsourcers or vendors for breach of contract?
6. Are we in danger of being sued because of problems with our software?
7. How many companies are involved in our supply chains?
8. What are we doing to keep our software useful and valuable?
9. What are our worst current software systems in terms of defects per function point?
10. What are we doing to improve our worst software systems?

Question Set 4: Employee Satisfaction and Demographic Questions

1. Do we have the best software people in our industry?
2. What is our current breakdown between software employees and contractors?
3. How many software personnel do we employ this year?
4. How many will we need 5 years in the future; how many in 10 years?
5. Are there any critical skill shortages that we're having trouble recruiting?
6. Are our software personnel satisfied or dissatisfied with us?
7. Is our voluntary attrition rate higher or lower than our competition?
8. How do we stack up to the competition in terms of benefits and salaries?
9. What advantages or disadvantages would we get from outsourcing our software?
10. What are the pros and cons of offshore outsourcing?

Question Set 5: Software Economic Impact Questions

1. How much software do we own right now in terms of both custom and COTS?
2. What is the replacement cost of the software that we own?
3. What is the taxable value of the software that we own?
4. What is our annual growth rate in software ownership?
5. How many of our software projects are canceled and don't get finished?
6. What are the main reasons for our canceled projects?
7. How much have we spent on canceled projects over the past 5 years?
8. How much have we spent on software warranty repairs over the past 5 years?
9. How many of our applications need immediate replacement?
10. How big is our backlog of applications we can't get built?

Question Set 6: Competitive Analysis Questions

1. Which companies in our industry are best in software?
2. Which companies should we benchmark with?
3. If we spent more on software, what should we spend it on?
4. If we spent less on software, what harm would it do to us?

5. How does our software productivity compare to competition using function points?
6. How many of our software projects run late and exceed their budgets?
7. How does our software quality compare to our competition using function points?
8. What are our main strengths in the software domain?
9. What are our main weaknesses in software?
10. Are we training our people so they stay current with new technologies?

These 60 questions are not the only important questions about software, but they are definitely topics that C-level executives need to understand in 2017.

Answers to the 60 Software Questions

The following general answers to the 60 key software questions are taken from among the Namcook Analytics LLC client base. Of course, what is important for any specific company is not general knowledge, but how that company in particular would answer the 60 questions for their own internal software organizations and portfolios.

The ranges of possible answers are very broad and vary widely by industry and also by company size. Insurance and banking, for example, were early adopters of automation and tend to have large populations of software professionals. Other industries, and in particular, conglomerates, may have answers that vary by location and business function.

Nonetheless, software is an important enough management topic so that the senior executives and CEOs of all medium and large enterprises should have a basic understanding of the economic impact of software on corporate operations.

Readers and also top executives should know that software has had one of the worst collections of bad metrics and bad measurement practices of any industry in human history. Many common metrics violate standard economic principles and yield results that distort reality. Cost per defect penalizes quality. Lines of code (LOC) penalize modern programming languages. Technical debt only covers 17% of the total cost of poor quality.

Function point metrics combined with the IBM-developed metrics of defect potentials and DRE are the best choices for understanding both software economics and software quality.

The *only* software metrics that allow quality and productivity to be measured with 1% precision are the 11 primary software metrics and 11 supplemental metrics.

Primary Software Metrics for High Precision

1. Application size in function points including requirements creep
2. Size of reusable materials (design, code, documents, etc.)

3. Activity-based costs using function points for normalization
4. Work hours per month including both paid and unpaid overtime
5. Work hours per function point by activity
6. Function points per month
7. Defect potentials using function points (requirements, design, code, document, and bad-fix defect categories)
8. DRE, percentage of defects removed before release
9. Delivered defects per function point
10. COQ including prerelease and postrelease bug repairs
11. Total cost of ownership (TCO) for development for at least 3 years

Supplemental Software Metrics for High Precision

1. Software project taxonomy (nature, scope, class type)
2. Occupation groups (business analysts, programmers, testers, managers, Quality Assurance (QA), etc.)
3. Team experience levels (expert to novice)
4. Software nonfunctional requirements using Software Non-functional Assessment Process (SNAP) metrics or equivalent
5. Capability Maturity Model Integrated (CMMI) level (1–5)
6. Development methodology used on application (Agile, Team Software Process (TSP), etc.)
7. Programming language(s) used on application (Ruby, Java, mixed, etc.)
8. Complexity levels (problem, data, code complexity)
9. User effort for internal applications
10. Documents produced (type, pages, words, illustrations, etc.)
11. Meeting and communication costs (client meetings, team meetings, etc.)

All C-level executives should understand productivity and quality results in terms of work hours per function point and function points per month. They should understand defect potentials in terms of defects per function point, and they should understand their company's ranges of DRE. They should also understand the distortions and errors caused by older metrics such as LOC, cost per defect, and technical debt.

The new metric of SNAP points for nonfunctional requirements is just being developed and released by the International Function Point Users Group (IFPUG), so there is very little empirical data on SNAP results, as of early 2017. However, corporate metric teams should stay current with SNAP progress and consider adding it to local measures.

Answers to the Current "Hot Topic" Questions

The topics of SOA and cloud applications have been sweeping the industry. SOA applications are loosely coupled collections of software packages that can provide features and share data. In a sense, SOA elevates software reusability from small individual modules up to the level of complete applications. Reusable modules are typically less than 10 function points in size, while SOA features range between 1000 and 10,000 function points in size. Cloud applications are usually stored remotely in the cloud and are available for corporate use from any location.

The goal of SOA is to allow quick construction of useful applications from large collections of existing software. However, as of 2017, SOA is still partly experimental and requires some special skills and methods that are not yet common. SOA is promising but not yet a fully operational technology. Migration to SOA costs several million dollars, but the potential savings once SOA is achieved would be more than $10,000,000 per application by speeding up development and delivery of business functions.

The next hot topics is that of the ITIL and its impact on SLAs. The ITIL approach attempts to professionalize customer service, incident management, availability of software, and all the associated record-keeping surrounding incidents or bugs. This is synergistic with SLAs.

ITIL also includes sample SLAs and suggested time intervals for topics such as defect reporting and repair. Since customer support and maintenance have been somewhat behind software development in technical sophistication, ITIL is a step in the right direction. In general, the ITIL approach seems to speed up incident processing by perhaps 20%. A caveat is that ITIL works best for companies with state-of-the-art quality control. If too many bugs are released, incidents will be numerous and change processing sluggish regardless of ITIL.

The next topic concerns ERP packages such as those offered by SAP, Oracle, JD Edwards, Peopleware, and similar companies. The ERP business is based on the fact that corporations are large and multifaceted. Each operating unit built their own applications, but few of these used the same data formats or were mutually supporting. The ERP concept is that a single umbrella application could be used by all business units. The single application would facilitate shared data and common information exchange.

Of course, the ERP packages are very large: some top 300,000 function points in size. They support multiple business functions such as purchasing, customer resource management, finance, marketing, and sales.

Although the theoretical basis of ERP is sound, in practice, these large applications are quite complex, very difficult to install and customize, and have a large number of latent bugs or defects. Also, as big as they are, ERP packages only cover about one-third of all corporate business needs, so companies still need to develop scores of customized applications that link to ERP packages in various ways. ERP is powerful, but also somewhat complicated to deploy safely.

In every industry that is more than 50 years of age, there are more people working on fixing and repairing existing products than there are people building new products. Software is no exception. For almost every kind of software except Web applications, maintenance personnel comprises about 52% of total software workers, while development personnel comprises about 48%.

The life expectancy of applications is directly proportional to size. Applications larger than 10,000 function points in size often are used for more than 10 years. Applications larger than 100,000 function points in size may be used for more than 25 years. Unfortunately, software does not age gracefully. As software ages, structure decays and maintenance becomes more difficult.

A useful response to the geriatric problems of aging software is that of renovation. Using a combination of tools and manual methods, aging software can be restructured, error-prone modules (EPM) can be surgically eliminated, dead code can be excised, and the applications put in shape for continued usage.

The next interesting topic is that of commercial off-the-shelf software or COTS as it is called. Surprisingly, corporate portfolios consist of about 50% COTS packages and 50% internal or custom applications. The major COTS packages include operating systems, office suites, ERP packages, and single-purpose applications such as those supporting manufacturing, sales, or other business functions.

Unfortunately, the COTS vendors are not much better in quality control than in-house development groups. This means that many COTS applications, and especially large applications such as Windows 8, Windows 10, or ERP packages, are delivered with hundreds or even thousands of latent bugs.

Another emerging hot topic concerns open-source applications such as Linux, Firefox, and a host of others. The main difference between COTS packages and open-source software is that COTS packages are marketed by specific companies with customer support groups and other business contact points. Open-source software, on the other hand, is provided by somewhat amorphous volunteer organizations or sometimes the software just floats around the Web. New features and maintenance for open-source applications are performed by volunteers, most of whom choose to remain anonymous.

While the idea of getting useful software for free via an open-source arrangement is attractive and growing fast, there are some issues that need to be considered. The issues include: (1) uncertain maintenance in the case of bugs or defects; (2) possible liabilities for users in the event of patent violations in open-source applications; and (3) unknown security levels with possible Trojans and spyware embedded in the open-source software.

As of 2017, open-source packages comprise less than 5% of the software used by business and government organizations. The most successful open-source application is Firefox, the Web browser. This tool is probably used by at least 40% of Windows users, and has features that make it quite attractive. The Linux operating system also has a few corporate and government users, but it is far from common.

The next hot topic deals with big data and cloud-based software. The term "big data" refers to huge databases that are so large that they strain conventional database applications. These large data sets are used for predictive analytics, and some aspects of big data are critical for exploring new markets and new technical advances.

The term "cloud computing" refers to applications or services stored on the Internet and available as needed for use with local applications or business processes. Some of the services available from the cloud (which is another term for the Internet) can speed up local development by providing reusable materials.

Another hot topic concerns the well-known Software Engineering Institute (SEI), which is a nonprofit organization funded by the U.S. Department of Defense. The SEI has developed an interesting five-point evaluation scale called the "capability maturity model" or CMM for short. There is also a newer CMMI or "capability maturity model integration."

The SEI CMM scoring system rates companies based on patterns of answers to about 150 questions. The most primitive level, where organizations typically have trouble with software cost and schedule overruns, poor quality, and cancellations is Level 1, which is termed the Initial level using the SEI capability.

As of 2017, about 50% of all organizations are found at this level. The approximate distribution of the five levels is shown in Table 3.1.

The empirical evidence of the benefits of moving up the SEI CMM scale has been controversial, but as of 2017, the benefits from higher quality and productivity levels are now solid. In general, it requires about a year to move from level to level. Ascending to each higher level may cost as much as $5000–$10,000 per capita for the entire software population. The bigger the company the higher the cost.

There is some overlap between the various levels in terms of both quality and productivity. However, as empirical data accumulates, it appears that the higher levels 2–5 often have better quality and sometimes better productivity than level 1.

Because the CMM and CMMI were developed in response to the problems of large systems in large corporations, these approaches are most widely used for applications larger than 10,000 function points in size. The defense industry is more likely to use CMM than any other, although usage is quite common in the systems and embedded domains too.

The last of the emerging hot topics concerns Agile development, which is sweeping the industry. The history of the Agile methods is not as clear as the history of

Table 3.1 Distribution of Results on the SEI Capability Maturity Model

CMMI Level	*Frequency (%)*	*Meaning*
1=Initial	50	Seriously inadequate methods and tools
2=Repeatable	10	Some rigor in processes and methods
3=Defined	30	Good enough to develop reusable materials
4=Managed	3	Very good in most software topics
5=Optimizing	7	State of the art in all software topics

the CMMI, because the Agile methods are somewhat diverse. However, in 2001, the famous "Agile manifesto" was published. This provided the essential principles of Agile development. That being said, there are quite a few Agile variations that include extreme programming (XP), crystal development, adaptive software development, feature-driven development, and several others.

Some of the principal beliefs found in the Agile manifesto include the following.

- Working software is the goal, not documents
- Working software is the primary measure of success
- Close and daily contact between developers and clients are necessary
- Face-to-face conversation is the best form of communication
- Small self-organizing teams give the best results
- Quality is critical, so testing should be early and continuous

The Agile methods and the CMMI are all equally concerned about three of the same fundamental problems:

1. Software requirements always change.
2. Fixing software bugs is the most expensive software activity in history.
3. High quality leads to high productivity and short schedules.
   However, the Agile method and the CMMI approach draw apart on two other fundamental problems:
4. Paperwork is the second most expensive software activity in history.
5. Without careful measurements, continuous progress is unlikely.

The Agile methods take a strong stand that paper documents in the form of rigorous requirements and specifications are too slow and cumbersome to be effective. In the Agile view, daily meetings with clients are more effective than written specifications. In the Agile view, daily team meetings or "Scrum" sessions are the best way of tracking progress, as opposed to written status reports. The CMM and CMMI do not fully endorse this view.

The CMMI takes a strong stand that measurements of quality, productivity, schedules, costs, etc. are a necessary adjunct to process improvement and should be done well. In the view of the CMMI, without data that demonstrate effective progress, it is hard to prove that a methodology is a success or not. The Agile methods do not fully endorse this view. In fact, one of the notable gaps in the Agile approach is any quantitative quality or productivity data that can prove the success of the methods.

Poor quantitative data for Agile is partly due to poor metrics. Story points have no ISO standards or certification and vary by over 400% from company to company. Velocity, burn up, burn down, and other Agile metrics have no benchmarks. Agile should have adopted function points.

Namcook uses function points for Agile and also converts sprint data into a standard chart of accounts, which allows side-by-side comparisons between Agile and other methods.

For small projects below about 1000 function points in size, the Agile approaches have proven to have high productivity rates and fairly good quality levels too. Although Agile can be scaled up a bit, the more recent disciplined Agile delivery method is a better choice than the original Agile above 1000 function points. However, small applications below 500 function points in size are by far the most common for Agile methods.

As of 2017, Agile averages defect potentials of about 3.5 bugs per function point combined with a DRE level of about 92%. This is better than waterfall (5 bugs per function point; 85% DRE), but not as good as Rational Unified Process (RUP) or TSP or other "quality strong" methodologies that top 96% in DRE.

Answers to the Security, Quality, and Governance Questions

Cyber-attacks and cybersecurity are important modern problems. Every large company should have internal security experts, and should bring in top security consultants for applications that process or contain valuable data.

Cybersecurity goes beyond software itself and includes topics such as proper vetting of employees and contractors, physical security of data centers and software office space, and possible use ethical hackers for critical applications. It also includes software security inspections and "penetration testing" of critical applications to ensure that they resist hacking and data theft attempts.

Governance is a modern issue made more serious by the Sarbanes–Oxley law of 2004 that can bring criminal charges against executives that fail to properly govern important financial applications. Governance is a specialized topic but includes accurate cost and progress tracking, very thorough quality control and quality measures, and the use of effective test case design methods followed by test coverage analysis.

Answers to the Software Usage, Value, and User Satisfaction Questions

Depending upon the industry, from less than 35% to more than 95% of U.S. white collar workers have daily hands-on usage of computers and software. Some of the industries with the highest levels of daily use include insurance, finance, telecommunications, and of course the computer industry itself. Daily usage of software is often inversely correlated with position level: Executives may use software less than technical staff and lower management.

Most software users feel that software helps their job-related tasks, but that software is often difficult to learn and sometimes of less than optimal quality. On the basis of surveys of commercial-grade software in the United States, about 70% of users are fairly well satisfied with both quality and service, but 30% are not. For internal Management Information System (MIS) applications, about 65% of users are satisfied with quality and service, but 35% are not. Basic complaints include poor quality, slow service, and software that is not easy to use. For internal MIS software, slow development schedules top the list of complaints.

Starting more than 20 years ago, a new subindustry appeared of companies and products that could provide "geriatric care" for aging software. It is now possible to analyze the structure of existing software and automatically restructure it. Reverse engineering and reengineering are newer technologies for analyzing and converting aging software applications into modern versions. These renovations or geriatric tools are proving to be fairly effective.

The leading companies in many industries have started to reverse the alarming growth in software maintenance and enhancement costs by using these techniques. Some industry leaders are now spending less than 20% of their annual software budgets on maintenance and enhancement of aging software, while laggards may be spending as much as 70%. If your company's maintenance and enhancement costs exceed 50% of your total software budget, something may be seriously wrong in your enterprise.

The area of maintenance has been one of the most successful targets of the outsource community. Some maintenance outsource vendors are often almost twice as productive as the companies whose software they are maintaining. This is due in part to having trained maintenance personnel, and in part to having excellent tool sets. Also, maintenance outsource groups are usually not splitting their time between new development projects and maintenance of legacy applications.

Answers to the Employee Satisfaction and Demographic Questions

The growth of software professionals has been extremely large since the industry began in the 1950s. Although the growth rate is now declining, it is still positive: The overall growth of software professionals in the United States appears to be rising at a rate of 2%–5% per year. This is down from the 13% of the early 1980s, and the 20% of the 1970s.

The percentage that software professionals constitute of total employment varies with industry: In banking and insurance, from 7% to 15% of all employees can be in the software functions. Large banks and insurance companies, for example, can employ as many as 5000 software professionals in a single location, and they may have more than one such location! Overall, software professionals in the United States may approximate 2%–5% of the total workforce.

Many other kinds of companies also have high software populations: telephone operating companies, manufacturing, and defense, for example.

There are now more than 150 different occupation groups associated with software in large corporations and government agencies: programmers, analysts, quality assurance specialists, database administrators, technical writers, measurement specialists, estimating specialists, human factors specialists, and many more are now part of software functions.

If your total software staff is larger than 1000 professionals, and your company cannot identify the occupation groups employed, then something is wrong in your software and human resource organizations.

Large enterprises find that they achieve the best results from judicious usage of software specialists: Maintenance specialists and testing specialists are among the first to be needed, as are database administrators. Software, like medicine and engineering, is too vast a domain for generalists to be expert in all its phases.

As of 2017, a number of software specialties have become so popular that shortages exist in the United States. Indeed, for a few of the specialties, it may even be necessary to pay "signing bonuses" similar to that received by professional athletes. Some of the software specialties in short supply as of early 2017 include:

- Agile coaches
- SAP specialists (or Oracle or other ERP packages)
- Orchestrators for combining SOA applications
- Scrum masters for Agile applications
- Certified testing specialists
- Certified function point specialists
- Ethical hackers for security and cyber-defense work

If your voluntary attrition rate of software professionals is at 3% or higher, there may be something troubling about your software organization. If your company does not carry out annual opinion surveys, does not have a dual salary plan that provides equivalent compensation for technical staffs and managers, or does not have other modern human resource practices in place, you may not be able to attract and keep good professional staffs.

One critical topic that every executive should have current data on is that of the numbers and costs of canceled software projects. The kind of canceled projects that are of concern are those due to excessive cost and schedule overruns, which are severe enough to reduce the value of the software to negative levels.

There are of course many business reasons for canceling a software project; for example, selling a portion of a business, or buying a company that already owns software similar to a project your company is developing. However, most software projects that are terminated are canceled because they are so late and so much more expensive than planned that they are no longer valuable. This topic is perhaps the

Table 3.2 U.S. Software Project Outcomes by Size of Project

	Probability of Schedule Outcomes				
	Early (%)	*On Time (%)*	*Delayed (%)*	*Canceled (%)*	*Sum (%)*
1FP	14.68	83.16	1.92	0.25	100
10FP	11.08	81.25	5.67	2	100
100FP	6.06	74.77	11.83	7.33	100
1,000FP	1.24	60.76	17.67	20.33	100
10,000FP	0.14	28.03	23.83	48.00	100
100,000FP	0.00	13.67	21.33	65.00	100.00
Average	5.53	56.94	13.71	23.82	100.00

chief complaint that CEOs have about software: large software projects often run late and overrun their budgets.

Table 3.2 shows the approximate U.S. distribution of project outcomes, sorted by size into six size plateaus from 1 to 100,000 function points.

Because software is important but has been difficult to control, an increasing number of companies are evaluating outsourcing or turning over software development and maintenance to companies that specialize in software.

While outsourcing may be advantageous, not every outsource agreement is satisfactory. The approximate distribution of domestic outsource results after 2 years is shown in Table 3.3.

From process assessments performed within several large outsource companies, and analysis of projects produced by outsource vendors, our data indicate slightly

Table 3.3 Approximate Distribution of U.S. Outsource Results after 24 Months

Results	*Percentage of Outsource Arrangements (%)*
Both parties generally satisfied	65
Some dissatisfaction by client or vendor	17
Dissolution of agreement planned	10
Litigation between client and contractor probable	6
Litigation between client and contractor in progress	2

better-than-average productivity and quality control approaches when compared to the companies and industries who engaged the outsource vendors. These results are for development projects. For maintenance projects, the data indicate that maintenance outsourcing is between 25% and 100% more productive.

However, our data are still preliminary and need refinement, as of 2017. Our main commissioned research in the outsource community has been with clients of the largest outsource vendors in the United States such as Accenture, EDS, Perot, Lockheed, Computer Associates (CA), Computer Sciences Corporation (CSC), and others in this class. There are a host of smaller outsource vendors and contractors where we have encountered only a few projects, or sometimes none at all since our clients have not utilized their services.

For international outsourcing, the results are more difficult to assemble because they can vary from country to country and contract to contract. However, outsource agreements with countries such as India, Ukraine, the Philippines, Russia, and China seem to have more or less the same distribution of results as do the U.S. outsource contracts, although the margin of error is quite high.

Answers to the Software Economic Impact Questions

A large bank or insurance company can expect to own as many as 5,000,000 function points. A major multinational manufacturing concern or an international conglomerate can own upward of 7,000,000 million function points scattered across dozens of software labs and locations.

Even a small manufacturing company with less than 500 total employees can expect to own up to 500,000 function points. (Note: Function points are becoming the de facto standard for measuring software. A function point is a synthetic metric composed of the weighted totals of the inputs, outputs, inquiries, interfaces, and files that users have identified as significant in a software application.) The growth rate of corporate portfolios is an annual increase of 5%–7% in new and changed function points.

A typical distribution in a manufacturing company would be to build about 40% of the function points they use, lease about 30%, and buy about 30%. The purchased functions are often personal computer applications, while the leased and developed applications are often for mainframes or mini computers.

Banks and insurance companies may lease or purchase more than 50% of the function points they utilize, and build less than 30%. This is because of the ready availability of commercial banking and insurance packages. Smaller companies buy or lease much more software than they build; very small companies may purchase 100% of all their software functions.

Also, many companies are migrating toward integrated software solutions such as the well-known SAP, Oracle, and other ERP software packages that can replace more than 50% of in-house applications within many companies.

Replacement costs vary significantly, but a rough rule of thumb for the United States is that each function point would cost from $200 to $3500 to replace and the 2015 average is roughly $1200.

The less expensive replacements would be the simpler MIS applications, and the more expensive ones would be complex systems applications that have many special characteristics.

The life expectancy of software is proportional to its volume: Systems of more than 5000 function points in size will run for 10–15 years in many cases. Therefore, a rational replacement strategy must be based not only on the future business of your company, but also on the size and class variances of your existing portfolio. Following are the approximate useful life expectancies of software by size in function points.

Size	*Useful Life (Years)*
100	2
1,000	5
10,000	10
100,000	25

Large and aging legacy applications are an endemic problem of all major companies as of 2017. Much more research is needed in cost-effective renovation and data migration strategies to solve the problems of aging and decaying software applications. Some of these are written in obsolete programming languages with few skilled modern programmers, and sometimes without even working compilers available!

Answers to the Competitive Analysis Questions

Software expenses in the United States currently constitute about 3%–15% of the sales volumes of typical enterprises. There are major variations by industry, however. Obviously, for software companies such as Microsoft, CA, Oracle, and the like, software expenses approach 100% of their sales volumes.

For non-software industries, insurance and banking often have the largest expenses of general businesses, although telecommunications companies, defense companies, and computer manufacturers also have the high percentages of software expenses. Both the manufacturing and service companies may be software intensive, so many businesses will also have a high percentage of software costs.

In data collected from Japan, Europe, and the United States, it appears that the optimal level of capital investment per software engineer will be about

$5000–$10,000 for a powerful workstation, but even more for software tools and support: $35,000 in round numbers. Historically, software development has been undercapitalized. For software, like all other industries, there is a direct correlation between capital investment and overall productivity.

One of the most surprising recent findings about software productivity is that there is a strong correlation between office space and overall performance. Here, in the United States, software professionals with more than 78 square feet of noise-free office space are in the high quartile in overall performance, while those with less than 44 square feet of open office space or crowded, noisy cubicles are in the low quartile in overall performance.

U.S. productivity for MIS software averages about 8–10 function points per staff month. In general, large projects of more than 1000 function points will have much lower rates—in the range of 2–6 function points per staff month.

Small projects of less than 100 function points in size can often exceed rates of 20 function points per staff month. The overall range of software in the United States is from a fraction of a function point per staff month to a high of about 140 function points per staff month. Systems software and military software averages are much lower than MIS averages, due to the greater complexity of those applications and, in the case of military software, the enormous volume of paperwork required by military specifications.

U.S. software quality for MIS projects is an average of about 3–5 defects per function point, with perhaps 75%–90% of those defects being found prior to delivery of the software to its users. An average of about 1 defect per function point can still be present in MIS software at delivery.

Systems software and military software both have higher defect potentials, but also higher DREs—systems software averages more than 95% in defect removal, and military software can average more than 98%. To achieve an excellent reputation for high software quality, strive to keep potential defects below 2.5 per function point, and to achieve DREs in excess of 97%, attempt for a delivered defect rate of less than 0.075 defects per function point.

There is a strong correlation between high levels of user satisfaction and low levels of delivered defects, which might be expected. However, there is also a strong and surprising correlation between low levels of delivered defects and short development schedules. Since finding and fixing bugs is the most expensive task in all software, leading-edge companies that utilize sophisticated combinations of defect prevention and defect removal techniques can achieve very short schedules and very high quality levels simultaneously!

Just as a medical doctor cannot tell a specific patient about medical problems without a diagnosis, it is not possible to tell a CEO about specific strengths or weaknesses without careful study. However, common software strengths found in the United States include experience levels and capabilities of the staff, adequacy of basic tools, use of high-level languages, and use of structured methods for development and enhancements.

Common software weaknesses in the United States include excessive schedule pressure, management skills that rank below technical staff skills, failure to use adequate pretest reviews and inspections, and generally lower levels of capital equipment and support than appears desirable.

A number of software research and consulting organizations specialize in performing software diagnostic studies; for example, the Namcook Analytics LLC, the Quality/Productivity Management Group (Q/P), Davids' Consulting Group, the SEI, and a number of other organizations can carry out software assessments and/or benchmark studies.

Now that the questions and answers have been stated, there is one significant final point: If your company has a large software population, but answers to these questions are difficult or impossible to ascertain, you may need to consider either outsourcing, or reengineering of your software organization so that topics such as the ones in these questions can be dealt with. You may even need new executive level positions reporting directly to the CEO and charged with leading software functions into the modern era, such as creation of a vice president of software engineering or a vice president of software quality assurance.

Twenty-Five Quantitative Software Engineering Targets

Progress in the software industry has resembled a drunkard's walk, with both improvement and regressions occurring at the same time. For example, agile is an improvement for small projects, but pair programming is a regression and an expensive one to boot.

One reason for the lack of continued progress is the dismal lack of effective measurements. The two oldest and most widely used metrics, cost per defect and LOC, don't show progress and actually distort reality. Cost per defect penalizes quality, and LOC penalizes high-level languages.

Function point metrics have improved software measurement accuracy, but are not without problems of their own. For one thing, there are too many function point "clones" such as COSMIC, FISMA, and NESMA over and above the original function point by the IFPUG. This chapter uses IFPUG function points. It also uses supplemental quality metrics such as DRE.

This portion of the chapter provides 25 tangible goals that should be achievable within 5 years, assuming a starting year of 2018. Many of the goals are expressed in terms of function point metrics, which are the only widely used metrics that are valid for showing economic productivity and also for showing quality without serious errors.

Following is a collection of 25 goals or targets for software engineering progress developed by Namcook Analytics LLC for the 5 years between 2018 and 2023. Some of these goals are achievable now in 2017 but not many companies have achieved them. Some have already been achieved by a small selection of leading companies.

Unfortunately, less than 5% of the United States and global companies have achieved any of these goals, and less than 0.1% have achieved most of them. None of the author's clients have achieved every goal.

The author suggests that every major software-producing company and government agency have their own set of 5-year targets, using the current list as a starting point.

1. *Raise DRE from <90% to >99.5%.* This is the most important goal for the industry. It cannot be achieved by testing alone, but requires pretest inspections and static analysis. Automated proofs and automated analysis of code are useful. Certified test personnel and mathematical test case design using techniques such as design of experiments are useful. The DRE metric was developed by IBM circa 1970 to prove the value of inspections. It is paired with the "defect potential" metric discussed in the next paragraph. DRE is measured by comparing all bugs found during development to those reported in the first 90 days by customers. The current U.S. average is about 90%. Agile is about 92%. Quality strong methods such as RUP and TSP usually top 96% in DRE. Only a few top companies using a full suite of defect prevention, pretest defect removal, and formal testing with mathematically designed test cases and certified test personnel can top 99% in DRE. The upper limit of DRE circa 2015 was about 99.6%. DRE of 100% is theoretically possible but has not been encountered on more than about 1 project out of 10,000.
2. *Lower software defect potentials from >4 per function point to <2 per function point.* The phrase defect potentials was coined in IBM circa 1970. Defect potentials are the sum of bugs found in all deliverables: (1) requirements, (2) architecture, (3) design, (4) code, (5) user documents, and (6) bad fixes. Requirements and design bugs often outnumber code bugs. Presently, defect potentials can top 6 per function point for large systems in the 10,000–function point size range. Achieving this goal requires effective defect prevention such as joint application design (JAD), quality function deployment (QFD), requirements modeling, certified reusable components, and others. It also requires a complete software quality measurement program. Achieving this goal also requires better training in common sources of defects found in requirements, design, and source code. The most effective way of lowering defect potentials is to switch from custom designs and manual coding, which are intrinsically error prone. Construction for certified reusable components can cause a very significant reduction in software defect potentials.
3. *Lower COQ from >45% of development to <15% of development.* Finding and fixing bugs has been the most expensive task in software for more than 50 years. A synergistic combination of defect prevention and pretest inspections and static analysis are needed to achieve this goal. The probable sequence would be to raise DRE from today's average of below 90% up to 99%. At the

same time, defect potentials can be brought down from today's averages of more than 4 per function point to less than 2 per function point. This combination will have a strong synergistic impact on maintenance and support costs. Incidentally, lowering COQ will also lower technical debt. But as of 2017 technical debt is not a standard metric and varies so widely it is hard to quantify.

4. *Reduce average cyclomatic complexity from >25 to <10.* Achieving this goal requires careful analysis of software structures, and of course, it also requires measuring cyclomatic complexity for all modules. Since cyclomatic tools are common and some are open source, every application should use them without exception.
5. *Raise test coverage from <75% to >98.5% for risks, paths, and requirements.* Achieving this goal requires using mathematical design methods for test case creation such as using design of experiments. It also requires measurement of test coverage. It also requires predictive tools that can predict numbers of test cases based on function points, code volumes, and cyclomatic complexity. The author's Software Risk Master (SRM) tool predicts test cases for 18 kinds of testing and therefore can also predict probable test coverage.
6. *Eliminate EPM in large systems.* Bugs are not randomly distributed. Achieving this goal requires careful measurements of code defects during development and after release with tools that can trace bugs to specific modules. Some companies such as IBM have been doing this for many years. EPM are usually less than 5% of total modules but receive more than 50% of total bugs. Prevention is the best solution. Existing EPM in legacy applications may require surgical removal and replacement. However, static analysis should be used on all identified EPM. In one study, a major application had 425 modules. Of these, 57% of all bugs were found in only 31 modules built by one department. Over 300 modules were zero-defect modules. EPM are easy to prevent but difficult to repair once they are created. Usually surgical removal is needed. EPM are the most expensive artifacts in the history of software. EPM is somewhat like the medical condition of smallpox; i.e., it can be completely eliminated with "vaccination" and effective control techniques. EPM often top 3 defects per function and remove less than 80% prior to release. They also tend to top 50 in terms of cyclomatic complexity. Higher defect removal via testing is difficult due to the high cyclomatic complexity levels.
7. *Eliminate security flaws in all software applications.* As cybercrime becomes more common, the need for better security is more urgent. Achieving this goal requires use of security inspections, security testing, and automated tools that seek out security flaws. For major systems containing valuable financial or confidential data, ethical hackers may also be needed.
8. *Reduce the odds of cyber-attacks from >10% to <0.1%.* Achieving this goal requires a synergistic combination of better firewalls, continuous antivirus checking with constant updates to viral signatures, and also increasing the

immunity of software itself by means of changes to basic architecture and permission strategies. It may also be necessary to rethink hardware and software architectures to raise the immunity levels of both.

9. *Reduce bad-fix injections from >7% to <1%.* Not many people know that about 7% of attempts to fix software bugs contain new bugs in the fixes themselves, commonly called "bad fixes." When cyclomatic complexity tops 50, the bad-fix injection rate can soar to 25% or more. Reducing bad-fix injection requires measuring and controlling cyclomatic complexity, using static analysis for all bug fixes, testing all bug fixes, and inspections of all significant fixes prior to integration.
10. *Reduce requirements creep from >1.5% per calendar month to <0.25% per calendar month.* Requirements creep has been an endemic problem of the software industry for more than 50 years. While prototypes, agile embedded users, and JAD are useful, it is technically possible to also use automated requirements models to improve requirements completeness. The best method would be to use pattern matching to identify the features of applications similar to the one being developed. A precursor technology would be a useful taxonomy of software application features, which did not actually exist in 2017 but could be created with several months of concentrated study.
11. *Lower the risk of project failure or cancellation on large 10,000–function point projects from >35% to <5%.* Cancellation of large systems due to poor quality, poor change control, and cost overruns that turn ROI from positive to negative is an endemic problem of the software industry, and totally unnecessary. A synergistic combination of effective defect prevention and pretest inspections and static analysis can come close to eliminating this far too common problem. Parametric estimation tools can predict risks, costs, and schedules with greater accuracy that ineffective manual estimates are also recommended.
12. *Reduce the odds of schedule delays from >50% to <5%.* Since the main reasons for schedule delays are poor quality and excessive requirements creep, solving some of the earlier problems in this list will also solve the problem of schedule delays. Most projects seem on time until testing starts, when huge quantities of bugs begin to stretch out the test schedule to infinity. Defect prevention combined with pretest static analysis can reduce or eliminate schedule delays. This is a treatable condition and it can be eliminated within 5 years.
13. *Reduce the odds of cost overruns from >40% to <3%.* Software cost overruns and software schedule delays have similar root causes; i.e., poor quality control and poor change control combined with excessive requirements creep. Better defect prevention combined with pretest defect removal can help cure both these endemic software problems. Using accurate parametric estimation tools rather than optimistic manual estimates are also useful in lowering cost overruns.

14. *Reduce the odds of litigation on outsource contracts from >5% to <1%.* The author of this chapter has been an expert witness in 12 breach of contract cases. All these cases seem to have similar root causes that include poor quality control, poor change control, and very poor status tracking. A synergistic combination of early sizing and risk analysis prior to contract signing plus effective defect prevention and pretest defect removal can lower the odds of software breach of contract litigation.
15. *Lower maintenance and warranty repair costs by >75% compared to 2017 values.* Starting in about 2000, the number of U.S. maintenance programmers began to exceed the number of development programmers. IBM discovered that effective defect prevention and pretest defect removal reduced delivered defects to such low levels that maintenance costs were reduced by at least 45% and sometimes as much as 75%. Effective software development and effective quality control have a larger impact on maintenance costs than on development. It is technically possible to lower software maintenance for new applications by over 60% compared to current averages. By analyzing legacy applications and removing EPM plus some refactoring, it is also possible to lower maintenance costs by legacy software by about 25%. Technical debt would be reduced as well, but technical debt is not a standard metric and varies widely so it is hard to quantify. Static analysis tools should routinely be run against all active legacy applications.
16. *Improve the volume of certified reusable materials from <15% to >85%.* Custom designs and manual coding are intrinsically error prone and inefficient, no matter what methodology is used. The best way of converting software engineering from a craft to a modern profession would be to construct applications from libraries of certified reusable material; i.e., reusable requirements, design, code, and test materials. Certification to near zero-defect levels is a precursor, so effective quality control is on the critical path to increasing the volumes of certified reusable materials. All candidate reusable materials should be reviewed, and code segments should also be inspected and have static analysis runs. Also, reusable code should be accompanied by reusable test materials, and supporting information such as cyclomatic complexity and user information.
17. *Improve average development productivity from <8 function points per month to >16 function points per month.* Productivity rates vary based on application size, complexity, team experience, methodologies, and several other factors. However, when all projects are viewed in aggregate, average productivity is below 8 function points per staff month. Doubling this rate needs a combination of better quality control and much higher volumes of certified reusable materials: probably 50% or more.
18. *Improve work hours per function point from >16.5 to <8.25.* Goals 17 and 18 are essentially the same but use different metrics. However, there is one important difference. Work hours per month will not be the same in every

country. For example, a project in the Netherlands with 116 work hours per month will have the same number of work hours as a project in China with 186 work hours per month. But the Chinese project will need fewer calendar months than the Dutch project due to the more intense work pattern.

19. *Improve maximum productivity to >100 function points per staff month for 1000 function points.* In early 2017, productivity rates for 1000 function points range from about 5–12 function points per staff month. It is intrinsically impossible to top 100 function points per staff month using custom designs and manual coding. Only construction from libraries of standard reusable component can make such high productivity rates possible. However, it would be possible to increase the volume of reusable materials. The precursor needs are a good taxonomy of software features, catalogs of reusable materials, and a certification process for adding new reusable materials. It is also necessary to have a recall method in case the reusable materials contain bugs or need changes.
20. *Shorten average software development schedules by >35% compared to 2017 averages.* The most common complaint of software clients and corporate executives at the CIO and CFO levels is that big software projects take too long. Surprisingly, it is not hard to make them shorter. A synergistic combination of better defect prevention, pretest static analysis and inspections, and larger volumes of certified reusable materials can make significant reductions in schedule intervals. In today's world, raising software application size in function points to the 0.4 power provides a useful approximation of schedule duration in calendar months. But current technologies are sufficient to lower the exponent to the 0.37 power. Raising 1000 function points to the 0.4 power indicates a schedule of 15.8 calendar months. Raising 1000 function points to the 0.37 power shows a schedule of only 12.9 calendar months. This shorter schedule is made possible by using effective defect prevention augmented by pretest inspections and static analysis. Reusable software components could lower the exponent down to the 0.3 power or 7.9 calendar months. Schedule delays are rampant today, but they are treatable conditions that can be eliminated.
21. *Raise maintenance assignment scopes from <1500 function points to >5000 function points.* The metric "maintenance assignment scope" refers to the number of function points that one maintenance programmer can keep up and running during a calendar year. This metric was developed by IBM in the 1970s. The current range is from <300 function points for buggy and complex software to >5000 function points for modern software released with effective quality control. The current U.S. average is about 1500 function points. This is a key metric for predicting maintenance staffing for both individual projects and also for corporate portfolios. Achieving this goal requires effective defect prevention, effective pretest defect removal, and effective testing using modern mathematically based test case design methods. It also requires low

levels of cyclomatic complexity. Static analysis should be run on all applications during development, and on all legacy applications as well.

22. *Replace today's static and rigid requirements, architecture, and design methods with a suite of animated design tools combined with pattern matching.* When they are operating, software applications are the fastest objects yet created by the human species. When being developed, software applications grow and change on a daily basis. Yet every single design method is static and consists either of text such as story points or very primitive and limited diagrams such as flowcharts or Unified Modeling Language (UML) diagrams. The technology for creating a new kind of animated graphical design method in full color and also in three dimensions existed in 2014. It is only necessary to develop the symbol set and begin to animate the design process.
23. *Develop an interactive learning tool for software engineering based on massively interactive game technology.* New concepts are occurring almost every day in software engineering. New programming languages are coming out on a weekly basis. Software lags medicine and law and other forms of engineering in having continuing education. But live instruction is costly and inconvenient. The need is for an interactive learning tool with a built-in curriculum-planning feature. It is technically possible to build such a tool today. By licensing a game engine, it would be possible to build a simulated software university where avatars could both take classes and also interact with one another.
24. *Develop a suite of dynamic, animated project planning and estimating tools that will show growth of software applications.* Today, the outputs of all software estimating tools are static tables augmented by a few graphs. But software applications grow during development at more than 1% per calendar month and they continue to grow after release at more than 8% per calendar year. It is obvious that software planning and estimating tools need dynamic modeling capabilities that can show the growth of features over time. They should also show the arrival (and discovery) of bugs or defects entering from requirements, design, architecture, code, and other defect sources. The ultimate goal, which is technically possible today, would be a graphical model that shows application growth from the first day of requirements through 25 years of usage.
25. *Introduce licensing and board certification for software engineers and specialists.* It is strongly recommended that every reader of this chapter also read Paul Starr's book *The Social Transformation of American Medicine.* This book won a Pulitzer Prize in 1982. Starr's book shows how the American Medical Association was able to improve academic training, reduce malpractice, and achieve a higher level of professionalism than other technical fields. Medical licenses and board certification of specialists were a key factor in medical progress. It took over 75 years for medicine to reach the current professional status, but with Starr's book as a guide, software could do the same within 10 years. This is outside the 5-year window of this chapter, but the process started in 2015.

Note that the function point metrics used in this chapter refer to function points as defined by the IFPUG. Other function points such as COSMIC, FISMA, NESMA, unadjusted, etc. can also be used but would have different quantitative results.

The technology stack available in 2017 is already good enough to achieve each of these 20 targets, although few companies have done so. Some of the technologies associated with achieving these 20 targets include but are not limited to the following.

Technologies Useful in Achieving Software Engineering Goals

- Use early risk analysis, sizing, and both quality and schedule/cost estimation before starting major projects, such as Namcook's SRM.
- Use parametric estimation tools rather than optimistic manual estimates. All the parametric tools COCOMO, CostXpert, ExcelerPlan, KnowledgePlan, SEER, SLIM, SRM, and True Price will produce better results for large applications than manual estimates, which become progressively optimistic as applications size grows larger.
- Use effective defect prevention such as JAD and QFD.
- Use pretest inspections of major deliverables such as requirements, architecture, design, code, etc.
- Use both text static analysis and source code static analysis for all software. This includes new applications and 100% of active legacy applications.
- Use the SANs Institute list of common programming bugs and avoid them all.
- Use the FOG and FLESCH readability tools on requirements, design, etc.
- Use mathematical test case design such as design of experiments.
- Use certified test and quality assurance personnel.
- Use function point metrics for benchmarks and normalization of data.
- Use effective methodologies such as agile and XP for small projects; RUP and TSP for large systems. Hybrid methods are also effective such as agile combined with TSP.
- Use automated test coverage tools.
- Use automated cyclomatic complexity tools.
- Use parametric estimation tools that can predict quality, schedules, and costs. Manual estimates tend to be excessively optimistic.
- Use accurate measurement tools and methods with at least 3% precision.
- Consider applying automated requirements models, which seem to be effective in minimizing requirements issues.
- Consider applying the new Software Engineering Methods and Theory (SEMAT) method that holds promise for improved design and code quality. SEMAT comes with a learning curve and so reading the published book is necessary prior to use.

It is past time to change software engineering from a craft to a true engineering profession. It is also past time to switch from partial and inaccurate analysis of software results to results with high accuracy for both predictions before projects start and measurements after projects are completed.

The 25 goals shown above are positive targets that companies and government groups should strive to achieve. But "software engineering" also has a number of harmful practices that should be avoided and eliminated. Some of these are bad enough to be viewed as professional malpractice. Following are six hazardous software methods, some of which have been in continuous use for more than 50 years without their harm being fully understood.

Six Hazardous Software Engineering Methods to Be Avoided

1. *Stop trying to measure quality economics with "cost per defect."* This metric always achieves the lowest value for the buggiest software, so it penalizes actual quality. The metric also understates the true economic value of software by several hundred percent. This metric violates standard economic assumptions and can be viewed as professional malpractice for measuring quality economics. The best economic measure for COQ is "defect removal costs per function point." Cost per defect ignores the fixed costs for writing and running test cases. It is a well-known law of manufacturing economics that if a process has a high proportion of fixed costs, the cost per unit will go up. The urban legend that it costs 100 times as much to fix a bug after release than before is not valid; the costs are almost flat if measured properly.
2. *Stop trying to measure software productivity with LOC metrics.* This metric penalizes high-level languages. This metric also makes noncoding work such as requirements and design invisible. This metric can be viewed as professional malpractice for economic analysis involving multiple programming languages. The best metrics for software productivity are work hours per function point and function points per staff month. Both of these can be used at activity levels and also for entire projects. These metrics can also be used for noncode work such as requirements and design. LOC metrics have limited use for coding itself, but are hazardous for larger economic studies of full projects. LOC metrics ignore the costs of requirements, design, and documentation that are often larger than the costs of the code itself.
3. *Stop measuring "design, code, and unit test" or DCUT.* Measure full projects including management, requirements, design, coding, integration, documentations, all forms of testing, etc. DCUT measures encompass less than 30% of the total costs of software development projects. It is professionally embarrassing to measure only part of software development projects.

4. *Be cautious of "technical debt."* This is a useful metaphor but not a complete metric for understanding quality economics. Technical debt omits the high costs of canceled projects, and it excludes both consequential damages to clients and also litigation costs and possible damage awards to plaintiffs. Technical debt only includes about 17% of the true costs of poor quality. COQ is a better metric for quality economics.
5. *Avoid "pair programming."* Pair programming is expensive and less effective for quality than a combination of inspections and static analysis. Do read the literature on pair programming, and especially the reports by programmers who quit jobs specifically to avoid pair programming. The literature in favor of pair programming also illustrates the general weakness of software engineering research, in that it does not compare pair programming to methods with proven quality results such as inspections and static analysis. It only compares pairs to single programmers without any discussion of tools, methods, inspections, etc.
6. *Stop depending only on testing* without using effective methods of defect prevention and effective methods of pretest defect removal such as inspections and static analysis. Testing by itself without pretest removal is expensive and seldom tops 85% in DRE levels. A synergistic combination of defect prevention, pretest removal such as static analysis and inspections can raise DRE to >99% while lowering costs and shortening schedules at the same time.

The software engineering field has been very different from older and more mature forms of engineering. One of the main differences between software engineering and true engineering fields is that software engineering has very poor measurement practices and far too much subjective information instead of solid empirical data.

This portion of this chapter suggests a set of 25 quantified targets that if achieved would make significant advances in both software quality and software productivity. But the essential message is that poor software quality is a critical factor that needs to get better to improve software productivity, schedules, costs, and economics.

References and Readings

Abran, A. and Robillard, P.N.; Function point analysis: An empirical study of its measurement processes; *IEEE Transactions on Software Engineering*; 22, 12; 1996; pp. 895–910.

Austin, R.D.; *Measuring and Managing Performance in Organizations*; Dorset House Press, New York; 1996; ISBN 0-932633-36-6; 216 pages.

Black, R.; *Managing the Testing Process: Practical Tools and Techniques for Managing Hardware and Software Testing*; Wiley; Hoboken, NJ; 2009; ISBN-10 0470404159; 672 pages.

Bogan, C.E. and English, M.J.; *Benchmarking for Best Practices*; McGraw Hill, New York; 1994; ISBN 0-07-006375-3; 312 pages.

Cohen, L.; *Quality Function Deployment: How to Make QFD Work for You*; Prentice Hall, Upper Saddle River, NJ; 1995; ISBN 10: 0201633302; 368 pages.

Crosby, P.B.; *Quality Is Free*; New American Library, Mentor Books, New York; 1979; 270 pages.

Curtis, B., Hefley, W.E., and Miller, S.; *People Capability Maturity Model*; Software Engineering Institute, Carnegie Mellon University, Pittsburgh, PA; 1995.

Department of the Air Force; *Guidelines for Successful Acquisition and Management of Software Intensive Systems*; Volumes 1 and 2; Software Technology Support Center, Hill Air Force Base, UT; 1994.

Gack, G.; *Managing the Black Hole: The Executive's Guide to Software Project Risk*; Business Expert Publishing, Thomson, GA; 2010; ISBN10: 1-935602-01-9.

Gack, G.; Applying Six Sigma to software implementation projects; 1996; http://software.isixsigma.com/library/content/c040915b.asp.

Gilb, T. and Graham, D.; *Software Inspections*; Addison-Wesley, Reading, MA; 1993; ISBN 10: 0201631814.

Humphrey, W.S.; *Managing the Software Process*; Addison-Wesley Longman, Reading, MA; 1989.

Jacobsen, I., Griss, M., and Jonsson, P.; *Software Reuse-Architecture, Process, and Organization for Business Success*; Addison-Wesley Longman, Reading, MA; 1997; ISBN 0-201-92476-5; 500 pages.

Jacobsen, I. et al.; *The Essence of Software Engineering: Applying the SEMAT Kernel*; Addison-Wesley Professional; Reading, MA; 2013.

Jones, C.; *A Ten-Year Retrospective of the ITT Programming Technology Center*; Software Productivity Research, Burlington, MA; 1988.

Jones, C.; *Assessment and Control of Software Risks*; Prentice Hall; Englewood Cliffs, NJ; 1994; ISBN 0-13-741406-4; 711 pages.

Jones, C.; *Patterns of Software System Failure and Success*; International Thomson Computer Press, Boston, MA; 1995; 250 pages; ISBN 1-850-32804-8; 292 pages.

Jones, C.; *Software Quality: Analysis and Guidelines for Success*; International Thomson Computer Press, Boston, MA; 1997a; ISBN 1-85032-876-6; 492 pages.

Jones, C.; *The Economics of Object-Oriented Software*; Software Productivity Research, Burlington, MA; 1997b; 22 pages.

Jones, C.; *Becoming Best in Class*; Software Productivity Research; Burlington, MA; 1998; 40 pages.

Jones, C.; *Software Assessments, Benchmarks, and Best Practices*; Addison-Wesley Longman, Boston, MA; 2000 (due in May of 2000); 600 pages.

Jones, C.; *Estimating Software Costs*; 2nd edition; McGraw Hill; New York; 2007.

Jones, C.; *Applied Software Measurement*; 3rd edition; McGraw Hill; New York; 2008.

Jones, C.; *Software Engineering Best Practices*; McGraw Hill, New York; 2010.

Jones, C.; *The Technical and Social History of Software Engineering;* Addison-Wesley; Reading, MA; 2014.

Jones, C. and Bonsignour, O.; *The Economics of Software Quality*; Addison-Wesley, Reading, Boston, MA; 2011; ISBN 978-0-3-258220-9; 587 pages.

Kan, S. H.; *Metrics and Models in Software Quality Engineering*; 2nd edition; Addison-Wesley Longman, Boston, MA; 2003; ISBN 0-201-72915-6; 528 pages.

Keys, J.; *Software Engineering Productivity Handbook*; McGraw Hill, New York; 1993; ISBN 0-07-911366-4; 651 pages.

Love, T.; *Object Lessons*; SIGS Books, New York; 1993; ISBN 0-9627477 3-4; 266 pages.

Melton, A.; *Software Measurement*; International Thomson Press, London; 1995; ISBN 1-85032-7178-7.

Paulk M. et al.; *The Capability Maturity Model: Guidelines for Improving the Software Process*; Addison-Wesley, Reading, MA; 1995; ISBN 0-201-54664-7; 439 pages.

Radice, R. A.; *High Quality Low Cost Software Inspections*; Paradoxicon Publishing, Andover, MA; 2002; ISBN 0-9645913-1-6; 479 pages.

Royce, W. E.; *Software Project Management: A Unified Framework*; Addison-Wesley Longman, Reading, MA; 1998; ISBN 0-201-30958-0.

Shepperd, M.; A critique of cyclomatic complexity as a software metric; *Software Engineering Journal*; 3; 1988; pp. 30–36.

Strassmann, P.; *The Squandered Computer*; The Information Economics Press, New Canaan, CT; 1997; ISBN 0-9620413-1-9; 426 pages.

Stukes, S., Deshoretz, J., Apgar, H., and Macias, I.; Air force cost analysis agency software estimating model analysis; TR-9545/008-2; Contract F04701-95-D-0003, Task 008; Management Consulting & Research, Inc.; Thousand Oaks, CA 91362; September 30, 1996.

Thayer, R.H. (editor); *Software Engineering and Project Management*; IEEE Press, Los Alamitos, CA; 1988; ISBN 0 8186-075107; 512 pages.

Umbaugh, R. E. (editor); *Handbook of IS Management*; 4th edition; Auerbach Publications, Boston, MA; 1995; ISBN 0-7913-2159-2; 703 pages.

Weinberg, Dr. G.; *Quality Software Management: Volume 2 First-Order Measurement*; Dorset House Press, New York; 1993; ISBN 0-932633-24-2; 360 pages.

Wiegers, K.A.; *Creating a Software Engineering Culture*; Dorset House Press, New York; 1996; ISBN 0-932633-33-1; 358 pages.

Yourdon, E.; *Death March: The Complete Software Developer's Guide to Surviving "Mission Impossible" Projects*; Prentice Hall, Upper Saddle River, NJ; 1997; ISBN 0-13-748310-4; 218 pages.

Zells, L.; *Managing Software Projects: Selecting and Using PC-Based Project Management Systems*; QED Information Sciences, Wellesley, MA; 1990; ISBN 0-89435-275-X; 487 pages.

Zvegintzov, N.; *Software Management Technology Reference Guide*; Dorset House Press, New York; 1994; ISBN 1-884521-01-0; 240 pages.

Chapter 4

Metrics to Solve Problems and Improve Software Engineering Quality and Productivity

One of the main uses of a good software metrics and measurement program is to use the data to solve chronic problems and to improve software performance for both productivity and quality.

The software industry has been running blind for 60 years. Real economic progress is not visible when using either "lines of code (LOC) per month" or "cost per defect" because both these metrics distort reality and conceal economic progress.

Progress is not visible with agile metrics because neither story points nor velocity have any effective benchmarks circa 2017. Also, story points can't be used for the thousands of applications that don't utilize user stories.

A number of endemic software problems can be eliminated or greatly reduced by means of effective metrics and measurements. One of these problems is "wastage" or lost days spent on canceled projects or fixing bugs that should not exist. Another problem is to improve software quality, since finding and fixing bugs is the #1 cost driver for the software industry.

If these two problems are eliminated or reduced then software failures will diminish, software schedules and costs will be reduced, and post-release bugs will be cut back by over 50% compared to 2017 norms.

The good metrics for problem-solving include function points, defect potentials, and defect removal efficiency (DRE).

Table 4.1 Good Metrics Impact on Chronic Software Problems for Applications of a Nominal 1000 Function Points in Size

	Bad Metrics	*Good Metrics*	*Difference*
Canceled projects (%)	12.00	1.00	11.00
Schedule delays (%)	24.00	0.50	23.50
Average schedule months (1000 function points)	15.85	12.88	2.97
Average delay months	1.90	0	1.90
Cost overruns (%)	23.00	1.00	22.00
Average cost per function point	$1100.00	$737.00	$363.00
Defect potentials per function point	4.25	2.50	$1.75
DRE (%)	92.50	99.50	7.00
Delivered defects per function point	0.32	0.01	0.31
Delivered defects	319	13	306
High-severity defects	48	2	46
Security flaws	6	0	6

The bad metrics include LOC, cost per defect, story points, and use case points. Technical debt is not very good since it is undefined and varies from company to company and project to project.

The benefits of good metrics measures cover both productivity and quality and include, but are not limited to, the topics illustrated in Table 4.1 that illustrates a generic project of 1000 function points coded in the Java programming language.

As can be seen from Table 4.1, good metrics and good measures reveal critical problems and help in eliminating or reducing them.

Reducing Software Wastage

When the work patterns of software development and maintenance projects are analyzed, a surprising hypothesis emerges. Software quality is so poor that productivity is much lower than it should be. Poor quality shows up in three major software economic problems: (1) Canceled projects that are never released due to poor quality; (2) Schedule delays due to poor quality extending test duration;

(3) Excessive work on finding and fixing bugs, which often exceeds 60% of the total software efforts.

The amount of software efforts spent on software projects that will be canceled due to excessive error content appears to absorb more than 20% of the U.S. software work force. In addition, about 60% of the U.S. software engineering work time centers on finding and fixing errors, which might have been avoided. Finally, software schedules for major applications are about 25% longer than they should be due to poor quality expanding testing intervals.

Out of a full software engineering working year, only about 48 days are spent on code development. About 53 days are spent on finding and fixing bugs in current applications. About 16 days are spent on canceled projects. About 15 days are spent on inspections and static analysis. About 13 days are spent on bug repairs in legacy code, i.e., 97 days per year are essentially "wastage" spent on bugs that might have been prevented or removed inexpensively. No other major occupation appears to devote so much effort to canceled projects or to defect repairs as does software engineering.

Software is one of the most labor-intensive occupations of the 21st century (*The Technical and Social History of Software Engineering*; Jones, 2014). Software is also one of the most challenging business endeavors, since software projects are difficult to control and subject to a significant percentage of delays and outright cancelations. The primary reason for both software delays and cancelations is due to the large numbers of "bugs" or errors whose elimination can absorb more than 60% of the efforts on really large software projects.

When the errors in software schedules and cost estimates are analyzed carefully, it can be seen that a major source of schedule slippage and cost overruns is the fact that the applications have so many bugs that they don't work or can't be released (*The Economics of Software Quality*; Jones and Bonsignour, 2011). A famous example of this phenomenon can be seen in the 1-year delay in opening the Denver Airport due to errors in the software controlling the luggage handling system. Problems with excessive bugs or errors have also caused delays in many software product releases, even by such well-known companies as IBM and Microsoft.

Canceled projects, schedule delays, and cost overruns all have a common origin: excessive defects that might be prevented, combined with labor-intensive defect removal methods such as testing and manual inspections.

Software engineering is a very labor-intensive occupation. A key reason for the high labor content of software applications is because these applications are very complex and hence very error-prone.

The large number of severe errors or "bugs" in software applications has several unfortunate effects on the software industry:

1. A substantial number of software projects are canceled due to high error rates.
2. Much of the development work of software engineering is defect removal.
3. Much of the maintenance work of software engineering is defect repair.

Using data gathered during Namcook's software assessments and benchmark studies, Table 4.1 shows the approximate number of software "projects" that were undertaken in the United States during the calendar year 2015. A software "project" is defined as the total effort assigned to developing or enhancing a specific software application.

Table 4.2 2017 Distribution of Software Development Efforts

Activities	*Work Days*	*Percentage of Time (%)*
Regular weekends	104	28.49
Testing and defect repairs	53	14.52
New code development	48	13.15
Meetings and status tracking	24	6.58
Producing paper documents	21	5.75
Vacations and personal time	20	5.48
Days spent on canceled projects	16	4.38
Pretest inspections/static analysis	15	4.11
Bug repairs in legacy code	13	3.56
Travel	12	3.29
Training and classes	10	2.74
Slack time between assignments	10	2.74
Sick leave	10	2.74
Public holidays	9	2.47
Total	365	100.00
Size in function points	1,000	
Size in Java statements	53,000	
Staff (development, test, management, etc.)	7	
Schedule in calendar months	15.85	
Work hours per function point	20.23	
Function points per month	6.52	
Costs per function point	$1,532.42	

(*Continued*)

Table 4.2 (*Continued*) 2017 Distribution of Software Development Efforts

Activities	*Work Days*	*Percentage of Time (%)*
Total costs for project	$1,532,423	
Defect potentials per function point	3.50	
Defect potential	3,500	
DRE (%)	92.50	
Delivered defects per function point	0.26	
Delivered defects	263	
High-severity defects	39	
Security flaws	6	

Notes: Assumes 1,000-function point projects, Agile development, 0% reuse, Java; Assumes 132 h per month, monthly costs of $10,000, CMMI 3, average skills.

Table 4.2 shows how a typical software engineer spends a calendar year. The background data for Table 4.2 comes from interviews and benchmarks carried out among the author's clients, and also among software personnel at IBM and ITT, since the author worked at both companies.

The tables include code development, canceled projects, defect repairs, and neutral activities such as training and holidays. Weekends are included because some countries use 6-day weeks but United States does not.

Note that days spent finding and fixing bugs or working on canceled projects total to 97 annual work days, while code development is only 48 annual work days. Worse, Agile is one of the newer and better methodologies. If the project had been done in Waterfall, even more days would have been spent in finding and fixing bugs.

It is obvious that as of 2017, software engineering is out of kilter. Bug repairs constitute a far larger proportion of the work year than is satisfactory. It is also obvious that custom designs and manual coding are intrinsically error-prone and expensive.

What would be the results if instead of developing software as unique projects using manual labor, it were possible to construct software applications using libraries of about 85% standard and certified reusable components instead of 0% reuse?

Table 4.3 uses the same format as that of Table 4.2 but makes the assumption that the application is constructed from a library that allows 85% of the code to be in the form of certified reusable modules rather than hand coding and custom development (*Software Engineering Best Practices*; Jones, 2012).

Table 4.3 2027 Distribution of Software Development Efforts

Activities	*Work Days*	*Percentage of Time (%)*
Regular weekends	104	28.49
Testing and defect repairs	8	2.18
New code development	119	32.60
Meetings and status tracking	24	6.58
Producing paper documents	21	5.75
Vacations and personal time	20	5.48
Days spent on canceled projects	2	0.66
Pretest inspections/static analysis	2	0.62
Bug repairs in legacy code	13	3.56
Travel	12	3.29
Training and classes	10	2.74
Slack time between assignments	10	2.74
Sick leave	10	2.74
Public holidays	9	2.47
Total	365	100.00
Size in function points	1,000	
Size in Java statements	53,000	
Staff (development, test, management, etc.)	4	
Schedule in calendar months	4.71	
Work hours per function point	6.14	
Function points per month	21.50	
Costs per function point	$465.11	
Total costs for project	$465,105	
Defect potentials per function point	0.53	
Defect potential	525	

(Continued)

expensive. Therefore, excellent quality control is the first stage in a successful reuse program.

The need for being close to zero defects and formal certification adds about 20% to the costs of constructing reusable artifacts and about 30% to the schedules for construction. However, using certified reusable materials subtracts over 80% from the costs of construction and can shorten schedules by more than 60%. The more frequently the materials are reused, the greater their cumulative economic value.

Achieving Excellence in Software Quality Control

In addition to moving toward higher volumes of certified reusable components, it is also obvious from the huge costs associated with finding and fixing bugs that the software industry needs much better quality control than was common in 2015.

Excellent Quality Control

Excellent software projects have rigorous quality control methods that include formal estimation of quality before starting, full defect measurement and tracking during development, and a full suite of defect prevention, pretest removal, and test stages. The combination of low defect potentials and high DRE is what software excellence is all about.

The most common companies that are excellent in quality control are usually the companies that build complex physical devices such as computers, aircraft, embedded engine components, medical devices, and telephone switching systems (*Software Engineering Best Practices*; Jones, 2010). Without excellence in quality, these physical devices will not operate successfully. Worse, failure can lead to litigation and even criminal charges. Therefore, all companies that use software to control complex physical machinery tend to be excellent in software quality.

Examples of organizations noted as excellent for software quality in alphabetical order include Advanced Bionics, Apple, AT&T, Boeing, Ford for engine controls, General Electric for jet engines, Hewlett Packard for embedded software, IBM for systems software, Motorola for electronics, NASA for space controls, the Navy for surface weapons, Raytheon, and Siemens for electronic components.

Companies and projects with excellent quality control tend to have low levels of code cyclomatic complexity and high test coverage, i.e., test cases cover >95% of paths and risk areas.

These companies also measure quality well and all know their DRE levels. (Any company that does not measure and know their DRE is probably below 85% in DRE.)

Excellence in software quality also uses pretest inspections for critical materials (i.e., critical requirements, design, architecture, and code segments). Excellence in quality also implies 100% usage of pretest static analysis for all new modules, for significant changes to modules, and for major bug repairs. For that matter, static analysis is also valuable during the maintenance of aging legacy applications.

Excellent testing involves certified test personnel, formal test case design using mathematical methods such as design of experiments and a test sequence that includes at least (1) unit test, (2) function test, (3) regression test, (4) performance test, (5) usability test, (6) system test, (7) Beta or customer acceptance test. Sometimes additional tests such as supply chain or security are also included.

Excellent quality control has DRE levels between about 97% for large systems in the 10,000-function point size range and about 99.6% for small projects <1,000 function points in size.

A DRE of 100% is theoretically possible but is extremely rare. The author has only noted DRE of 100% in 10 projects out of a total of about 25,000 projects examined. As it happens, the projects with 100% DRE were all compilers and assemblers built by IBM and using >85% certified reusable materials. The teams were all experts in compilation technology and of course a full suite of pretest defect removal and test stages were used as well.

Average Quality Control

In the present-day world, Agile is the new average and indeed a long step past waterfall development. Agile development has proven to be effective for smaller applications below 1000 function points in size. Agile does not scale up well and is not a top method for quality. Indeed both Tables 4.1 and 4.2 utilize Agile since it was very common in 2015.

Agile is weak in quality measurements and does not normally use inspections, which has the highest DRE of any known form of defect removal. Inspections top 85% in DRE and also raise testing DRE levels. Among the author's clients that use Agile, the average value for DRE is about 92–94%. This is certainly better than the 85–90% industry average, but not up to the 99% actually required to achieve optimal results.

[Methods with stronger quality control than Agile include personal software process (PSP), team software process (TSP), and the Rational Unified Process (RUP), which often top 97% in DRE (*Applied Software Measurement*; Jones, 2008).]

Some but not all agile projects use "pair programming" in which two programmers share an office and a work station and take turns coding while the other watches and "navigates." Pair programming is very expensive but only benefits quality by about 15% compared to single programmers. Pair programming is much less effective in finding bugs than formal inspections, which usually bring three to five personnel together to seek out bugs using formal methods. Critical inspection combined with static analysis has higher defect removal than pair programming at costs below 50% of pair programming.

Agile is a definite improvement for quality compared to waterfall development, but is not as effective as the quality-strong methods of TSP and the RUP.

Average projects usually do not predict defects by origin and do not measure DRE until testing starts, i.e., requirements and design defects are underreported and sometimes invisible.

A recent advance since 1984 in software quality control now frequently used by average as well as advanced organizations is that of static analysis. Static analysis tools can find about 55–65% of code defects, which is much higher than most forms of testing.

Many test stages such as unit test, function test, regression test, etc., are only about 35% efficient in finding code bugs or in finding one bug out of three. This explains why 6–10 separate kinds of testing are needed.

The kinds of companies and projects that are "average" would include internal software built by hundreds of banks, insurance companies, retail and wholesale companies, and many government agencies at federal, state, and municipal levels.

Average quality control has DRE levels from about 85% for large systems up to 96% for small and simple projects.

Poor Quality Control

Poor quality control is characterized by weak defect prevention and almost a total omission of pretest defect removal methods such as static analysis and formal inspections. Poor quality control is also characterized by inept and inaccurate quality measures that ignore front-end defects in requirements and design. There are also gaps in measuring code defects. For example, most companies with poor quality control have no idea how many test cases might be needed or how efficient various kinds of test stages are.

Companies with poor quality control also fail to perform any kind of up-front quality predictions so they jump into development without a clue as to how many bugs are likely to occur and what are the best methods for preventing or removing these bugs. Testing is usually by untrained, uncertified developers without using any formal test case design methods.

One of the main reasons for the long schedules and high costs associated with poor quality is the fact that so many bugs are found when testing starts that the test interval stretches out to two or three times longer than planned (*Estimating Software Costs*; Jones, 2007).

Some of the kinds of software that are noted for poor quality control include the Obamacare website, municipal software for property tax assessments, and software for programmed stock trading, which has caused several massive stock crashes. Indeed, government software projects tend to have more poor quality projects than corporate software by a considerable margin. For example, the author has worked as an expert witness in lawsuits for poor quality for more state government software failures than any other industrial segment.

Table 4.6 Distribution of DRE for 1000 Projects

DRE (%)	*Projects*	*Percentage (%)*
>99.00	10	1.00
95–99	120	12.00
90–94	250	25.00
85–89	475	47.50
80–85	125	12.50
<80.00	20	2.00
Totals	1000	100.00

Poor quality control is often below 85% in DRE levels. In fact, for canceled projects or those that end up in litigation for poor quality, the DRE levels may drop below 80%, which is low enough to be considered professional malpractice. In litigation where the author has been an expert witness, DRE levels in the low 80% range have been the unfortunate norm.

Table 4.6 shows the ranges in DRE noted from a sample of 1000 software projects. The sample included systems and embedded software, Web projects, cloud projects, information technology projects, and also defense and commercial packages.

As is evident, high DRE does not occur often. This is unfortunate because projects that are above 95% in DRE have shorter schedules and lower costs than projects below 85% in DRE. The software industry does not measure either quality or productivity well enough to know this.

However, the most important economic fact about high quality is as follows: *Projects >97% in DRE have shorter schedules and lower costs than projects <90% in DRE*. This is because projects that are low in DRE have test schedules that are at least twice as long as projects with high DRE due to omission of pretest inspections and static analysis!

Table 4.7 shows DRE for four application size plateaus and for six technology combinations.

It can be seen that DRE varies by size and also by technology stack. A combination of inspections, static analysis, and formal testing has the highest DRE values for all sizes.

The DRE metric was first developed by IBM circa 1970. It is normally calculated by measuring all defects found prior to release to customers, and then customer-reported defects for the first 90 days of usage (*The Economics of Software Quality*; Jones and Bonsignour, 2012).

As of 2017, DRE measures are being used by dozens of high-technology companies, but are not widely used by government agencies or other industry segments such as banks, insurance, and manufacturing. However, DRE is one of the most

Table 4.7 Software DRE

Cases		*100 DRE (%)*	*1,000 DRE (%)*	*10,000 DRE (%)*	*100,000 DRE (%)*	*Average (%)*
1	Inspections, static analysis, formal testing	99.60	98.50	97.00	96.00	97.78
2	Inspections, formal testing	98.00	97.00	96.00	94.50	96.38
3	Static analysis, formal testing	97.00	96.00	95.00	93.50	95.38
4	Formal testing	93.00	91.00	90.00	88.50	90.63
5	Informal testing	87.00	85.00	83.00	80.00	83.75
	Average	94.92	93.50	92.20	90.50	92.78

Notes: DRE is total removal before release; Size is expressed in terms of IFPUG function points 4.3.

useful of all quality measures. Namcook suggests that all companies and government agencies use "defect potentials" and "DRE" on all projects. The phrase defect potentials also originated in IBM circa 1970 and is the sum total of bugs that originate in requirements, architecture, design, code, documents, and "bad fixes" or new bugs found in bug repairs themselves.

Formal testing implies certified test personnel and mathematical test case design such as use of "design of experiments." Informal testing implies untrained uncertified developers.

Table 4.8 shows the schedules in calendar months for the same combinations of application size plateaus and also for technology combinations.

As is evident from Table 4.8, high quality does not add time to development. High quality shortens schedules because the main reason for schedule delays is too many bugs when testing starts, which stretches out test schedules by weeks or months (*Estimating Software Costs*; Jones, 2007).

(Note: Namcook Analytics' Software Risk Master (SRM) tool can predict the results of any combination of defect prevention methods, pretest defect removal methods, and testing stages. SRM can also predict the results of 57 development methods, 79 programming languages, and the impact of all 5 levels of the capability maturity model integrated (CMMI).)

Since one of the major forms of "wastage" involves canceled projects, Table 4.8 shows the impact of high quality on project cancelation rates. Since many canceled projects are above 10,000 function points in size, only data for large systems are shown in Table 4.9.

Table 4.8 Software Schedules Related to Quality Control

Cases		*100*	*1,000*	*10,000*	*100,000*	*Average*
1	Inspections, static analysis, formal testing	5.50	12.88	30.20	70.79	29.84
2	Inspections, formal testing	5.75	13.80	33.11	79.43	33.03
3	Static analysis, formal testing	5.62	13.34	31.62	74.99	31.39
4	Formal testing	6.03	14.79	36.31	89.13	36.56
5	Informal testing	6.31	15.85	39.81	100.00	40.49
	Average	5.84	14.13	34.21	82.87	34.26

Notes: Size is expressed in terms of IFPUG function points 4.3; Schedule is expressed in terms of calendar months.

From Table 4.9, it can be noted that poor quality control and lack of pretest inspections and static analysis are major causes of canceled projects.

Table 4.10 shows the impact of quality control on software outsource litigation. The author has been an expert witness in 15 breach of contract cases and has provided data to other testifying experts in about 50 other cases. In fact, data from the author's book *The Economics of Software Quality*; *Jones and Bonsignour, 2011* is

Table 4.9 Impact of Software Quality on Cancelation

Cases		*100*	*1,000*	*10,000 (%)*	*100,000 (%)*	*Average (%)*
1	Inspections, static analysis, formal testing			7.00	14.00	10.50
2	Inspections, formal testing			12.00	24.00	18.00
3	Static analysis, formal testing			16.00	32.00	24.00
4	Formal testing			43.00	57.00	50.00
5	Informal testing			72.00	83.00	77.50
	Average			30.00	42.00	36.00

Notes: Size is expressed in terms of IFPUG function points 4.3; Projects <10,000 function points are seldom canceled and not shown.

Table 4.10 Impact of Software Quality on Litigation

Cases		*100*	*1,000*	*10,000 (%)*	*100,000 (%)*	*Average (%)*
1	Inspections, static analysis, formal testing			2.00	5.00	3.50
2	Inspections, formal testing			4.00	12.00	8.00
3	Static analysis, formal testing			6.00	17.00	11.50
4	Formal testing			12.00	24.00	18.00
5	Informal testing			24.00	48.00	36.00
	Average			9.60	21.20	15.40

Notes: Size is expressed in terms of IFPUG function points 4.3; Data show % of outsource projects in breach of contract litigation; Most breach of contract litigation occurs >10,000 function points.

used in many lawsuits, sometimes by both sides, because there is no other useful published source of quality data.

As can be seen, poor quality control on outsourced projects has an alarmingly high probability of ending up in court either because the projects are canceled or because of major errors after deployment that prevent the owners from using the software as intended.

Table 4.11 shows the kinds of industries that are most likely to be found in each of the five levels of quality control. In general, the industries with the best quality tend to be those that build complex physical devices such as medical equipment, aircraft, or telephone switching systems.

The industries that bring up the bottom with frequently poor quality control include state, municipal, and federal civilian government groups (military software is fairly good in quality control), and also stock trading software. For that matter tax software is not very good and there are errors in things like property tax and income tax calculations.

Table 4.11 shows trends but the data are not absolute. Some government groups are better than expected but not very many. Namcook has data on over 70 industry sectors.

The final Table 4.12 in this paper is taken from Chapter 7 of the author's book *The Economics of Software Quality* with Olivier Bonsignour, Jones 2011 as coauthor. This table shows the approximate distribution of excellent, average, and poor software quality by application size.

It can be observed that software quality declines as application size increases. This is due to the intrinsic complexity of large software applications, compounded

Table 4.11 Industries Noted at Each Software Quality Level

	Cases	*Industries*
1	Inspections, static analysis, formal testing	Medical devices, aircraft, telecom
2	Inspections, formal testing	Airlines, pharmaceuticals, defense
3	Static analysis, formal testing	Open source, commercial software, automotive
4	Formal testing	Banks/insurance, health care, outsourcers
5	Informal testing	Retail, stock trading, government (all levels)

Notes: High-tech industries have the best quality control at all sizes; Defense and avionics have good quality control; All forms of civilian government have poor quality control; Stock trading has poor quality control.

by the lack of standard certified reusable components. It is also compounded by the omission of pretest inspections and static analysis for many of the large applications above 10,000 function points in size.

Analysis of the work patterns of the software engineering world reveals a surprising fact. Much of the work of software engineering is basically "wasted" because it concerns either working on projects that will not be completed, or working on repairing defects that should not be present at all.

If the software community can be alerted to the fact that poor software engineering economics are due to poor quality, then we might be motivated to take defect prevention, defect removal, reusability, and risk analysis more seriously than they are taken today.

Table 4.12 Distribution of Software Projects by Quality Levels

Function Points	*Low Quality (%)*	*Average Quality (%)*	*High Quality (%)*
10	25.00	50.00	25.00
100	30.00	47.00	23.00
1,000	36.00	44.00	20.00
10,000	40.00	42.00	18.00
100,000	43.00	45.00	12.00
Average	34.80	44.50	18.25

Note: Sample=~25,000 software projects to 2017.

Metrics to Improve Software Quality

Since high quality leads to lower costs and shorter schedules, measuring and improving software quality should be the main goal of software engineering.

Software quality depends upon two important variables. The first variable is that of *defect potentials* or the sum total of bugs likely to occur in requirements, architecture, design, code, documents, and "bad fixes" or new bugs in bug repairs. Defect potentials are measured using function point metrics, since "LOC" cannot deal with requirements and design defects.

(This paper uses the International Function Point Users Group (IFPUG) function points version 4.3. The newer software nonfunctional assessment process (SNAP) metrics are only shown experimentally due to insufficient empirical quality data with SNAP as of 2016. However, an experimental tool is included for calculating SNAP defects.)

The second important measure is *DRE* or the percentage of bugs found and eliminated before release of software to clients.

The author's SRM estimating tool predicts defect potentials and DRE as standard quality outputs for all software projects.

Defect potentials and DRE are useful quality metrics developed by IBM circa 1973 and widely used by technology companies as well as by banks, insurance companies, and other organizations with large software staffs.

This combination of defect potentials using function points and DRE is the only accurate and effective measure for software quality. The "cost-per-defect metric" penalizes quality and makes buggy software look better than high-quality software. The "LOC" metric penalizes modern high-level languages. The LOC metric can't measure or predict bugs in requirements and design. The new technical debt metric only covers about 17% of the true costs of poor quality.

Knowledge of effective software quality control has major economic importance because for over 60 years, the #1 cost driver for the software industry has been the costs of finding and fixing bugs. Table 4.13 shows the 15 major cost drivers for software projects in 2016.

Table 4.13 illustrates an important but poorly understood economic fact about the software industry. About 4 of the 15 major cost drivers can be attributed specifically to poor quality. The poor quality of software is a professional embarrassment and a major drag on the economy of the software industry and for that matter a drag on the entire U.S. and global economies.

Poor quality is also a key reason for cost driver #2. A common reason for canceled software projects is because quality is so bad that schedule slippage and cost overruns turned the project ROI from positive to negative.

Note the alarming location of successful cyber-attacks in sixth place (and rising) on the cost-driver list. Since security flaws are another form of poor quality, it is obvious that high quality is needed to deter successful cyber-attacks.

Table 4.13 U.S. Software Costs in Rank Order

1. The cost of finding and fixing bugs
2. The cost of canceled projects
3. The cost of producing English words
4. The cost of programming or code development
5. The cost of requirements changes
6. The cost of successful cyber-attacks
7. The cost of customer support
8. The cost of meetings and communication
9. The cost of project management
10. The cost of renovation and migration
11. The cost of innovation and new kinds of software
12. The cost of litigation for failures and disasters
13. The cost of training and learning
14. The cost of avoiding security flaws
15. The cost of assembling reusable components

Poor quality is also a key factor in cost driver #12 or litigation for breach of contract. (The author has worked as an expert witness in 15 lawsuits. Poor software quality is an endemic problem with breach of contract litigation. In one case against a major ERP company, the litigation was filed by the company's own shareholders who asserted that the ERP package quality was so bad that it was lowering stock values!)

A chronic weakness of the software industry for over 50 years has been poor measurement practices and bad metrics for both quality and productivity. For example, many companies don't even start quality measures until late testing, so early bugs found by inspections, static analysis, desk checking, and unit testing are unmeasured and invisible.

If you can't measure a problem, then you can't fix the problem either. Software quality has been essentially unmeasured and therefore unfixed for 50 years. This paper shows how quality can be measured with high precision, and also how quality levels can be improved by raising DRE up above 99%, which is where it should be for all critical software projects.

Software defect potentials are the sum total of bugs found in requirements, architecture, design, code, and other sources of error. The approximate U.S. average for defect potentials is shown in Table 4.14 using IFPUG function points version 4.3.

Note that the phrase "bad fix" refers to new bugs accidentally introduced in bug repairs for older bugs. The current U.S. average for bad-fix injections is about 7%, i.e., 7% of all bug repairs contain new bugs. For modules that are high in cyclomatic complexity and for "error-prone modules (EPMs)," bad-fix injections

Table 4.14 Average Software Defect Potentials circa 2016 for the United States

• Requirements	0.70 defects per function point
• Architecture	0.10 defects per function point
• Design	0.95 defects per function point
• Code	1.15 defects per function point
• Security code flaws	0.25 defects per function point
• Documents	0.45 defects per function point
• Bad fixes	0.65 defects per function point
• Totals	4.25 defects per function point

can top 75%. For applications with low cyclomatic complexity, bad fixes can drop below 0.5%.

Defect potentials are of necessity, measured using function point metrics. The older "LOC" metric cannot show requirements, architecture, and design defects nor any other defect outside the code itself. (As of 2016, function points are the most widely used software metric in the world. There are more benchmarks using function point metrics than all other metrics put together.)

Because of the effectiveness of function point measures compared to older LOC measures, an increasing number of national governments are now mandating function point metrics for all software contracts. The governments of Brazil, Italy, Japan, Malaysia, and South Korea now require function points for government software. Table 4.15 shows the countries with rapid expansions in function point use.

To be blunt, any company or government agency in the world that does not use function point metrics does not have accurate benchmark data on either quality or productivity. The software industry has had poor quality for over 50 years and a key reason for this problem is that the software industry has not measured quality well enough to make effective improvements. Cost per defect and LOC both distort reality and conceal progress. They are harmful rather than being helpful in improving either quality or productivity.

LOC reverses true economic productivity and makes assembly language seem more productive than Objective C. Cost per defect reverses true quality economics and makes buggy software look cheaper than high-quality software. These distortions of economic reality have slowed software progress for over 50 years.

The U.S. industries that tend to use function point metrics and therefore understand software economics fairly well include automotive manufacturing, banks, commercial software, insurance, telecommunications, and some public utilities.

For example, Bank of Montreal was one of the world's first users of function points after IBM placed the metric in the public domain; Ford has used function point metrics

Table 4.15 Countries Expanding Use of Function Points 2016

1	Argentina	
2	Australia	
3	Belgium	
4	Brazil	Required for government contracts
5	Canada	
6	China	
7	Finland	
8	France	
9	Germany	
10	India	
11	Italy	Required for government contracts
12	Japan	Required for government contracts
13	Malaysia	Required for government contracts
14	Mexico	
15	Norway	
16	Peru	
17	Poland	
18	Singapore	
19	South Korea	Required for government contracts
20	Spain	
21	Switzerland	
22	Taiwan	
23	The Netherlands	
24	United Kingdom	
25	United States	

for fuel injection and navigation packages; Motorola has used function points for smartphone applications; AT&T has used function points for switching software; IBM has used function points for both commercial software and also operating systems.

The U.S. industries that do not use function points widely and hence have no accurate data on either software quality or productivity include the Department of Defense, most state governments, the U.S. Federal government, and most universities (which should understand software economics but don't seem to).

Although the Department of Defense was proactive in endorsing the Software Engineering Institute (SEI) CMMI, it lags the civilian sector in software metrics and measurements. For that matter, the SEI itself has not yet supported function point metrics nor pointed out to clients that both LOC and cost per defect distort reality and reverse the true economic value of high-quality and high-level programming languages.

It is interesting that the author had a contract from the U.S. Air Force to examine the benefits of ascending to the higher CMMI levels because the SEI itself had no quantitative data available.

Although the Department of Defense itself lags in function point use, some of the military services have used function points for important projects. For example, the U.S. Navy has used function points for shipboard gun controls and cruise missile navigation.

If a company or government agency wants to get serious in improving quality, then the best and only effective metrics for achieving this are the combination of defect potentials in function points and DRE.

DRE is calculated by keeping accurate counts of all defects found during development. After release, all customer-reported bugs are included in the total. After 90 days of customer usage, DRE is calculated. If developers found 900 bugs and customers reported 50 bugs in the first 3 months, then DRE is 95%.

Obviously, bug reports don't stop cold after 90 days, but the fixed 90-day interval provides an excellent basis for statistical quality reports.

The overall range in defect potentials runs from about 2 per function point to more than 7 per function point. Factors that influence defect potentials include team skills, development methodologies, CMMI levels, programming languages, and defect prevention techniques such as joint application design (JAD) and quality function deployment (QFD).

Some methodologies such as TSP are "quality strong" and have low defect potentials. Agile is average for defect potentials. Waterfall is worse than average for defect potentials.

Table 4.16 shows the U.S. ranges for defect potentials circa 2017.

NOTE: The author's SRM estimating tool predicts defect potentials as a standard output for every project estimated.

Defect potentials obviously vary by size, with small projects typically having low defect potentials. Defect potentials rise faster than size increases, with large systems above 10,000 function points having alarmingly high defect potentials.

Table 4.16 U.S Average Ranges of Defect Potentials circa 2017

Defect Origins	*Best*	*Average*	*Worst*
Requirements	0.34	0.70	1.35
Architecture	0.04	0.10	0.20
Design	0.63	0.95	1.58
Code	0.44	1.15	2.63
Security flaws	0.18	0.25	0.40
Documents	0.20	0.45	0.54
Bad fixes	0.39	0.65	1.26
Total	2.22	4.25	7.96

Note: Defects per IFPUG 4.3 function point.

Table 4.17 shows U.S. ranges in defect potentials from small projects of 1 function point up to massive systems of 100,000 function points.

As can be seen, defect potentials go up rapidly with application size. This is one of the key reasons why large systems fail so often and also run late and over budget.

Table 4.18 shows the overall U.S. ranges in DRE by applications size from a size of 1 function point up to 100,000 function points. As can be seen, DRE goes down as size goes up.

Table 4.19 is a somewhat complicated table that combines the results of Tables 4.5 and 4.6; i.e., both defect potentials and DRE ranges are now shown together on the same table. Note that as size increases defect potentials also increase, but DRE comes down.

Table 4.17 Software Defect Potentials per Function Point by Size

Function Points	*Best*	*Average*	*Worst*
1	0.60	1.50	2.55
10	1.25	2.50	4.25
100	1.75	3.25	6.13
1,000	2.14	4.75	8.55
10,000	3.38	6.50	12.03
100,000	4.13	8.25	14.19
Average	2.21	4.31	7.95

Note: Defects per IFPUG 4.3 function point.

Table 4.18 U.S. Software Average DRE Ranges by Application Size

Function Points	*Best (%)*	*Average (%)*	*Worst (%)*
1	99.90	97.00	94.00
10	99.00	96.50	92.50
100	98.50	95.00	90.00
1,000	96.50	94.50	87.00
100,00	94.00	89.50	83.50
100,000	91.00	86.00	78.00
Average	95.80	92.20	86.20

Table 4.19 Software Defect Potentials and DRE Ranges by Size

Function Points		*Best*	*Average*	*Worst*
	Defect potential	0.60	1.50	2.55
	DRE (%)	99.90	97.00	94.00
1	Delivered defects	0.00	0.05	0.15
	Defect potential	1.25	2.50	4.25
	DRE (%)	99.00	96.00	92.50
10	Delivered defects	0.01	0.10	0.32
	Defect potential	1.75	3.50	6.13
	DRE (%)	98.50	95.00	90.00
100	Delivered defects	0.03	0.18	0.61
	Defect potential	2.14	4.75	8.55
	DRE (%)	96.50	94.50	87.00
1,000	Delivered defects	0.07	0.26	1.11
	Defect potential	3.38	6.50	12.03
	DRE (%)	94.00	89.50	83.50
10,000	Delivered defects	0.20	0.68	1.98
	Defect potential	4.13	8.25	14.19
	DRE (%)	91.00	86.00	78.00
100,000	Delivered defects	0.37	1.16	3.12

Best-case results are usually found for software controlling medical devices or complex physical equipment such as aircraft navigation packages, weapons systems, operating systems, or telecommunication switching systems. These applications are usually large and range from about 1,000 to over 100,000 function points in size. Large complex applications require very high DRE levels in order for the physical equipment to operate safely. They normally use pretest inspections and static analysis and usually at least 10 test stages.

Average-case results are usually found among banks, insurance companies, manufacturing, and commercial software. These applications are also on the large size and range from 1,000 to more than 10,000 function points. Here too high levels of DRE are important since these applications contain and deal with confidential data. These applications normally use pretest static analysis and at least eight test stages.

Worst-case results tend to show up in litigation for canceled projects or for lawsuits for poor quality. State, municipal, and civilian Federal government software projects, and especially large systems such as taxation, child support, and motor vehicles are often in the worst-case class.

It is an interesting point that every lawsuit where the author has worked as an expert witness has been for large systems >10,000 function points in size. These applications seldom use either pretest inspections or static analysis and sometimes use only six test stages.

While function point metrics are the best choice for normalization, it is also important to know the actual numbers of defects that are likely to be present when software applications are delivered to customers. Table 4.20 shows data from Table 4.7 only expanded to show total numbers of delivered defects.

Here too it is painfully obvious that defect volumes go up with application size. However, Table 4.21 shows all severity levels of delivered defects. Only about 1% of delivered defects will be in high-severity class 1 and only about 14% in severity class 2. Severity class 3 usually has about 55% while severity 4 has about 30%.

Table 4.20 U.S. Average Delivered Defects by Application Size

Function Points	*Best*	*Average*	*Worst*
1	0	0	1
10	0	1	3
100	3	18	61
1,000	75	261	1,112
10,000	2,028	6,825	19,841
100,000	3,713	11,550	31,218
Average	970	3,109	8,706

Table 4.21 IBM Defect Severity Scale (1960 – 2017)

Severity 1	Software does not work at all
Severity 2	Major features disabled and inoperative
Severity 3	Minor bug that does not prevent normal use
Severity 4	Cosmetic errors that do not affect operation
Invalid	Defects not correctly reported such as hardware problems
Duplicate	Multiple reports of the same bug
Abeyant	Unique defects found by only one client that cannot be duplicated

This classification of defect severity levels was developed by IBM circa 1960. It has been used for over 50 years by thousands of companies for hundreds of thousands of software applications.

It is obvious that valid high-severity defects of severities 1 and 2 are the most troublesome for software projects.

DRE is a powerful and useful metric. Every important project should measure DRE and should top 99% in DRE, but only a few do.

As defined by IBM circa 1973, DRE is measured by keeping track of all bugs found internally during development, and comparing these to customer-reported bugs during the first 90 days of usage. If internal bugs found during development total 95 and customers report 5 bugs in the first 3 months of use, then the DRE is 95%.

Another important quality topic is that of "error-prone modules" (EPM) also discovered by IBM circa 1970. IBM did a frequency analysis of defect distributions and was surprised to find that bugs are not randomly distributed, but clump in a small number of modules. For example, in the IBM IMS database application, there were 425 modules. About 300 of these were zero-defect modules with no customer-reported bugs.

About 57% of all customer-reported bugs were noted in only 31 modules out of 425. These tended to be high in cyclomatic complexity and also had failed to use pretest inspections. Table 4.22 shows approximate results for EPM in software by application size.

EPM was discovered by IBM but unequal distribution of bugs was also noted by many other companies whose defect-tracking tools can highlight bug reports by modules. For example, EPM was confirmed by AT&T, ITT, Motorola, Boeing, Raytheon, and other technology companies with detailed defect-tracking systems.

EPM tends to resist testing, but is fairly easy to find using pretest static analysis, pretest inspections, or both. EPMs are treatable, avoidable conditions, and should not be allowed to occur in modern software circa 2016. The presence of EPM is a sign of inadequate defect quality measurements and inadequate pretest defect removal activities.

Table 4.22 Distribution of "EPM" in Software

Function Points	*Best*	*Average*	*Worst*
1	0	0	0
10	0	0	0
100	0	0	0
1,000	0	2	4
10,000	0	18	49
100,000	0	20	120
Average	0	7	29

The author had a contract from the U.S. Air Force to examine the value of ascending to the higher levels of the CMMI. Table 4.23 shows the approximate quality results for all five levels of the CMMI.

Table 4.23 was based on the study by the author commissioned by the U.S. Air Force. Usage of the CMMI is essentially limited to military and defense software. Few civilian companies use the CMMI and the author has met several CIO's from large companies and state governments that have never even heard of SEI or the CMMI.

Software defect potentials and DRE also vary by industry. Table 4.24 shows a sample of 15 industries with higher than average quality levels out of a total of 75 industries where the author has data.

There are also significant differences by country. Table 4.24 shows a sample of 15 countries with better than average quality out of a total of 70 countries where the author has data.

Table 4.23 Software Quality and the SEI Capability Maturity

CMMI Level	*Defect Potential per Function Point*	*DRE (%)*	*Delivered Defects per Function Point*	*Delivered Defects*
SEI CMMI 1	4.50	87.00	0.585	1,463
SEI CMMI 2	3.85	90.00	0.385	963
SEI CMMI 3	3.00	96.00	0.120	300
SEI CMMI 4	2.50	97.50	0.063	156
SEI CMMI 5	2.25	99.00	0.023	56

Note: Model integrated (CMMI) for 2500 function points.

Table 4.24 Software Quality Results by Industry

Industry		*Defect Potentials per Function Point 2017*	*DRE 2017 (%)*	*Delivered Defects per Function Point 2017*
Best Quality				
1	Manufacturing—medical devices	4.60	99.50	0.02
2	Manufacturing—aircraft	4.70	99.00	0.05
3	Government—military	4.70	99.00	0.05
4	Smartphone/tablet applications	3.30	98.50	0.05
5	Government—intelligence	4.90	98.50	0.07
6	Software (commercial)	3.50	97.50	0.09
7	Telecommunications operations	4.35	97.50	0.11
8	Manufacturing—defense	4.65	97.50	0.12
9	Manufacturing—telecommunications	4.80	97.50	0.12
10	Process control and embedded	4.90	97.50	0.12
11	Manufacturing—pharmaceuticals	4.55	97.00	0.14
12	Professional support—medicine	4.80	97.00	0.14
13	Transportation—airlines	5.87	97.50	0.15
14	Manufacturing—electronics	4.90	97.00	0.15
15	Banks—commercial	4.15	96.25	0.16

Countries such as Japan and India tend to be more effective in pretest defect removal operations and to use more certified test personnel than those lower down the table. Although not shown in Table 4.25, the United States ranks as country #19 out of the 70 countries from which the author has data.

Table 4.26 shows quality comparison of 15 software development methodologies (this table is cut down from a larger table of 80 methodologies that will be published in the author's next book).

Table 4.25 Samples of Software Quality by Country

Countries		Defect Potential per FP 2016	DRE 2016 (%)	Delivered Defects per Function Point 2016
Best Quality				
1	Japan	4.25	96.00	0.17
2	India	4.90	95.50	0.22
3	Finland	4.40	94.50	0.24
4	Switzerland	4.40	94.50	0.24
5	Denmark	4.25	94.00	0.26
6	Israel	5.00	94.80	0.26
7	Sweden	4.45	94.00	0.27
8	Netherlands	4.40	93.50	0.29
9	Hong Kong	4.45	93.50	0.29
10	Brazil	4.50	93.00	0.32
11	Singapore	4.80	93.40	0.32
12	United Kingdom	4.55	93.00	0.32
13	Malaysia	4.60	93.00	0.32
14	Norway	4.65	93.00	0.33
15	Taiwan	4.90	93.30	0.33

Table 4.27 shows the details of how DRE operates. Table 4.27 must of course use fixed values but there are ranges for every row and column for both pretest and test methods.

There are also variations in the numbers of pretest removal and test stages used. Table 4.27 illustrates the maximum number observed.

The data in Table 4.27 is originally derived from IBM's software quality data collection, which is more complete than most companies. Other companies have been studied as well. Note that requirements defects are among the most difficult to remove since they are resistant to testing.

To consistently top 99% in DRE, the minimum set of methods needed include most of the following.

In other words, a series of about 13 kinds of defect removal activities are generally needed to top 99% in DRE consistently. Testing by itself without inspections or static analysis usually is below 90% in DRE.

Table 4.26 Comparisons of 15 Software Methodologies

	Methodologies	*Defect Potential per FP 2016*	*DRE 2016 (%)*	*Delivered Defects per Function Point 2016*
Best Quality				
1	Reuse-oriented (85% reusable materials)	1.30	99.50	0.007
2	Pattern-based development	1.80	99.50	0.009
3	Animated, 3-D, full color design development	1.98	99.20	0.016
4	TSP+PSP	2.35	98.50	0.035
5	Container development (65% reuse)	2.90	98.50	0.044
6	Microservice development	2.50	98.00	0.050
7	Model-driven development	2.60	98.00	0.052
8	Microsoft SharePoint development	2.70	97.00	0.081
9	Mashup development	2.20	96.00	0.088
10	Product Line engineering	2.50	96.00	0.100
11	DevOps development	3.00	94.00	0.180
12	Pair programming development	3.10	94.00	0.186
13	Agile+scrum	3.20	92.50	0.240
14	Open-source development	3.35	92.00	0.268
15	Waterfall development	4.60	87.00	0.598

Of course, some critical applications such as medical devices and weapons systems use many more kinds of testing. As many as 18 kinds of testing have been observed by the author. This paper uses 12 kinds of testing since these are fairly common on large systems >10,000 function points in size, which is where quality is a critical factor.

Note that DRE includes bugs that originate in architecture, requirements, design, code, documents, and "bad fixes" or new bugs in bug repairs themselves. All bug origins should be included since requirements and design bugs often outnumber code bugs.

Table 4.27 Summary of Software Defect Removal for High Quality

Pretest Removal
1. Formal inspections (requirements, design, code, etc.)
2. Code static analysis
3. Automated requirements modeling
4. Automated correctness proofs
Test Removal
1. Unit test (manual/automated)
2. Function test
3. Regression test
4. Integration test
5. Performance test
6. Usability test
7. Security test
8. System test
9. Field or acceptance test

Note that the defect potential for Table 4.28 is somewhat lower than the 4.25 value shown in Tables 4.1 through 4.3. This is because those tables include all programming languages and some have higher defect potentials than Java, which is used for Table 4.28.

Code defect potentials vary by language with low-level languages such as assembly and C having a higher defect potential than high-level languages such as Java, Objective C, C#, Ruby, Python, etc.

Note: The letters "IV and V" in Table 4.28 stand for "independent verification and validation." This is a method used by defense software projects but it seldom occurs in the civilian sector. The efficiency of IV and V is fairly low and the costs are fairly high.

DRE measures can be applied to any combination of pretest and testing stages. Table 4.28 shows 7 pretest DRE activities and 12 kinds of testing—19 forms of defect removal in total. This combination would only be used on large defense systems and also on critical medical devices. It might also be used on aircraft navigation and avionics packages. In other words, software that might cause injury or death to humans if quality lags are the most likely to use both DRE measures and sophisticated combinations of pretest and test removal methods.

As of 2017, the U.S. average for DRE is only about 92.50%. This is close to the average for Agile projects.

The U.S. norm is to use only static analysis before testing and six kinds of testing: unit test, function test, regression test, performance test, system test, and acceptance test. This combination usually results in about 92.50% DRE.

Table 4.28 Software Quality and DRE

	Application size in function points					1,000	
	Application language					Java	
	Source lines per Function Point					53.33	
	Source LOC					53,330	
	Pretest Defect Removal Methods	*Architecture Defects per Function Point*	*Requirements Defects per Function Point*	*Design Defects per Function Point*	*Code Defects per Function Point*	*Document Defects per Function Point*	*Totals*
	Defect potentials per function point	0.25	1.00	1.15	1.30	0.45	4.15
	Defect potentials	250	1,000	1,150	1,300	450	4,150
1	*Requirement inspection (%)*	5.00	87.00	10.00	5.00	8.50	26.52
	Defects discovered	13	870	115	65	38	1,101
	Bad-fix injection	0	26	3	2	1	33
	Defects remaining	237	104	1,032	1,233	411	3,016
2	*Architecture inspection (%)*	**85.00**	10.00	10.00	2.50	12.00	13.10

(Continued)

Table 4.28 (*Continued*) Software Quality and DRE

	Pretest Defect Removal Methods	*Architecture Defects per Function Point*	*Requirements Defects per Function Point*	*Design Defects per Function Point*	*Code Defects per Function Point*	*Document Defects per Function Point*	*Totals*
	Defects discovered	202	10	103	31	49	395
	Bad-fix injection	6	0	3	1	1	12
	Defects remaining	30	93	925	1,201	360	2,609
3	*Design inspection* (%)	10.00	14.00	**87.00**	7.00	16.00	36.90
	Defects discovered	3	13	805	84	58	963
	Bad-fix injection	0	0	24	3	2	48
	Defects remaining	26	80	96	1,115	301	1,618
4	*Code inspection* (%)	12.50	15.00	20.00	**85.00**	10.00	62.56
	Defects discovered	3	12	19	947	30	1,012
	Bad-fix injection	0	0	1	28	1	30
	Defects remaining	23	67	76	139	270	575
5	*Code Static Analysis* (%)	2.00	2.00	7.00	**55.00**	3.00	15.92

(Continued)

Table 4.28 (*Continued*) Software Quality and DRE

	Pretest Defect Removal Methods	*Architecture Defects per Function Point*	*Requirements Defects per Function Point*	*Design Defects per Function Point*	*Code Defects per Function Point*	*Document Defects per Function Point*	*Totals*
	Defects discovered	0	1	5	76	8	92
	Bad-fix injection	0	0	0	2	0	3
	Defects remaining	23	66	71	60	261	481
6	*IV and V* (%)	10.00	12.00	**23.00**	7.00	18.00	16.16
	Defects discovered	2	8	16	4	47	78
	Bad-fix injection	0	0	0	0	1	2
	Defects remaining	20	58	54	56	213	401
7	*SQA review* (%)	10.00	17.00	17.00	12.00	12.50	30.06
	Defects discovered	2	10	9	7	27	54
	Bad-fix injection	0	0	0	0	1	3
	Defects remaining	18	48	45	49	185	344
	Pretest defects removed	232	952	1,105	1,251	265	3,805

(Continued)

Table 4.28 (*Continued*) Software Quality and DRE

	Pretest Defect Removal Methods	*Architecture Defects per Function Point*	*Requirements Defects per Function Point*	*Design Defects per Function Point*	*Code Defects per Function Point*	*Document Defects per Function Point*	*Totals*
	Pretest efficiency (%)	92.73	95.23	96.12	96.24	58.79	91.69
Test Defect Removal Stages							
		Architecture	*Requirements*	*Design*	*Code*	*Document*	*Total*
1	*Unit testing (manual)* (%)	2.50	4.00	7.00	**35.00**	10.00	11.97
	Defects discovered	0	2	3	17	19	41
	Bad-fix injection	0	0	0	1	1	1
	Defects remaining	18	46	41	31	166	301
2	*Function testing* (%)	7.50	5.00	22.00	**37.50**	10.00	13.63
	Defects discovered	1	2	9	12	17	41
	Bad-fix injection	0	0	0	0	0	1
	Defects remaining	16	43	32	19	149	259

(Continued)

Table 4.28 (*Continued*) Software Quality and DRE

		Architecture	*Requirements*	*Design*	*Code*	*Document*	*Total*
3	*Regression testing (%)*	2.00	2.00	5.00	**33.00**	7.50	7.84
	Defects discovered	0	1	2	6	11	20
	Bad-fix injection	0	0	0	0	0	1
	Defects remaining	16	43	30	13	138	238
4	*Integration testing (%)*	6.00	20.00	22.00	**33.00**	15.00	17.21
	Defects discovered	1	9	7	4	21	41
	Bad-fix injection	0	0	0	0	1	1
	Defects remaining	15	34	23	8	116	196
5	*Performance testing (%)*	14.00	2.00	**20.00**	18.00	2.50	6.07
	Defects discovered	2	1	5	2	3	12
	Bad-fix injection	0	0	0	0	0	0
	Defects remaining	13	33	19	7	113	184

(Continued)

Table 4.28 (*Continued*) Software Quality and DRE

		Architecture	*Requirements*	*Design*	*Code*	*Document*	*Total*
6	*Security testing (%)*	12.00	15.00	**23.00**	8.00	2.50	7.71
	Defects discovered	2	5	4	1	3	14
	Bad-fix injection	0	0	0	0	0	0
	Defects remaining	11	28	14	6	110	169
7	*Usability testing (%)*	12.00	17.00	15.00	5.00	**48.00**	36.42
	Defects discovered	1	5	2	0	53	62
	Bad-fix injection	0	0	0	0	2	2
	Defects remaining	10	23	12	6	56	106
8	*System testing (%)*	16.00	12.00	18.00	12.00	**34.00**	24.81
	Defects discovered	2	3	2	1	19	26
	Bad-fix injection	0	0	0	0	1	1
	Defects remaining	8	20	10	5	36	79

(Continued)

Table 4.28 (*Continued*) Software Quality and DRE

		Architecture	*Requirements*	*Design*	*Code*	*Document*	*Total*
9	*Cloud testing* (%)	10.00	5.00	13.00	10.00	**20.00**	13.84
	Defects discovered	1	1	1	1	7	11
	Bad-fix injection	0	0	0	0	0	0
	Defects remaining	7	19	8	5	29	69
10	*Independent testing* (%)	12.00	10.00	11.00	10.00	**23.00**	15.81
	Defects discovered	1	2	1	0	7	11
	Bad-fix injection	0	0	0	0	0	0
	Defects remaining	6	17	8	4	22	57
11	*Field (Beta) testing* (%)	14.00	12.00	14.00	12.00	**34.00**	20.92
	Defects discovered	1	2	1	1	7	12
	Bad-fix injection	0	0	0	0	0	0
	Defects remaining	6	15	6	4	14	45

(Continued)

Table 4.28 (*Continued*) Software Quality and DRE

		Architecture	*Requirements*	*Design*	*Code*	*Document*	*Total*
12	*Acceptance testing (%)*	13.00	14.00	15.00	12.00	**24.00**	20.16
	Defects discovered	1	2	1	0	6	10
	Bad-fix injection	0	0	0	0	0	0
	Defects remaining	5	13	6	3	8	35
	Test defects removed	13	35	39	46	177	309
	Testing efficiency (%)	73.96	72.26	87.63	93.44	95.45	89.78
	Total defects removed	245	987	1,144	1,297	442	4,114
	Total bad-fix injection	7	30	34	39	13	123
	Cumulative removal (%)	98.11	98.68	99.52	99.75	98.13	99.13

(Continued)

Table 4.28 (*Continued*) Software Quality and DRE

		Architecture	*Requirements*	*Design*	*Code*	*Document*	*Total*
	Remaining defects	5	13	6	3	8	36
	High-severity defects	1	2	1	1	1	5
	Security defects	0	0	0	0	0	1
	Remaining defects per function point	0.0036	0.0102	0.0042	0.0025	0.0065	0.0278
	Remaining defects per K function points	3.63	10.17	4.23	2.46	6.48	27.81
	Remaining defects per KLOC	0.09	0.25	0.10	0.06	0.16	0.68

Note: The table represents high-quality defect removal operations.

Table 4.29 Quality and Security Flaws for 1000 Function Points

Defect Potentials per Function Point	*DRE (%)*	*Delivered Defects per Function Point*	*Delivered Defects*	*High Severity Defects*	*Security Flaw Defects*
2.50	99.50	0.01	13	1	0
3.00	99.00	0.03	30	3	0
3.50	97.00	0.11	105	10	1
4.00	95.00	0.20	200	21	3
4.25	92.50	0.32	319	35	4
4.50	92.00	0.36	360	42	6
5.00	87.00	0.65	650	84	12
5.50	83.00	0.94	935	133	20
6.00	78.00	1.32	1,320	206	34

If static analysis is omitted and only six test stages are used, DRE is normally below 85%. In this situation quality problems are numerous.

Note that when a full suite of pretest defect removal and test stages are used, the final number of defects released to customers often has more bugs originating in requirements and design than in code.

Due to static analysis and formal testing by certified test personnel, DRE for code defects can top 99.75%. It is harder to top 99% for requirements and design bugs since both resist testing and can only be found via inspections, or by text static analysis.

Software Quality and Software Security

Software quality and software security have a tight relationship. Security flaws are just another kind of defect potential. As defect potentials go up, so do security flaws; as DRE declines, more and more security flaws will be released.

Of course, security has some special methods that are not part of traditional quality assurance. One of these is the use of ethical hackers and another is the use of penetration teams that deliberately try to penetrate the security defenses of critical software applications.

Security also includes social and physical topics that are not part of ordinary software operations. For example, security requires careful vetting of personnel. Security for really critical applications may also require Faraday cages around computers to ensure that remote sensors are blocked and can't steal information from a distance or through building walls.

To provide an approximate set of values for high-severity defects and security flaws, Table 4.28 shows what happens when defect potentials increase and DRE declines. To add realism to this example, Table 4.29 uses a fixed size of 1000 function points. Delivered defects, high-severity defects, and security flaws are shown in whole numbers rather than defects per function point:

The central row in the middle of this table highlighted in *blue* shows approximate 2017 U.S. averages in terms of delivered defects, high-severity defects, and latent security flaws for 1000 function points. The odds of a successful cyber-attack would probably be around 15%.

At the safe end of the spectrum where defect potentials are low and DRE tops 99%, the number of latent security flaws is 0. The odds of a successful cyber-attack are very low at the safe end of the spectrum: probably below 1%.

At the dangerous end of the spectrum with high defect potentials and low DRE, latent security flaws top 20 for 1000 function points. This raises the odds of a successful cyber-attack to over 50%.

Software Quality and Technical Debt

Ward Cunningham introduced an interesting metaphor called "technical debt," which concerns latent defects present in software applications after deployment.

The idea of technical debt is appealing but unfortunately technical debt is somewhat ambiguous and every company tends to accumulate data using different methods and so it is hard to get accurate benchmarks.

In general, technical debt deals with the direct costs of fixing latent defects as they are reported by users or uncovered by maintenance personnel. However, there are other and larger costs associated with legacy software and also new software that are not included in technical debt.

1. Litigation against software outsource contractors or commercial software vendors by disgruntled users who sue for excessive defects.
2. Consequential damages or financial harm to users of defective software. For example, if the computerized brake system of an automobile fails and causes a serious accident, neither the cost of repairing the auto nor any medical bills for injured passengers are included in technical debt.
3. Latent security flaws that are detected by unscrupulous organizations and lead to data theft, denial of service, or other forms of cyber-attack are not included in technical debt either.

Technical debt is an appealing metaphor but until consistent counting rules become available, it is not a satisfactory quality metric. The author suggests that the really high cost topics of consequential damages, cyber-attacks, and litigation for poor quality should be included in technical debt or at least not ignored as they are in 2016.

Assume that a software outsource vendor builds a 10,000-function point application for a client for a cost of $30,000,000 and it has enough bugs to make the client unhappy. True technical debt or the costs of repairing latent defects found and reported by clients over several years after deployment might cost about $5,000,000.

However, depending upon what the application does, consequential damages to the client could top $25,000,000; litigation by the unhappy client might cost $5,000,000; severe cyber-attacks and data theft might cost $30,000,000: a total cost of $60,000,000 over and above the nominal amount for technical debt.

Of these problems, cyber-attacks are the most obvious candidates to be added to technical debt because they are the direct result of latent security flaws present in the software when it was deployed. The main difference between normal bugs and security flaws is that cyber criminals can exploit security flaws to do very expensive damages to software (and even hardware) or to steal valuable and sometimes classified information.

In other words, possible post-release costs due to poor quality control might approach or exceed twice the initial costs of development and 12 times the costs of "technical debt" as it is normally calculated.

SNAP Metrics for Nonfunctional Size

In 2012, the IFPUG organization developed a new metric for nonfunctional requirements. This metric is called "SNAP," which is sort of an acronym for "software nonfunctional assessment process." (No doubt future sociologists will puzzle over software-naming conventions.)

Unfortunately, the SNAP metric was not created to be equivalent to standard IFPUG function points. That means if you have 100 function points and 15 SNAP points you cannot add them together to create 115 total "points." This makes both productivity and quality studies more difficult because function point and SNAP work needs to be calculated separately.

Since one of the most useful purposes for function point metrics has been for predicting and measuring quality, the addition of SNAP metrics to the mix has raised the complexity of quality calculations.

Pasted below are the results of an experimental quality calculation tool developed by the author that can combine defect potentials and DRE for both function point metrics and the newer SNAP metrics (Table 4.30).

In real life, defect potentials go up with application size and DRE comes down with it. This experimental tool holds defect potentials and DRE as constant values. The purpose is primarily to experiment with the ratios of SNAP defects and with DRE against SNAP bugs.

Table 4.30 SNAP Software Defect Calculator

Size in function points		1,000	
Size in SNAP points		152	
Defect Origins	*Defects Per Function Point*	*Defects per SNAP*	*SNAP Percentage (%)*
Requirements	0.70	0.14	19.50
Architecture	0.10	0.02	15.50
Design	0.95	0.18	18.50
Source code	1.15	0.13	11.50
Security flaws	0.25	0.05	20.50
Documents	0.45	0.02	3.50
Bad fixes	0.65	0.12	18.50
Totals	4.25	0.65	15.23
Defect Origins	*Defect Potential*	*Defects Potential*	
Requirements	700	21	
Architecture	100	2	
Design	950	27	
Source code	1,150	20	
Security flaws	250	8	
Documents	450	2	
Bad fixes	650	18	
Totals	4,250	99	
Defect Origins	*Removal Percentage (%)*	*Removal Percentage (%)*	
Requirements	75.00	75.00	
Architecture	70.00	70.00	
Design	96.00	96.00	

(Continued)

Tsble 4.30 (*Continued*) SNAP Software Defect Calculator

Defect Origins	*Removal Percentage (%)*	*Removal Percentage (%)*	
Source code	98.00	98.00	
Security flaws	87.00	87.00	
Documents	95.00	95.00	
Bad fixes	78.00	78.00	
Average	85.57	85.57	
Defect Origins	*Delivered Defects*	*Delivered Defects*	
Requirements	175	5	
Architecture	30	1	
Design	38	1	
Source code	23	0	
Security flaws	33	1	
Documents	23	0	
Bad fixes	143	4	
Total	464	13	
Defect Origins	*Delivered per Function Point*	*Delivered per SNAP*	*SNAP Percentage (%)*
Requirements	0.175	0.034s	19.50
Architecture	0.030	0.005	15.50
Design	0.038	0.007	18.50
Source code	0.023	0.003	11.50
Security flaws	0.023	0.007	29.61
Documents	0.143	0.026	18.50
Bad fixes	0.464	0.082	17.75
Total	0.896	0.164	18.31

A great deal more study and more empirical data is needed before SNAP can actually become useful for software quality analysis. Currently, there is hardly any empirical data available on SNAP and software quality.

Economic Value of High Software Quality

One of the major economic weaknesses of the software industry due to bad metrics and poor measurements is a total lack understanding of the economic value of high software quality. If achieving high quality levels added substantially to development schedules and development costs it might not be worthwhile to achieve it. But the good news is that high software quality levels comes with shorter schedules and lower costs than average or poor quality!

These reductions in schedules and costs, or course, are due to the fact that finding and fixing bugs has been the #1 software cost driver for over 50 years. When defect potentials are reduced and DRE is increased due to pretest defect removal such as static analysis, then testing time and testing costs shrink dramatically.

Table 4.31 shows the approximate schedules in calendar months, the approximate effort in work hours per function point, and the approximate $ cost per function point that results from various combinations of software defect potentials and DRE.

The good news for the software industry is that low defect potentials and high DRE levels are the fastest and cheapest ways to build software applications!

The central row highlighted in *blue* shows approximate U.S. average values for 2016. This table also shows the "cost-per-defect" metric primarily to caution readers that this metric is inaccurate and distorts reality since it makes buggy applications look cheaper than high-quality applications.

A Primer on Manufacturing Economics and the Impact of Fixed Costs

The reason for the distortion of the cost-per-defect metric is because cost per defect ignores the fixed costs of writing test cases, running test cases, and for maintenance people are always needed whether bugs or reported or not.

To illustrate the problems with the cost-per-defect metric, assume you have data on four identical applications of 1000 function points in size. Assume for all four that writing test cases costs $10,000 and running test cases costs $10,000, so fixed costs are $20,000 for all four cases.

Now assume that fixing bugs costs exactly $500 each for all four cases. Assume Case 1 found 100 bugs, Case 2 found 10 bugs, Case 3 found 1 bug, and Case 4 had zero defects with no bugs found by testing. Table 4.32 illustrates both cost per defect and cost per function point for these four cases.

Table 4.31 Schedules, Effort, Costs for 1000 Function Points (Monthly Costs=$10,000)

Defect Potentials per Function Point	*DRE (%)*	*Delivered Defects per Function Point*	*Delivered Defects*	*Schedule Months*	*Work Hours per Function Point*	*Development Cost per Function Point*	*$ per Defect (Caution!)*
2.50	99.50	0.01	13	13.34	12.00	$909.09	$4,550.00
3.00	99.00	0.03	30	13.80	12.50	$946.97	$3,913.00
3.50	97.00	0.11	105	14.79	13.30	$1,007.58	$3,365.18
4.00	95.00	0.20	200	15.85	13.65	$1,034.09	$2,894.05
4.25	92.50	0.32	319	16.00	13.85	$1,050.00	$2,488.89
4.50	92.00	0.36	360	16.98	14.00	$1,060.61	$2,140.44
5.00	87.00	0.65	650	18.20	15.00	$1,136.36	$1,840.78
5.50	83.00	0.94	935	19.50	16.50	$1,250.00	$1,583.07
6.00	78.00	1.32	1,320	20.89	17.00	$1,287.88	$1,361.44

Table 4.32 Comparison of $ per Defect and $ per Function Point

	Case 1	*Case 2*	*Case 3*	*Case 4*
Fixed costs	$20,000	$20,000	$20,000	$20,000
Bug repairs	$50,000	$5,000	$500	$0
Total costs	$70,000	$25,000	$20,500	$20,000
Bugs found	100	10	1	0
$ per defect	$700	$2,500	$20,500	Infinite
$ per function point	$70.00	$25.00	$20.50	$20.00

As can be seen, the "cost-per-defect" metric penalizes quality and gets more expensive as defect volumes decline. This is why hundreds of refereed papers all claim that cost per defect goes up later in development.

The real reason that cost per defect goes up is not that the actual cost of defect repairs goes up, but rather fixed costs make it look that way. Cost per function point shows the true economic value of high quality and this goes down as defects decline.

Recall a basic law of manufacturing economics that "If a manufacturing process has a high percentage of fixed costs and there is a decline in the number of units produced, the cost per unit will go up." For over 50 years, the cost-per-defect metric has distorted reality and concealed the true economic value of high-quality software.

Some researchers have suggested leaving out the fixed costs of writing and running test cases and only considering the variable costs of actual defect repairs. This violates both economic measurement principles and also good sense.

Would you want a contractor to give you an estimate for building a house that only showed foundation and framing costs but not the more variable costs of plumbing, electrical wiring, and internal finishing? Software cost of quality (COQ) needs to include ALL the cost elements of finding and fixing bugs and not just a small subset of those costs.

The author has read over 100 refereed software articles in major journals such as IEEE Transactions, IBM Systems Journal, Cutter Journal, and others that parroted the stock phrase "It costs 100 times more to fix a bug after release than it does early in development."

Not even one of these 100 articles identified the specific activities that were included in the cost-per-defect data. Did the authors include test case design, test case development, test execution, defect logging, defect analysis, inspections, desk checking, correctness proofs, static analysis, all forms of testing, post-release defects, abeyant defects, invalid defects, duplicate defects, bad-fix injections, EPM, or any of the other topics that actually have a quantified impact on defect repairs?

Not even one of the 100 journal articles included such basic information on the work elements that comprised the "cost-per-defect" claims by the authors.

In medical journals this kind of parroting of a stock phrase without defining any of its elements would be viewed as professional malpractice. But the software literature is so lax and so used to bad data, bad metrics, and bad measures that none of the referees probably even noticed that the cost-per-defect claims were unsupported by any facts at all.

The omission of fixed costs also explains why "LOC" metrics are invalid and penalize high-level languages. In the case of LOC metrics requirements, design, architecture, and other kinds of noncode work are fixed costs, so when there is a switch from a low-level language such as assembly to a higher level language such as Objective C, the "cost per LOC" goes up.

Table 4.33 shows 15 programming languages with cost per function point and cost per LOC in side-by-side columns to illustrate that LOC penalizes high-level programming languages, distorts reality, and reverses the true economic value of high-level programming languages.

Recall that the standard economic definition for productivity for more than 200 years has been "Goods or services produced per unit of labor or expense." If a line of code is selected as a unit of expense, then moving to a high-level programming language will drive up the cost per LOC because of the fixed costs of noncode work.

Function point metrics, on the other hand, do not distort reality and are a good match to manufacturing economics and standard economics because they correctly show that the least expensive version has the highest economic productivity. LOC metrics make the most expensive version seem to have higher productivity than the cheapest, which of course violates standard economics.

Also, software has a total of 126 occupation groups. The only occupation that can be measured at with "LOC" is that of programming. Function point metrics, on the other hand, can measure the productivity of those employed in noncode occupations such as business analysts, architects, database designers, technical writers, project management, and everybody else.

The author is often asked questions such as "If cost per defect and LOC are such bad metrics why do so many companies still use them?"

The questioners are assuming, falsely, that if large numbers of people do something it must be beneficial. There is no real correlation between usage and benefits. Usually it is only necessary to pose a few counter questions:

> If obesity is harmful why are so many people overweight?
>
> If tobacco is harmful why do so many people smoke?
>
> If alcohol is harmful why are there so many alcoholics?

As will be shown later in this book, the number of users of the very harmful antipattern development methodology outnumber the users of the very beneficial

Table 4.33 Productivity Expressed Using Both LOC and Function Points

	Languages	*Size in LOC*	*Coding Work Hours*	*Total Work Hours*	*Total Costs*	*$ per Function Point*	*$ per LOC*
1	Application Generators	7,111	1,293	4,293	$325,222	$325.22	$45.73
2	Mathematica10	9,143	1,662	4,662	$353,207	$353.21	$38.63
3	Smalltalk	21,333	3,879	6,879	$521,120	$521.12	$24.43
4	Objective C	26,667	4,848	7,848	$594,582	$594.58	$22.30
5	Visual Basic	26,667	4,848	7,848	$594,582	$594.58	$22.30
6	APL	32,000	5,818	8,818	$668,044	$668.04	$20.88
7	Oracle	40,000	7,273	10,273	$778,237	$778.24	$19.46
8	Ruby	45,714	8,312	11,312	$856,946	$856.95	$18.75
9	Simula	45,714	8,312	11,312	$856,946	$856.95	$18.75
10	C#	51,200	9,309	12,309	$932,507	$932.51	$18.21
11	ABAP	80,000	14,545	17,545	$1,329,201	$1,329.20	$16.62
12	PL/I	80,000	14,545	17,545	$1,329,201	$1,329.20	$16.62
13	COBOL	106,667	19,394	22,394	$1,696,511	$1,696.51	$15.90
14	C	128,000	23,273	26,273	$1,990,358	$1,990.36	$15.55
15	Macro Assembly	213,333	38,788	41,788	$3,165,748	$3,165.75	$14.84

pattern-based development methodology. There is very poor correlation between value and numbers of users. Many harmful applications and other things have thousands of users.

The reason for continued usage of bad metrics is "cognitive dissonance," which is a psychological topic studied by Dr. Leon Festinger and first published in 1962. Today there is an extensive literature on cognitive dissonance.

Dr. Festinger studied opinion formation and found that once an idea is accepted by the human mind, it is locked in place and won't change until evidence against the idea is overwhelming. Then there will be an abrupt change to a new idea.

Cognitive dissonance has been a key factor for resistance to many new innovations and new scientific theories including the following:

- Resistance to the theories of Copernicus and Galileo.
- Resistance to Lister's and Semmelweis's proposals for sterile surgical procedures.
- Resistance to Alfred Wegener's theory of continental drift.
- Resistance to Charles Darwin's theory of evolution.
- British naval resistance to self-leveling shipboard naval cannons.
- Union and Confederate Army resistance to replacing muskets with rifles.
- Naval resistance to John Ericsson's inventions of screw propellers
- Naval resistance to John Ericsson's invention of ironclad ships.
- Army resistance to Christie's invention of military tank treads.
- Military and police resistance to Samuel Colt's revolvers (he went bankrupt.)
- Military resistance and the court martial of Gen. Billy Mitchell for endorsing air power.

Cognitive dissonance is a powerful force that has slowed down acceptance of many useful technologies.

Table 4.34 illustrates the use of function points for 40 software development activities. It is obvious that serious software economic analysis needs to use activity-based costs and not just use single-point measures or phase-based measures, neither of which can be validated.

In Table 4.34, over half of the activities are related to software quality. Out of a total of 40 activities, 26 of them are directly related to quality and defect removal. These 26 quality-related activities sum to 50.50% of software development costs while the actual coding amounts to only 30.58% of development costs.

The accumulated costs for defect-related activities were $8,670,476. The author is not aware of any other industry where defect-related costs sum to more than half of total development costs. This is due to the high error content of custom designs and manual coding, rather than construction of software from certified reusable components.

So long as software is built using custom designs and manual coding, defect detection and defect removal must be the major cost drivers of all software applications. Construction of software from certified reusable components would greatly

Table 4.34 Function Points for Activity-Based Cost Analysis for 10,000 Function Points

	Development Activities	*Work Hours per Function Point*	*Burdened Cost per Function Point*	*Project Cost*	*% of Total*
1	Business analysis	0.01	$0.42	$4,200	0.02
2	Risk analysis/sizing	0.00	$0.14	$1,400	0.01
3	Risk solution planning	0.00	$0.21	$2,100	0.01
4	Requirements	0.29	$23.33	$233,333	1.36
5	Requirement inspection	0.24	$19.09	$190,909	1.11
6	Prototyping	0.38	$30.00	$30,000	0.17
7	Architecture	0.05	$4.20	$42,000	0.24
8	Architecture inspection	0.04	$3.00	$30,000	0.17
9	Project plans/estimates	0.04	$3.00	$30,000	0.17
10	Initial design	0.66	$52.50	$525,000	3.06
11	Detail design	0.88	$70.00	$700,000	4.08
12	Design inspections	0.53	$42.00	$420,000	2.45
13	Coding	6.60	$525.00	$5,250,000	30.58
14	Code inspections	3.30	$262.50	$2,625,000	15.29
15	Reuse acquisition	0.00	$0.14	$1,400	0.01
16	Static analysis	0.01	$0.70	$7,000	0.04
17	COTS package purchase	0.01	$0.42	$4,200	0.02
18	Open-source acquisition	0.00	$0.21	$2,100	0.01
19	Code security audit	0.07	$5.25	$52,500	0.31
20	Independent verification and validation (IV and V)	0.01	$1.05	$10,500	0.06
21	Configuration control	0.03	$2.10	$21,000	0.12
22	Integration	0.02	$1.75	$17,500	0.10
23	User documentation	0.26	$21.00	$210,000	1.22

(Continued)

Table 4.34 (*Continued*) Function Points for Activity-Based Cost Analysis for 10,000 Function Points

	Development Activities	*Work Hours per Function Point*	*Burdened Cost per Function Point*	*Project Cost*	*% of Total*
24	Unit testing	1.06	$84.00	$840,000	4.89
25	Function testing	0.94	$75.00	$750,000	4.37
26	Regression testing	1.47	$116.67	$1,166,667	6.80
27	Integration testing	1.06	$84.00	$840,000	4.89
28	Performance testing	0.26	$21.00	$210,000	1.22
29	Security testing	0.38	$30.00	$300,000	1.75
30	Usability testing	0.22	$17.50	$175,000	1.02
31	System testing	0.75	$60.00	$600,000	3.49
32	Cloud testing	0.06	$4.38	$43,750	0.25
33	Field (Beta) testing	0.03	$2.63	$26,250	0.12
34	Acceptance testing	0.03	$2.10	$21,000	0.12
35	Independent testing	0.02	$1.75	$17,500	0.10
36	Quality assurance	0.18	$14.00	$140,000	0.82
37	Installation/training	0.03	$2.63	$26,250	0.15
38	Project measurement	0.01	$1.11	$11,053	0.06
39	Project office	0.24	$19.09	$190,909	1.11
40	Project management	1.76	$140.00	$1,400,000	8.15
	Cumulative results	21.91	$1,743.08	$17,168,521	100.00

increase software productivity and benefit the economics of not only software itself but of all industries that depend on software, which essentially means every industry in the world.

Table 4.34 shows the level of granularity needed to understand the cost structures of large software applications where coding is just over 30% of the total effort. Software management and C-level executives such as Chief Financial Officers (CFO) and Chief Information Officers (CIO) need to understand the complete set of activity-based costs, including costs by occupation groups such as business analysts and architects over and above programmers.

When you build a house, you need to know the costs of everything: foundations, framing, electrical systems, roofing, plumbing, etc. You also need to know the separate costs of architects, carpenters, plumbers, electricians, and of all the other occupations that work on the house.

Here too for large systems in the 10,000-function point size range, a proper understanding of software economics needs measurements of ALL activities and all occupation groups and not just coding programmers, whose effort is often <30% of the total effort for large systems.

Both LOC metrics and cost-per-defect metrics should probably be viewed as *professional malpractice* for software economic studies because they both distort reality and make bad results look better than good results.

It is no wonder that software progress resembles a drunkard's walk when hardly anybody knows how to measure either quality or productivity with metrics that make sense and match standard economics.

Software's Lack of Accurate Data and Poor Education on Quality and Cost of Quality

One would think that software manufacturing economics would be taught in colleges and universities as part of computer science and software engineering curricula, but universities are essentially silent on the topic of fixed costs, probably because the software faculty does not understand software manufacturing economics either. There are a few exceptions such as the University of Montreal, however.

The private software education companies and the professional associations are also silent on the topic of software economics and the hazards of cost per defect and LOC. It is doubtful if either of these sectors understands software economics well enough to teach it. They certainly don't seem to understand either function points or quality metrics such as DRE.

Even more surprising, some of the major software consulting groups with offices and clients all over the world are also silent on software economics and the hazards of both cost per defect and LOC. Gartner Group uses function points but apparently has not dealt with the impact of fixed costs and the distortions caused by the LOC and cost-per-defect metrics.

You would think that major software quality tool vendors such as those selling automated test tools, static analysis tools, defect-tracking tools, automated correctness proofs, or test-case design methods based on cause–effect graphs or design of experiments would measure defect potentials and DRE because these metrics could help to demonstrate the value of their products. Recall that IBM used defect potentials and DRE metrics to prove the value of formal inspections back in 1973.

But the quality companies are just as clueless as their clients when it comes to defect potentials and DRE and the economic value of high quality. They make vast claims of quality improvements but provide zero quantitative data. For example, only CAST Software that sells static analysis uses function points on a regular basis from among the major quality tool companies. But even CAST does not use defect potentials and DRE although some of their clients do.

You would also think that project management tool companies that market tools for progress and cost accumulation reporting and project dashboards would support function points and show useful economic metrics such as work hours per function point and cost per function point. You would also think that they would support activity-based costs.

However, most project management tools do not support either function point metrics or activity-based costs, although a few do support earned value and some forms of activity-based cost analysis. This means that standard project management tools are not useful for software benchmarks since function points are the major benchmark metric.

The only companies and organizations that seem to know how to measure quality and economic productivity are the function point associations such as COSMIC, FISMA, IFPUG, and NESMA; the software benchmark organizations such as ISBSG, David's Consulting, Namcook Analytics, TIMetricas, Q/P Management Group, and several others; and some of the companies that sell parametric estimation tools such as KnowledgePlan, SEER, SLIM, and the author's SRM. In fact, the author's SRM tool predicts software application size in a total of 23 metrics including all forms of function points plus story points, use case points, physical and logical codes, and a number of others. It even predicts bad metrics such as cost per defect and LOC primarily to demonstrate to clients why those metrics distort reality.

Probably not 1 reader out of 1000 of this paper has quality and cost measures that are accurate enough to confirm or challenge the data in this chapter's tables because software measures and metrics have been fundamentally incompetent for over 60 years. This kind of analysis can't be done with "cost per defect" or "LOC" because they both distort reality and conceal the economic value of software quality.

However, the comparatively few companies and fewer government organizations that do measure software costs and quality well using function points and DRE can confirm the results. The quality pioneers of Joseph Juran, W. Edwards Deming, and Phil Crosby showed that for manufactured products, quality is not only free it also saves time and money.

The same findings are true for software, only software has lagged all other industries in discovering the economic value of high software quality because software metrics and measures have been so bad that they distorted reality and concealed progress.

The combination of function point metrics and DRE measures can finally prove that high software quality, like the quality of manufactured products, lowers development costs and shortens development schedules. High quality also lowers maintenance costs, reduces the odds of successful cyber-attacks, and improves customer satisfaction levels.

Summary and Conclusions on Metrics for Problem-Solving

The combination of *defect potentials* and *DRE* measures provides software engineering and quality personnel with powerful tools for predicting and measuring all forms of defect prevention and all forms of defect removal.

Function points are the best metric for normalizing software defect potentials because function points are the only metrics that can handle requirements, design, architecture, and other sources of noncode defects.

This chapter uses IFPUG 4.3 function points. Other forms of function point metric such as COSMIC, FISMA, NESMA, etc., would be similar but not identical to the values shown here.

As of 2017, there is insufficient data on SNAP metrics to show defect potentials and DRE. However, it is suspected that nonfunctional requirements contribute to defect potentials in a significant fashion. There was insufficient data in 2016 to judge DRE values against nonfunctional defects.

Note that the author's SRM tool predicts defect potentials and DRE as standard outputs for all projects estimated.

For additional information on 25 methods of pretest defect removal and 25 forms of testing, see *The Economics of Software Quality*, Addison-Wesley, 2012 by Capers Jones and Olivier Bonsignour.

Improving Software Project Management Tools and Training

Software projects have been notoriously difficult to manage well. Cost overruns, schedule delays, and outright cancelations of software projects have been common occurrences for more than 50 years. Poor software quality when delivered remains a chronic issue. Although software project management tools are not a panacea, when used by capable managers they do simplify the complexity of software project management and raise the odds of successful completions of large software projects.

Software project management tools have many divisions and subcategories. However, they can be broadly classified into two major areas of focus: (1) Tools for planning and estimating software projects before they begin and (2) Tools for tracking and reporting software project status and costs while projects are underway.

Beneath this division are 12 topical areas and scores of project management activities. Note that the distinction between software projects and systems engineering projects is ambiguous. Also note that modern Software as a Service (SaaS) and cloud computing projects require interfaces among numerous software systems.

This section discusses the tasks of project management and the available tools that support those tasks. From the frequent failures of over 35% of large software projects and from analysis of the depositions and testimony of litigation for software failures, it is apparent that software project managers are often not well trained in managerial tasks, and often fail to utilize appropriate tools to support critical tasks such as estimating and progress tracking.

As of 2017, about 75% of significant software projects run late and about 50% are over budget. As already stated, about 35% of large systems in the 10,000-function point range are canceled and never delivered.

For those large software projects that are delivered, poor quality and high technical debt are endemic software problems. This is why corporate CEOs regard software as the least professional of any corporate operating units.

Before discussing the tools of software project management, it is appropriate to consider the many tasks that software project managers are likely to confront in carrying out their jobs. The two major divisions of software project management include planning and estimating before projects start, and tracking status, progress, and expenses while software projects are underway. Beneath this broad division there are a dozen major functional areas and scores of individual tasks as follows:

1. Software Project Sizing
    a. Size in function points of new development
    b. Size in function points of legacy components
    c. Size including the new SNAP metrics for nonfunctional features
    d. Size in LOC
    e. Size in database contents
    f. Numbers of screens
    g. Numbers of reports
    h. Numbers of interfaces
    i. Number of associated applications
    j. Numbers of files
    k. Prediction of requirements creep
    l. Prediction of requirements churn
    m. Prediction of requirements slippage to future releases
    n. Sprint contents for Agile development
2. Software Project Schedule Planning
    a. Project schedules
    b. Activity schedules
    c. Task schedules
    d. Overlap and concurrency of scheduled task
    e. Time splitting among key technical workers
    f. Sprint schedules for Agile projects
    g. Work breakdown analysis
    h. Critical paths
    i. Earned value
    j. Team assignments
    k. Individual assignments

  l. Requirements creep
  m. Requirements churn
  n. Requirements slip to future releases
  o. Requirements cancelation
3. Software Project Cost Estimating
  a. Estimating method selection
  b. Estimating tool selection
  c. Historical data analysis from similar projects
  d. Project cost estimating
  e. Phase cost estimating
  f. Activity cost estimating
  g. Task cost estimating
  h. Reusable material estimating
  i. Commercial-off-the-shelf (COTS) acquisition estimating
  j. Open-source acquisition estimating
  k. Project earned value calculation
  l. Creeping requirements cost prediction
  m. Churned requirements cost prediction
  n. Slipped requirements cost prediction
  o. Canceled requirements cost accumulation
  p. Complexity estimating
  q. Special estimates for SaaS links
4. Software Project Risk Estimating
  a. Sarbanes–Oxley risks for financial applications by large U.S. companies
  b. Governance criteria for key financial applications
  c. Participation in software process assessments
  d. Risks associated with low levels of capability maturity
  e. Risks of serious problems with COTS applications
  f. Risks of serious problems with open-source applications
  g. Risks of serious problems with cloud computing components
  h. Risks of serious problems with SaaS components
  i. Risks of serious problems with legacy components
  j. Risks of serious problems with hardware components
  k. Risks of serious problems with reused components
  l. Risks of theft of intellectual property
  m. Risks of schedule delays
  n. Risks of cost overruns
  o. Risks of project cancelation
  p. Risks of poor quality
  q. Risks of poor customer satisfaction
  r. Risks of loss of stakeholder or sponsor support
  s. Risks of poor team morale
  t. Risks of hardware changes during development

   u. Risks of software platform changes during development
   v. Risks of litigation for breach of contract
   w. Risks of litigation for intellectual property issues and patent issues
   x. Risks of loss of key personnel
   y. Risks of strikes by union personnel
   z. Risks of loss of key clients
   aa. Risks of unexpected competitive applications in same field
   bb. Risk of business failure due to recession
   cc. Risk of layoffs due to recession
   dd. Risks of natural disasters (flu epidemic, weather, etc.)
   ee. Potential security problems—prevention
   ff. Potential security problems—active defenses
   gg. Potential security problems—recovery
5. Software Project Quality Estimating
   a. Defect potentials for application under development
   b. Defect potentials for legacy components
   c. Defect potentials for COTS packages
   d. Defect potentials for reusable components
   e. Defect potentials for SaaS links and Cloud applications
   f. Defect potentials for open-source applications
   g. Defect potentials for hardware platforms
   h. Defect prevention methods
   i. Defect removal activities
   j. DRE for inspections
   k. DRE for static analysis
   l. DRE from independent verification and validation (IV and V)
   m. Code coverage from static analysis
   n. DRE for audits
   o. DRE for test stages (up to 20 forms of testing)
   p. Defect tracking tools and protocols
   q. Bad-fix injections
   r. EPM
   s. Bad test cases
   t. Cyclomatic and essential complexity
   u. Number of defect removal activities
   v. Number of test stages
   w. Number of test scripts
   x. Number of test cases per stage
   y. Test coverage of applications and features
   z. Human factors and usability defect removal
   aa. Nationalization and translation defect removal
   bb. User guide and HELP text defect removal
6. Software Project Outsource Contract Analysis

   a. Requirements scope in contract
   b. Requirements changes in contract
   c. COTS package responsibilities in contract
   d. Quality terms in contract
   e. Security criteria in contract
   f. Intellectual property and nondisclosure in contract
   g. Incentives for early completion
   h. Penalties for late completion
   i. Penalties for poor quality
   j. Required tracking and reporting
   k. Service-level agreements (SLAs) in contract
   l. Usage of Information Technology Infrastructure Library (ITIL)
7. Software Project COTS Acquisition
   a. COTS acquisition analysis
   b. Open-source acquisition analysis
   c. SaaS acquisition analysis
   d. Cloud computing acquisition analysis
   e. Due diligence prior to major acquisitions
   f. Warranty provisions by vendors
   g. Installation schedules
   h. Installation assistance from vendor or consultants
   i. Installation defects
   j. Learning curves
   k. Loss of performance during learning
8. Software Personnel Management
   a. Specialist selection
   b. Hiring of new employees
   c. Employee confidentiality agreements
   d. Training and education of employees
   e. Certification of employees (quality assurance, testing, function points, etc.)
   f. Appraisals of existing employees
   g. Promotions
   h. Demotions
   i. Terminations (voluntary)
   j. Terminations (involuntary)
   k. Awards
   l. Patents filed by employees
   m. Books and articles written by employees
9. Software Project Measurement, Tracking, and Control
   a. Financial tracking
   b. Earned value (EV) tracking
   c. Balanced scorecard tracking

   d. Goal-question metric tracking (GQM)
   e. Milestone tracking
   f. Issue and "red flag" tracking
   g. Risk tracking
   h. Quality and defect tracking
   i. Security vulnerability tracking
   j. Change request tracking
   k. Requirements creep tracking
   l. Requirements churn tracking
   m. Requirements slippage or cancelation tracking
   n. Complexity measurement
   o. Benchmark productivity data
   p. Benchmark quality data
   q. Coordination with other managers in-house
   r. Coordination with sponsors and stakeholders
   s. Coordination with related hardware or system managers
   t. Coordination with quality assurance managers
   u. Coordination with supply chain managers
   v. Coordination with SaaS and cloud organizations
   w. Coordination with user associations (if any)
10. Software Technology Acquisition
   a. Evaluation of proposed tools and methods
   b. Licensing patents and intellectual property
   c. Project planning tools
   d. Project estimating tools
   e. Project tracking tools
   f. Defect tracking tools
   g. Complexity analysis tools
   h. Requirements methodologies
   i. Design methodologies
   j. Development methodologies
   k. Training in new methodologies, if any
   l. Maintenance and enhancement methodologies
   m. Renovation of legacy applications prior to use
   n. Customer support methodologies
   o. Quality control methodologies
   p. Methodology management tools
   q. Development workbenches
   r. Maintenance workbenches
   s. Renovation workbenches
   t. Reverse engineering tools
   u. Reengineering tools
   v. Programming languages

11. Software Mergers and Acquisitions
    a. Due diligence for purchaser
    b. Analysis of patents and intellectual property in company to be acquired
    c. Analysis of major applications owned by company to be acquired
    d. Analysis of tools in company to be acquired
    e. Analysis of portfolio in company to be acquired
    f. Analysis of litigation against company to be acquired
    g. Analysis of litigation by company to be acquired
    h. Technology conversion after merger
    i. Portfolio consolidation after mergers
    j. Organization planning after merger
    k. Reorganization after merger
    l. Layoffs and downsizing after merger
12. Software Litigation Support
    a. Breach of contract litigation
    b. Intellectual property and patent litigation
    c. Employee confidentiality and noncompetition litigation
    d. Preparation of materials on plaintiff side
    e. Preparation of materials on defendant side
    f. Discovery documents if required
    g. Depositions if required
    h. Testimony if required

As of 2017, automated tools are available for many but not all of the tasks that software project managers are likely to perform. For common tasks such as schedule planning and cost estimating there are dozens of available tools. For less common tasks such as litigation support or due diligence prior to a merger or acquisition, few tools are available.

Software project management tools are evolving as new kinds of applications are developed. They are also evolving to support new kinds of software development methods such as Agile development, container development, mashups, DevOps, the RUP, and the TSP, among others.

Project Management Knowledge Acquisition

Academic training in the area of software project management is not as sophisticated as academic training in basic software engineering or even in other forms of engineering management such as electrical and aeronautical engineering. For example, there is a shortage of academic training even for important topics such as software cost estimation, software quality estimation, risk analysis, and milestone tracking. For more unusual topics, such as litigation support or acquisition of COTS software, academic training hardly exists at all.

Worse, most academic institutions are unaware of the hazards and dangers of common software metrics such as "LOC" and "cost per defect." The LOC metric makes requirements and design invisible and penalizes high-level languages. Cost per defect penalizes quality and makes buggy software look better than excellent software. Academia is for the most part unaware of these problems.

Because academic training tends to lag the state of the art, newer topics such as cloud computing, Saas, and even Agile development usually have no academic courses for perhaps 5–10 years after the technologies start use in industry. This is a normal situation because experience must be accumulated before it can be taught.

The lack of academic training for software project managers is compensated for in some degree by training and "bodies of knowledge" or BOK provided by a number of professional associations. The large associations that provide

Academy of Management (AOM)	www.aomonline.org
Agile Alliance	www.agilealliance.org
American Management Association (AMA)	www.amanet.org
American Society for Quality (ASQ)	www.asq.org
Cosmic function point users group	Cosmic-sizing.org
Finnish function point users group	www.totalmetrics.com
IEEE Computer Society	www.computer.org
Information Technology Metrics and Productivity Institute (ITMPI)	www.itmpi.org
International Function Point Users Group (IFPUG)	www.ifpug.org
International Project Management Association (IPMA)	www.ipma.org
International Organization for Standards (ISO)	www.iso.org
Netherlands function point users group	www.Nesma.org
Project Management Association of Japan	www.pmaj.or.jp
International Software Benchmark Standards Group (ISBSG)	www.isbsg.org
Project Management Institute (PMI)	www.pmi.org
Software Engineering Institute (SEI)	www.SEI.CMU.edu.
Society of Project Management (SPM)	www.spm-japan.jp
tiMetricas (Brazil)	www.Metrics.com/BR

useful information to software project managers can be found on the following websites.

There are also scores of national and local organizations that also provide information and seminars on software project management topics. Examples include the Australian Institute of Project Management & Australian Software Metrics Association & Software Quality Assurance (ASMA-SQA), the Brazilian Association for Project Management, the Netherlands Software Metrics Association (NESMA), and dozens of local chapters of the Software Process Improvement Network (SPIN) in the United States, Europe, South America, and Asia. A search of the Web will turn up at least 50 national software project management associations.

Several websites contain useful links to the project management literature. Among the commercial links can be found the previously mentioned ITMPI portal (www.ITMPI.org). Among academic links one of the most complete is that of Dave W. Farthing of the University of Glamorgan in the United Kingdom. This interesting portal has links to dozens of project management sites and the publishers of scores of project management books.

From discussions with project managers in large corporations, in-house training of managers in companies such as IBM, Microsoft, AT&T, Google, and CISCO is actually the most effective source of information for software project managers. Topics such as quality estimation, measurement, and risk analysis are often better covered via in-house training than by universities.

However, this form of training is only available for project managers in rather large corporations. Small companies must depend upon commercial training or upon professional associations, plus a few universities that do have curricula in software management topics.

In recent years, new forms of management training have started to take advantage of the Web. Almost every day there are webinars and podcasts on project management topics. In addition, many companies such as IBM, Rational (an IBM subsidiary), Microsoft, MicroFocus, Computer Aid Inc., the ITMPI, and scores of others offer online training on dozens of managerial topics.

In recent years new forms of software such as Saas, cloud computing, and the integration of open-source applications and commercial applications into private application has raised the complexity levels of project management. Some projects circa 2010 involve coordination among perhaps a dozen companies and more than 25 components whose development may be taking place in different countries.

Training in specific project management tools is also available from the vendors themselves. For example, fairly sophisticated training in software cost estimating is available from the vendors of commercial estimating tools such as KnowledgePlan, Price-S, SLIM, and SEER. Computer Aid Inc. and its subsidiary the ITMPI also provide training in a number of software management topics.

Overall, training of software project managers is a topic that needs to be improved. Quality control, security control, economic analysis, estimation, and measurement are all topics where additional training is needed for entry-level managers just starting their careers.

The History of Software Project Management Tools

Software project management is of course only one of the many different forms of project management, such as home construction projects, electrical engineering projects, or civil engineering projects such as building roads and bridges. Software project management does have some distinguishing attributes that seem to differ from other and more established forms of project management:

1. Because software is abstract and invisible to the eye, sizing software projects is more difficult than for many other kinds of project.
2. Because software engineering is not highly automated, manual effort by skilled knowledge workers plays a more important role in software projects than in many other kinds of project.
3. Because of the large volume of bugs or defects in all software deliverables, quality estimation and quality control are more important for software projects than for many other kinds of project.
4. Because viruses, worms, botnets, and denial-of-service attacks are endemic and sophisticated, security control is becoming an extremely important aspect of both software project management, software development, and software operation.
5. For large software projects in the 10,000-function point range or larger, outright failure, major cost overruns, and major schedule delays occur more frequently than for many other kinds of project.
6. Although changing requirements can occur for any kind of project, they occur with unusual frequency for software projects. Sometimes more than 50% of the delivered features of software projects are not identified until the project is already in process. Software requirements changes have been measured to occur at a rate of between 1% and 2% per calendar month.
7. Some new laws such as Sarbanes–Oxley require careful governance of key financial software in large companies, with threats of large fines or criminal charges if governance is absent or ineffective. These new legal liabilities have increased the complexity of management tasks for large projects in large companies.

Long before the computer era and long before software itself, project management was starting to be studied as a technology that needed formalization. Among the pioneers who dealt with project management issues include Henry Gantt whose famous Gantt charts circa 1917 facilitated schedule planning. Later in the 1950s, Willard Fazar's "program evaluation and review technique" (PERT) added rigor to isolating and identifying the critical paths which complex projects needed to follow. On the quality side, W. Edwards Deming revolutionized the statistical analysis of quality. Deming's work was initially adopted in Japan circa 1950 and later came to the United States when it became obvious that quality was a major factor in successful competition. Other pioneers in related topics include Walter Shewart (one of Deming's teachers), Phil Crosby of ITT, Joseph Juran, and Frederick Taylor.

In terms of software project management tools as opposed to the general-purpose project management tools, some of the pioneers include A.J. Albrecht and the invention of function point metrics circa 1973; Dr. Barry Boehm and the development of the famous "constructive cost model" or COCOMO circa 1981; Watts Humphrey and the development of the capability maturity model (CMM) circa 1983; and Peter Hill and the foundation of the ISBSG circa 1997.

The phrase "project management tools" has been applied to a large family of tools whose primary purpose is sophisticated scheduling for projects with hundreds or even thousands of overlapping and partially interdependent tasks. These tools are able to drop down to very detailed task levels, and can even handle the schedules of individual workers. However, for software, there are many other critical areas that also need the support of tools: quality prediction and measurement, requirements changes, and the earned value of ongoing projects. Software projects require additional capabilities to be under full management control.

There are a host of managerial functions that standard project management tools don't deal with in depth, such as sizing with function point metrics, contract management, personnel management, assessments, quality estimating, defect tracking, and the like.

The family of project management tools originated in the 1950s with the U.S. Navy and Air Force. Weapons systems and other complex projects were becoming so large that manual planning and control were no longer feasible. Since computers were starting to become more powerful, it was natural to want to use computer power as an aid for project managers.

Project management tools are an automated form of several management aids developed by the Navy for controlling large and complex weapons systems: the "PERT", critical path analysis, resource leveling, and the classic Gantt charts. Such general-purpose tools can of course be used for software projects, but in addition specialized tools are needed that support the unique requirements of software projects.

Project management tools did not originate for software, but rather for handling very complex scheduling situations where hundreds or even thousands of tasks need to be determined and sequenced, and where dependencies such as the completion of a task might affect the start of subsequent tasks. Project management tools are general purpose in nature, and can be applied to any kind of project: building an aircraft, constructing an office building, or a software project.

General-purpose project management tools have no built-in expertise regarding software, as do specialized software cost estimating tools. For example, if you wish to explore the quality and cost impact of an object-oriented programming language such as Objective C, a standard project management tool is not the right choice. By contrast, many software cost estimating tools have built-in tables of programming languages and will automatically adjust the estimate based on which language is selected for the application.

Since software cost estimating tools originated about 10 years after commercial project management tools, the developers of software cost estimating tools seldom tried to replicate project management functions such as construction of detailed PERT diagrams or critical path analysis. Instead, the cost estimating tools would export data so that the final schedule could be adjusted via the project management tool. Thus, bidirectional interfaces between software cost estimating tools and generic project management tools are now standard features in the commercial estimation market.

Usage Patterns of Software Project Management Tools

Observations made from depositions and testimony during software litigation for breach of contract reveals that software projects that end up in court for outright failure or major overruns usually do not utilize many project management tools. Estimates were often informal and tracking seriously deficient. By contrast, data gathered while performing assessments and benchmarks of successful software projects that combined high productivity and high quality indicate that many specialized project management tools were utilized for both predictions before the projects started and for tracking and measurement while they were underway.

Table 4.35 shows the approximate sizes of project management tool features measured in function points. The table also illustrates the patterns of tool usage noted on "lagging," "average," and "leading" projects. Information on the lagging projects is taken from information discovered during litigation. Information on the average and leading project is taken from benchmark and assessment studies by the author.

Table 4.35 illustrates two important aspects of software project management tool usage: (1) leading or sophisticated project managers uses more kinds of tools than do managers on lagging and average projects; (2) leading or sophisticated project managers use more of the features or select more powerful tools than do project managers on lagging or average projects.

The very significant use of project management tools on leading projects results in one overwhelming advantage: "No surprises." The number of on-time projects in the leading set is far greater than in the lagging set, and all measurement attributes (quality, schedules, productivity, etc.) are also significantly better.

Software project management tools are far from being a panacea, but advanced knowledge of software project size, potential numbers of defects or bugs, and schedules derived from similar historical projects can go a long way toward yielding a successful conclusion and eliminating the rather high odds of failure associated with large software projects.

Table 4.35 Numbers and Size Ranges of Software Project Management Tools

	Project Management Tools	*Lagging*	*Average*	*Leading*
1	Project schedule planning	1,000	1,250	3,000
2	Project parametric cost estimation			3,000
3	Statistical analysis			3,000
4	Methodology management		750	3,000
5	Reusable feature analysis			2,000
6	Quality estimation			2,000
7	Assessment support		500	2,000
8	Project office support		500	2,000
9	Project measurement			1,750
10	Portfolio analysis			1,500
11	Risk analysis			1,500
12	Resource tracking	300	750	1,500
13	Governance tools			1,500
14	Value analysis		350	1,250
15	Cost variance reporting	500	500	1,000
16	Personnel support	500	500	750
17	Milestone tracking		250	750
18	Budget support		250	750
19	Function point analysis		250	750
20	Backfiring: LOC to function point			300
21	Earned value analysis		250	300
22	Benchmark data collection			300
	Subtotal	1,800	4,600	30,000
	Tools	4	12	22

Note: Tool sizes are expressed in terms of IFPUG function points, version 4.2.

For large software projects in large corporations, the project will typically be supported by a "Project Office." IBM was a pioneer in the utilization of project offices and first deployed them on versions of the main IBM operating system in the early 1970s.

Typically, a project office will be formed for a major software application in the 100,000-function point size range. Applications of this large size have hundreds of development personnel and dozens of subordinate project managers under at least three higher levels of management. A typical project office for such a large system would contain perhaps half a dozen staff to as many as 10 personnel equipped with a variety of estimating and planning tools, plus historical data from similar projects. The project office would assist the individual project managers in cost estimating, quality estimating, schedule estimating, and other predictive tasks. They would also monitor rates of progress, bug reports, earned value, costs, and other status data as the project was underway.

Recent Evolution of Software Project Management Tools

Project management tools, like other forms of commercial software, continue to evolve and add new features. The following paragraphs summarize some of the recent observations of project management tool capabilities noted in 2008.

Software Project Sizing: Sizing software projects has been a difficult activity for project managers. Sizing in terms of "LOC" primarily involved either comparison to similar projects or guess work. Adding to the difficulty, there are more than 2500 programming languages in existence and some of these lack any rules for counting code. Further, many software applications utilize several programming languages at the same time. More important, for large software projects the costs and schedules associated with producing paper documents and finding and repairing defects costs more than the code itself and takes more time during the schedule. These noncoding tasks cannot be estimated or measured using LOC metrics.

Function point metrics have proven to be effective for sizing noncoding work and also programming, so function points are a good choice for life cycle economic analysis of software projects. Sizing in terms of function point metrics can be done with accuracy from requirements, but by the time requirements are available several iterations of cost estimates are normally required. Also, function point analysis has been somewhat slow and expensive. Certified function point analysts count function points only at a rate of perhaps 400 function points per day. Thus, the long schedules and high costs of counting function point metrics have been a barrier.

As of 2010, several new forms of high-speed function point tools have been announced. Relativity Technologies has announced a tool for sizing legacy applications called "Function Point Analyzer" or FPA by parsing source code in selected languages. The assertion is that this new tool provides counts with an accuracy equivalent to normal function point analysis, but is able to accomplish this in a matter of minutes rather than a matter of days or weeks. Other companies such as CAST, Relativity Technologies, and Total Metrics also are involved with high-speed function points. There are even open-source function point tools available, such as the "Early and Quick" function point tool available from Italy under a Creative Common license.

Several other forms of high-speed function point sizing are also under development. Other high-speed function point methods include the following: (1) Pattern matching against historical projects; (2) Automated prediction of function points from requirements and specifications; (3) Data mining of legacy source code to recreate synthetic specifications; (4) Automated source code analysis to improve mathematical conversion from source code statements to equivalent function point metrics.

The last method of mathematical conversion between source code and function points has existed since 1975. The common name for this method is "backfiring." The first ratios of source code to function points were developed by A.J. Albrecht and his colleagues at IBM during the development of the original function point metric. As of 2017, published ratios are available for more than 700 programming languages and dialects out of around 3000 total programming languages. However, the accuracy of backfiring is low due to variations in code-counting methods and variations in individual programming styles. The backfire method is most effective when based on automated counts of logical source code statements, as opposed to counts of physical lines.

A problem with function point sizing as of 2017 is the fact that there are several different kinds of function point metrics. The function point metric defined by the IFPUG is the most widely used. But there are also COSMIC function points, Mark II function Points, web-object points, Story points, and several national variants from the Netherlands and Finland, plus usage of mathematical conversion from LOC to function points called "backfiring." All of these variants produce different size predictions, and as of 2017 there are no proven conversion rules from one variation to another.

(Of course, there have been no conversion rules between the variations in counting LOC, and this problem is now more than 60 years old. About a third of the literature that uses LOC metrics is based on counts of physical lines; another third is based on counts of logical statements; and the final third reports data in terms of LOC but fails to identify whether physical or logical lines were the bases of the count. For some languages there can be more than a 500% difference in size based on whether the counts are derived from physical or logical LOC. Other ambiguous code-counting topics include whether data definitions should be included; comments; and blank lines between paragraphs.)

Another problem with function point sizing is that conventional sizing by certified experts is slow and expensive. The practical upper limit is about 10,000 function points. For really massive applications such as large ERP packages and large defense applications that may approach 300,000 function points in size, normal function point analysis would take years. This is one of the economic motives for developing faster and cheaper alternatives to standard function point analysis.

Another new sizing topic circa 2017 deals with the impact of technical and architectural factors on application size. While user-requested functions are measured with function point metrics, part of the size of applications is also due to architectural and technical factors and the "layers" that the application will contain.

An extension of this concept was developed by IFPUG in 2012 for counting "nonfunctional requirements." These are things such as security or government mandates that must be in the software whether the users want them or not. A new metric was developed for nonfunctional requirements called SNAP. This term is based on the very awkward phrase "software nonfunctional assessment process."

The gist of the SNAP concept is to count function points and SNAP points separately and accumulate costs separately as well. Since having two costs is counter to standard cost analysis, the author and all other parametric estimation companies add nonfunctional costs to functional costs and use function points for a single overall aggregate metric.

If an application has 1,000 function points at a cost of $1,000 per function and 200 SNAP points at a cost of $2,000 per SNAP point, the total cost for the project is $1,400,000.

Since the nonfunctional requirements would not exist at all except for user requirements, the author asserts that the total cost of the project can be measured using function points and the result is $1400 per function point. This same logic is used by all other parametric cost estimating tools.

This is no different than aggregating the costs of home construction for foundations, framing, roofing, plumbing, etc., into the single metric of "cost per square foot." Obviously, houses are not built one square foot at a time, but the single metric is useful for comparing different kinds of home construction.

(Incidentally, the author is aware of a 5,000 square foot home that costs $40,000,000 or $8,000 per square foot. Some of the nonfunctional costs for this home included using helicopters to bring in construction equipment and the initial work crews; building an airport since there was no road to the construction site; building a private 5-mile road and also a private 5-mile utility line. Another nonfunctional cost factor was building a heated 200-foot dome around the home so work could continue during winter (the home was in Maine). A few other high-cost items included gold plumbing fixtures, hand-carved moldings, and custom hand-carved doors, and custom triple pane-insulated hurricane-proof glass for all windows. The home materials included exotic woods from South America and triple backup electric generators. The owner was a hedge-fund billionaire. This was only a summer home and not a year-round residence.)

The equivalent of nonfunctional requirements for home construction in Rhode Island where the author lives would be the building code mandates of requiring hurricane-proof windows for new homes within half a mile of the ocean and requiring a special septic system for new homes within 100 ft of an aquifer. These items add about $50 per square foot to any new homes that are required to have them.

Other sizing topics are waiting more research and more empirical data: sizing the volume databases and sizing value other than financial value (i.e., customer satisfaction, enterprise prestige) are two topics that need more research.

Software Project Schedule Planning: There are dozens of commercial tools, in-house tools, and even open-source freeware tools that can perform critical path analysis of software projects, as well as other kinds of projects. Examples of such tools include Microsoft Project, Artemis Views, and TimeLine. However, for software projects schedule planning can also be performed by specialized software estimating tools such as COCOMO II, KnowledgePlan, Price-S, SEER, and SLIM in alphabetical order. There are even newer forms of tools about to be released circa 2010 that can handle schedules. For example, the Advanced Management Insight (AMI) tool by Computer Aid Inc. integrates schedule planning, risk analysis, cost estimates, and tracking of resources, defects, and costs. This tool also supports interfaces with other tools, and even allows plugging in "cartridges" of specialized functions such as quality prediction or function point analysis.

Overall, a detailed work breakdown structure that reaches the level of specific tasks and individual assignments is the most effective, although complex and not easy to achieve. This level of granularity can only be provided by automated tools: manual methods are too labor-intensive and also too difficult to update.

As of 2017, schedule estimating remains a difficult challenge for software project managers. Above 10,000 function points in size, about two-thirds of all software projects run late, exceed their budgets, or are canceled because the delays and overruns have degraded the ROI to a negative number.

Two critical factors tend to make software schedules longer than initially anticipated. One of these factors is the growth of creeping requirements, which averages more than 1% per calendar month. The second is inadequate defect prevention combined with failure to use early defect removal activities such as inspections or static analysis. Detailed study of software slippage indicates that testing is the main activity where schedule slippage tends to be severe.

Software Cost Estimating: Software cost estimation can be traced back to the first IBM software cost estimating tool in 1973 and also to the first commercial estimating tool, which was the Price-S tool originally marketed by the RCA corporation. These pioneering tools were soon followed by other commercial software cost estimating tools such as Checkpoint, COCOMO, CostXpert, ExcelerPlan, Estimacs, ISBSG, SEER, SLIM, SoftCost, SPQR/20, and dozens of others.

These specialized software cost estimating tools were often superior to manual cost estimates, and they also included schedule and quality estimates as well. Because the software estimating tools appeared about 10 years after general-purpose project management tools, they support bidirectional data transfer with general purpose tools such as Microsoft Project or Artemis Views.

Another advantage of the commercial software cost estimating tools is that they contain built-in knowledge of topics that affect software projects, which are lacking in general-purpose tools. Examples of these factors include the specific programming languages to be used, the CMMI level of the development organization, the methodology to be used such as Agile development, RUP, TSP, or Extreme programming, and the structure or complexity of the application.

The commercial software estimating tools also have built-in knowledge of key topics such as the DRE levels of various kinds of inspection and testing. The first commercial estimating tool that could predict defect quantities and DRE levels was SPQR/20, which entered the market in 1985. This was also the first commercial estimating tool based on function point metrics, although it predicted LOC as well.

Large software projects often spend more money on the production of paper documents such as requirements, specifications, plans, user manuals, etc., than they do on the production of the source code itself. Estimating these costs lagged for many years because LOC metrics could neither measure nor predict paperwork. However, modern commercial software estimating tools can predict the sizes and costs for more than 50 kinds of paper documents, which collectively can accumulate more than 35% of the total development cost of a large software project. A few software cost estimating tools can even predict the costs of nationalization, or converting some documents into foreign languages for international sales.

Software Project Risk Estimating: The major forms of risk for software projects include outright cancelation due to failure to reach operational stability, cost overruns, schedule overruns, poor quality, security vulnerabilities, difficult human interfaces, and sluggish performance. Since software project failures are common, it can be stated that software risk assessment and risk predictions are either inadequate or not performed with rigor. There are also business risks such as indictment under the Sarbanes–Oxley law for inadequate governance of key financial applications. In the modern world, there are also increasing severe risks from hackers, viruses, and denial of service attacks.

As of 2017, there are some risk prediction features in commercial estimating tools such as KnowledgePlan, SLIM, and SEER. For example, the author's SRM identifies about 60 risks out of the total 210 risks known to impact software projects; this is probably the largest number of software risks identified.

There are also some stand-alone risk models as well as risk estimation in integrated tools such as the AMI by Computer Aid, Inc. However, automated risk estimation is not as sophisticated as other forms of predictive modeling as of 2017. Currently, risk analysis by trained risk experts is perhaps the most effective form of risk abatement.

Research and prototypes of risk analysis tools are a sign that future kinds of risk analysis may improve the situation. In particular, security vulnerabilities need better predictive capabilities in the modern world where viruses, spyware, and denial-of-service attacks are daily events that affect thousands of companies and millions of software applications. As outsourcing becomes more and more common, litigation avoidance is also a risk that needs more attention than it has received to date.

Another emerging kind of risk prediction is that of predicting the odds of litigation for contract projects. Indeed, there is a working prototype of a litigation risk tool that also predicts the probable costs for the plaintiff and defendant in breach of contract litigation.

Software Quality Estimating: Quality estimation for software was not done well before the invention of function point metrics because LOC could not be used for either measuring or predicting defects in requirements, design, user documents, and other noncode items. Once function point metrics were applied to the task of quality prediction, it was soon discovered that errors in requirements and design outnumbered coding errors. Commercial software estimating tools that could predict the full spectrum of software defects from all major sources (requirements, design, code, user documents, and "bad fixes" or secondary defects) only became available in 1985 with the SPQR/20 commercial estimation tool. As of 2017, most of the modern commercial software estimation tools can also predict quantities of defects in the applications under development.

Here too the author's SRM tool predicts defect potentials (bugs in requirements, design, code, documents, and bad fixes). SRM also predicts the DRE of inspections, static analysis, and various test stages. SRM also predicts post-release defect repair costs and technical debt.

However, in today's world of cloud computing, SaaS, open-source components, and COTS packages, quality estimation needs to expand beyond a single application. Quality estimation needs to encompass all components of modern systems, both in-house and acquired from external sources. This is an area that needs more data and additional research as of 2014.

Another important aspect of quality estimation is that of DRE, or predicting the percentages of defects that will be found by various kinds of review, inspection, and test stage. Historical data gathered since the 1970s indicate that formal design and code inspections have the highest DRE and alone can top 85%. Static analysis tools also can top 85% in terms of DRE. However, static analysis tools as of 2014 only support about 25 programming languages out of more than 2500 that are known to exist.

Most forms of testing such as unit test, new function test, regression test, etc., are <35% efficient or only find about one bug out of three. Some of the newer forms of testing such as automated unit test can approach static analysis or top 50% in DRE.

A sequence of formal requirement and design inspections, followed by static analysis of code, followed by six to eight different testing stages can top 98% in cumulative DRE. Since the industry average for DRE has stayed almost constant at about 85% for more than 30 years, companies and projects that top 95% are to commended. As of 2010, several commercial estimating tools can predict defect potentials and DRE levels.

Some commercial estimating tools can also predict the specific sequence of inspections, static analysis, and test stages that should be used to achieve optimal levels of DRE. Numbers of test cases and test coverage can also be predicted for many forms of testing such as unit test (manual or automated), regression test, new function test, performance test, system test, etc.

Since most forms of testing are <35% efficient in finding bugs, it is obvious that a multistage series of inspections, static analysis, and tests will be needed to achieve high levels of DRE. The good news about this fact is that projects that approximate 95% in cumulative DRE have shorter schedules and lower costs than "average" projects with only about 85% cumulative DRE. Maintenance costs are also reduced. Phil Crosby's aphorism that "Quality is Free" has not only been proven, but has also been measured on many successful software projects.

Because the costs and schedules associated with finding and fixing bugs is the most significant cost element of large software projects and the primary factor leading to cancelation and delay, quality prediction is a critical element of successful cost and schedule prediction. Poor quality is a frequent cause of litigation too.

Many specific software quality methods such as Six Sigma for software, QFD, formal design and code inspections, static analysis, automated testing, EPM removal, and JAD can now be modeled and predicted using automated tools. The most interesting results come from side-by-side comparisons that show a specific approach such as Six Sigma compared against an identical project that did not use the same approach.

Software Project Outsource Contract Analysis: Since many outsource contracts are canceled and about 5% end up in court for breach of contract, it can be stated that the available tools for predicting the results of software outsource contracts need improvement. There are some specialized and proprietary tools available from consulting companies that assist clients in dealing with outsource vendors. Also, normal commercial software cost estimating tools can be useful in predicting topics such as schedules, costs, and quality levels for specific projects. However, contract analysis as of 2017 does not have much in the way of sophisticated project management tools.

In recent years, newer topics have found their way into outsource agreements, such as SLA. The ITIL also has features that may be included in outsource agreements. Military and government projects are generally required to use earned value analysis, which is common in the government software domain.

What often occurs in litigation involving canceled projects are side-by-side comparisons of the project as it was actually performed against an alternate version that would have used better methods of quality control and dealt with requirement changes. This feature of producing side-by-side comparisons of alternate approaches should become a standard part of predictive tools uses for cost, quality, and schedule planning purposes.

Software contracts often are incomplete or ambiguous in dealing with requirements changes and also with quality control. All contracts need language and sections that define how changes in requirements will be funded and what kind of schedule relief will be provided. They should also deal specifically with quality, such as including clauses that require at least 95% DRE levels on the part of the developing company.

Software Project COTS Acquisition: Most corporations and government agencies lease or purchase more than half of the software they use on a daily basis. Operating systems, office suites, ERP packages, financial and accounting packages, human resource packages, and a host of others comprise a large portion of total corporate portfolios. There are very few tools available that can assist in making good selections of commercial software or open-source applications. Open-source applications such as Firefox and Open Office or Sun's office suite are finding increasing usage in the corporate world, due in part to the high cost of Microsoft Office.

Software vendors are reluctant to publish data on defects in their commercial packages. They are also reluctant to publish size data in either function point or LOC form. As a result, it is hard to ascertain basic information such as after-market defect levels, installation problems, and other issue. There are some specialized tools available for specific tasks such as estimating the deployment of ERP packages, but these are usually proprietary tools owned by consulting companies that specialize in ERP deployment. Overall, this is a major gap in project management tools circa 2017. However, the author's SRM tool does predict ERP deployment and also training of ERP users.

As it happens, open-source applications are frequent users of static analysis tools. Somewhat surprisingly, the quality levels of open-source software compares not unfavorably with the quality levels of some commercial applications from vendors such as Microsoft, Oracle, SAP, Computer Associates, etc.

Historical data on the numbers of bugs and installation problems in major commercial software packages such as Microsoft Vista, of the ERP packages from Oracle, SAP, and other vendors is not available to ordinary consumers and clients, even though the vendors actually have such data available internally.

Some of the data for ERP predictions come from litigation filed against these vendors, so the predictions also include learning curves by users, installation defects, and even predictions of consequential damages or the harm that bugs in commercial packages can cause to client companies. For example, a serious bug in a commercial financial package caused a client company to restate past year earnings and therefore lose credit with its bank.

Software Personnel Management: Strictly speaking, personnel management is not the same as project management. However, in day-to-day work, software project managers are also software personnel managers and so they need to be involved in both kinds of activities. As of 2017, there are dozens of commercial tools for dealing with personnel issues such as hiring, skills inventories, appraisals, compensation, and other human resource issues. This is one of the most extensively automated areas of the entire spectrum of management work.

Software Project Measurement, Tracking, and Control: There are scores of commercial and in-house tools for collecting resource and cost data for software projects. Often these cost- and resource-tracking tools are part of corporate-wide tool suites, so they operate for other kinds of projects besides software projects. However, there are some common gaps and omissions in resource and cost tracking. The most common omission is measuring unpaid overtime. Software personnel tend to have rather energetic work habits, and often work on weekends or at home. However, accurate measurement of unpaid overtime is rare. Failure to record unpaid overtime tends to artificially inflate productivity rates.

Another gap in resource tracking concerns the work of part-time specialists such as quality assurance, database administration, and technical writers. Since these workers may charge time to several projects or even be classified as overhead, sometimes their effort on specific projects is missing or inaccurate.

Yet another kind of gap is the work performed by users when they participate in requirements analysis or are assigned to project teams as may occur when Agile development is used. Indeed, the Agile approach brings a new level of complexity to tracking due to having the applications divided into sprints or short development intervals, each of which is semi-independent and devoted to specific features.

From on-site studies where software project personnel were interviewed and asked to compare their actual effort to the results of project tracking systems, it was noted that the "leakage" from many corporate resource tracking systems was in excess of 35%. That is, more than one-third of the actual effort was not recorded. Unpaid overtime and the work of users and specialists were the most common omissions. However, some companies also failed to track managerial effort and only reported technical effort by the programming team or software engineers. Leakage of more than 50% was noted in some cases.

An interesting form of project tracking has been developed by the Shoulders Corporation for keeping track of object-oriented projects. This method uses a three-dimensional (3-D) model of software objects and classes using Styrofoam balls of various sizes that are connected by dowels to create a kind of mobile. The overall structure is kept in a visible location viewable by as many team members as possible. The mobile makes the status instantly visible to all viewers. Color-coded ribbons indicate status of each component, with different colors indicating design complete, code complete, documentation complete, and testing complete (gold). There are also ribbons for possible problems or delays. This method provides almost instantaneous visibility of overall project status. The same method has been automated using a 3-D modeling package, but the physical structures are easier to see and have proven more useful on actual projects. The Shoulders Corporation method condenses a great deal of important information into a single visual representation that nontechnical staff can readily understand.

From documents produced during the discovery phase of software breach of contract litigation and from depositions of the project managers on the side of the defendant, poor tracking of progress is a very common problem. Almost always, the project teams are aware of problems that might cause delays, cancelation, or poor quality. But this information somehow is left out of project status reports so it never reaches executives or clients until it is too late to take corrective action. In a few cases, the omissions were actually deliberate, which is professional malpractice. Even when the omissions are accidental, the problem of poor status tracking is a serious endemic problem for software project managers.

While the earned value method is a step toward solving inaccurate status reports, it is not perfect. There are still issues with delays and quality even when earned value methods are in use.

Software Technology Acquisition: A weak link in the chain of project management tasks is that of technology acquisition. With more than 2500 programming languages available, more than 40 named design methods, and more than 30 forms of development methods are not easy to be sure that the languages, tools, and methods selected for any given project are in fact the best available.

There are many important questions that are not easy to answer: when should Agile methods be used, and when are Agile methods inappropriate? What about the RUP or the TSP? Is level 3 on the CMMI sufficient for a specific project, or would level 5 give better results? What kind of benefits will occur from using the TSP and Personal Software Projects (PSP)? What value does lean Six Sigma provide to specific projects? Will object-oriented programming languages be suitable for a specific project, or should some other language be used? Should QFD be used? Will automated testing be beneficial or not? If formal inspections are used, will they lower testing costs? Should static analysis be used in place of code inspections? Are use cases more effective than other forms of specification? These are samples of questions that need accurate empirical data.

There are few, if any, tools available for technology selection that match specific projects with the best choices for methodologies, languages, or design techniques. The growing volume of benchmark data collected by the ISBSG is starting to provide better data based on historical projects. However, more data and more measured projects are still needed. While the ISBG benchmark collection has exceeded 5000 projects circa 2017, it is not an exact match to what the industry really does. For example, since function point metrics are almost never used above 10,000 function points, the ISBSG data has no examples of massive applications in the 100,000-function point size range (such as Windows 7, Microsoft Office, SAP, Oracle, etc.) This is not the fault of ISBSG—the problem lies in the high costs and slow speed of function point analysis.

Some of the commercial software estimation tools provide a side-by-side capability that allows the same project to be estimated with varying assumptions, such as changing programming languages, changing design methods, or changing CMM levels. These side-by-side estimates can be helpful, but not every language or method is supported. However, the use of side-by-side modeling to predict the results of alternate scenarios is definitely a feature that should be added to all software project-predictive tools.

Software Mergers and Acquisitions: Some software managers can spend an entire career without being involved with mergers and acquisitions. Other managers may be involved as often as several times a year. Some of the issues with mergers and acquisitions are those of joining the portfolios of the two companies, merging the software organizations, and selecting the software tools and methods that will be used after the merger is complete, assuming the two companies use different approaches for development and maintenance. There are no standard tools available for supporting mergers and acquisitions, although some commercial estimating tools and personnel management tools can be helpful.

Because the technologies deployed by merging partners are often different, here too side-by-side comparisons of one set of methods against another would be useful.

Software Litigation Support: Software litigation comes in many varieties such as breach of contract, patent violations, tax litigation, theft of intellectual property, and violations of noncompetition agreements. For breach of contract litigation, expert witnesses are often asked to predict what a failing project would have looked like if state-of-the-art methods had been used, i.e., what would the schedule and costs have been had the project been done well instead of done poorly. Questions about the cost and quality impacts of requirements changes often occurs in litigation too. For tax litigation, reconstructing the original development cost of software at a given point in time is a topic that occurs often, as well as questions about what it would cost to build a new version today.

A combination of software cost estimating tools, software quality estimating tools, and general-purpose critical path tools are often used by expert witnesses in software litigation. For cases involving patent violations or theft of intellectual property, specialized tools that analyze the structure of software applications and produce graphs of that structure may be used. If two different software applications have common structures, it can be assumed that one version may have been copied from another version.

For many kinds of litigation, it is important to compare what actually took place against an alternate scenario of what might have taken place had more suitable methods been deployed. Therefore, side-by-side comparisons are an important aspect of litigation support.

Because many lawsuits are filed due to poor quality or the presence of major bugs that prevented the applications from being used successfully, quality prediction and quality measurement are important aspects of litigation support.

Litigation is a topic that is seldom taught by universities to computer science students, software engineering students, or even MBA candidates. However, litigation is common enough in the software world so that information on litigation costs, probable duration, and disruption of staff time needs to be included in software project management tools.

As of 2017, SRM can predict litigation costs and schedules for breach of contract suits for both the plaintiff and the defendant. But the prototype does not deal with patent violations, theft of intellectual property, criminal hacking, violations of employment agreements, or other forms of litigation.

The Costs and Value of Software Project Management Tools

The cost of software project management tools ranges from free (for open-source tools or in-house tools) to more than $10,000 per seat for some commercial software cost estimating tools. Assuming a normal mix of in-house tools, open-source tools, and commercial tools, the acquisition cost of a suite of software project management tools would cost about $0.50 per function point using the size of the tools in function points as the base. Thus, to equip an individual software project manager with a suite of tools totaling 30,000 function points would cost about $15,000 for one manager. Because of discounts and multiseat licenses, the effective cost per manager in a large company or a large project would probably be about $5000 for a suite of project management tools that encompass estimating, planning, risk analysis, quality control, and tracking of costs and progress.

The average cost to build a successful software application in the 10,000-function point size range is about $1,000 per function point or $10,000,000. However, unsuccessful projects that are canceled usually accrue costs of more than $1500 per function point and are about 12 months late when they are canceled. In other words, canceled software projects in the 10,000-function point range cost about $15,000,000 and obviously are a write-off without any positive value.

If the failing project becomes subject to breach of contract litigation, the damage costs can be much higher than the cost of the software itself. But litigation damages are difficult to predict ahead of time.

A 10,000-function point project would normally have about 8 project managers and perhaps 64 development personnel. The costs of fully equipping the software project managers on 10,000-function point projects with state-of-the-art software cost and quality estimating tools is not very expensive: At an average cost of $5000 per manager and 8 managers, the total cost would be perhaps $40,000.

Thus, an investment of $40,000 in state-of-the-art project management tools and quality control tools has the potential to keep the project within bounds and hence eliminate a $15,000,000 failure. If you subtract the $10,000,000 cost of a successful project from the potential cost of a canceled project, the savings due to effective project management with a powerful tool suite would be about $5,000,000. Thus, an investment of $40,000 in software project management tools could yield savings of $5,000,000, which is a ROI of $125 for every dollar expended. This prediction is only an approximation, but the basic principles are probably valid.

The presence of a suite of project management tools is not, by itself, the main differentiating factor between successful and unsuccessful software projects. The primary reason for the differences noted between failing and successful projects is that the project managers who utilize a full suite of management tools are usually better trained and have a firmer grasp of the intricacies of software development than the managers who lack adequate management tools.

Bringing a large software project to a successful conclusion is a very difficult task that is filled with complexity. The managers who can deal with this complexity recognize that some of the cost and resource scheduling calculations exceed the ability of manual methods. They also recognize that state-of-the-art quality control is a necessary prerequisite for success.

Managers on failing projects, on the other hand, tend to have a naïve belief that project planning and estimating are simple enough to be done using rough rules of thumb and manual methods. Further, when projects begin to show stress and experience problems, the managers on failing projects often do not know how to take corrective actions, and may just ignore the problems until it is far too late to solve them. Worse, they may even conceal the problems from executives and clients, which is professional malpractice. Last, managers on failing projects (as noted in litigation) do not understand the economic value of quality control and hence encounter massive delays and cost overruns when testing begins.

Effective cost estimating, quality control, change management, and progress tracking are the four keys to successful software project management, and to successful software projects as well. Inadequate estimates, poor change control, poor quality control, and poor progress tracking are sure to bring either massive overruns or outright failure to large software projects in the 10,000-function point size range.

The Future of Software Project Management Tools

Software project management circa 2017 is facing a number of future challenges even before successfully solving historical challenges. For the past 50 years, the average size of software projects has been increasing. Since risk of failure and risk of poor quality correlate directly with size, larger projects require more care in predicting quality and controlling quality than small projects. This has long been a weakness of both software engineering and software management.

New forms of software such as the service-oriented architecture (SOA) running on new kinds of platforms such as "cloud computing" are rapidly approaching. New development methods such as Agile development, Extreme Programming (XP), RUP, and TSP need specialized estimating and planning methods.

Security vulnerabilities have long been troublesome, and in the future will become even more troublesome than in the past. Thus, project planning and estimating are becoming more complex as the industry evolves.

As of 2017, many software applications are closely linked to hardware, so integration of software and systems project management tools are needed. Modern cloud computing and SOA applications may involve dozens of companies and scores of features concurrently, and this is a very complex management problem for which no fully effective tools have been released.

Although hundreds of commercial project management tools were available in 2010, many of these dealt only with a small subset of overall management tasks. For example, a software project management tool that can deal with schedule assignments, but which cannot deal with quality, is not an effective solution because bugs or defects are the primary source of schedule slippage.

Further, current project management tools often do not interface well or share data well with other tools, so that it is difficult to create a consolidated picture of every important topic. What would benefit software project management tools for the next generation would be better ease of sharing data, and an improved ability to link together sets of specialized functions. In other words, apply the principles of SOA to the domain of project management tools.

From examining scores of canceled projects and hundreds of costs and schedule overruns, it is obvious that a few basic topics need to be estimated and controlled in the future much better than they have been in the past.

Quality control is the prime weakness of software projects. Excessive numbers of bugs and inadequate methods of defect removal are associated with every project failure and all litigation involving breach of contract.

Requirements changes are endemic in the software world. It is not possible to stop requirements from changing, so the solution is to deal with changing requirements in the most effective and efficient manner. Canceled projects or those with major overruns tended to be swamped by unexpected changes for which no schedule or cost relief had been envisioned.

Tracking of progress on software projects has been a weak link. Project tracking omissions or outright concealment of true status tends to occur in breach of contract lawsuits. When problems are noted, they should be highlighted and addressed, not concealed and ignored.

Furthermore, gaps in historical cost data are common. Unpaid overtime, managerial costs, and the work of part-time specialists such as quality assurance and database administration are often missing or incomplete in software project cost tracking systems. This is especially true for in-house projects, but some omissions such as unpaid overtime even occur for outsourced applications.

Accurate plans and estimates are needed before projects start. But it often happens that even accurate plans and estimates are rejected by clients or top executives and replaced by impossible schedule demands. Accurate estimation by itself is not a total solution, but accurate estimation supported by accurate historical data is likely to be believed. Therefore, accurate benchmarks of historical projects are needed in larger volumes than have been available in the past.

The shared goals of the software project management community should be to improve the professionalism of software project management. We need to be able to develop large software projects with lower defect potentials, higher levels of DRE, lower costs, and shorter schedules than has been accomplished in the past. The technologies for achieving these goals are starting to become available, but improvements in academic preparation for software project managers and improvements in selecting the best methods are still needed.

References and Readings on Software Issues

Beck, K.; *Test-Driven Development*; Addison-Wesley, Boston, MA; 2002; ISBN 10: 0321146530; 240 pages.

Black, R.; *Managing the Testing Process: Practical Tools and Techniques for Managing Hardware and Software Testing*; Wiley; Hoboken, NJ; 2009; ISBN-10 0470404159; 672 pages.

Chelf, B. and Jetley, R.; *Diagnosing Medical Device Software Defects Using Static Analysis*; Coverity Technical Report, San Francisco, CA; 2008.

Chess, B. and West, J.; *Secure Programming with Static Analysis*; Addison-Wesley, Boston, MA; 2007; ISBN 13: 978-0321424778; 624 pages.

Cohen, L.; *Quality Function Deployment: How to Make QFD Work for You*; Prentice Hall, Upper Saddle River, NJ; 1995; ISBN 10: 0201633302; 368 pages.

Crosby, P.B.; *Quality Is Free*; New American Library, Mentor Books, New York; 1979; 270 pages.

Everett, G.D. and McLeod, R.; *Software Testing*; John Wiley & Sons, Hoboken, NJ; 2007; ISBN 978-0-471-79371-7; 261 pages.

Festinger, Dr. L.; *A Theory of Cognitive Dissonance*; Stanford University Press, 1962.

Gack, G.; *Managing the Black Hole: The Executive's Guide to Software Project Risk*; Business Expert Publishing, Thomson, GA; 2010; ISBN10: 1-935602-01-9.

Gack, G.; Applying Six Sigma to software implementation projects; http://software.isixsigma.com/library/content/c040915b.asp.

Gilb, T. and Graham, D.; *Software Inspections*; Addison-Wesley, Reading, MA; 1993; ISBN 10: 0201631814.

Hallowell, D.L.; Six Sigma software metrics, part 1; http://software.isixsigma.com/library/content/03910a.asp.

International Organization for Standards; ISO 9000/ISO 14000; http://www.iso.org/iso/en/iso9000-14000/index.html.

Jones, C.; *Critical Problems in Software Measurement*; Information Systems Management Group; 1993; ISBN 1-56909-000-9; 195 pages.

Jones, C.; *Assessment and Control of Software Risks*; Prentice Hall, Englewood Cliffs, NJ; 1994; ISBN 0-13-741406-4; 711 pages.

Jones, C.; *Patterns of Software System Failure and Success*; International Thomson Computer Press, Boston, MA; 1995; 250 pages; ISBN 1-850-32804-8; 292 pages.

Jones, C.; *Software Quality: Analysis and Guidelines for Success*; International Thomson Computer Press, Boston, MA; 1997; ISBN 1-85032-876-6; 492 pages.

Jones, C.; The economics of object-oriented software; SPR Technical Report; Software Productivity Research, Burlington, MA; April 1997; 22 pages.

Jones, C.; *Software Project Management Practices: Failure versus Success*; Crosstalk; Hill Air Force Base, Utah; 2004.

Jones, C.; *Software Estimating Methods for Large Projects*; Crosstalk, Hill Air Force Base, Utah; 2005.

Jones, C.; *Estimating Software Costs*; 2nd edition; McGraw-Hill, New York; 2007; 700 pages.

Jones, C.; *Applied Software Measurement*; 3rd edition; McGraw-Hill, New York; 2008; ISBN 978=0-07-150244-3; 662 pages.

Jones, C.; *Software Engineering Best Practices*; McGraw-Hill, New York; 2010; ISBN 978-0-07-162161-8; 660 pages.

Jones, C.; *Software Engineering Best Practices*; McGraw Hill, New York; 2012; ISBN 10007162161X.

Jones, C.; *The Technical and Social History of Software Engineering*; Addison-Wesley; Reading, MA; 2014.

Jones, C.; *Software Risk Master (SRM) Tutorial*; Namcook Analytics LLC, Narragansett, RI; 2015a.

Jones, C.; *Software Defect Origins and Removal Methods*; Namcook Analytics LLC; Narragansett, RI; 2015b.

Jones, C.; *The Mess of Software Metrics*; Namcook Analytics LLC, Narragansett, RI; 2015c.

Jones, C. and Bonsignour, O.; *The Economics of Software Quality*; Addison-Wesley, Boston, MA; 2011; ISBN 978-0-13-258220-9; 587 pages.

Kan, S.H.; *Metrics and Models in Software Quality Engineering*; 2nd edition; Addison-Wesley Longman, Boston, MA; 2003; ISBN 0-201-72915-6; 528 pages.

Land, S.K; S. D. B; Walz, J.Z.; *Practical Support for Lean Six Sigma Software Process Definition: Using IEEE Software Engineering Standards*; Wiley-Blackwell; Hoboken, NJ; 2008; ISBN 10: 0470170808; 312 pages.

Mosley, D. J.; *The Handbook of MIS Application Software Testing*; Yourdon Press, Prentice Hall; Englewood Cliffs, NJ; 1993; ISBN 0-13-907007-9; 354 pages.

Myers, G.; *The Art of Software Testing*; John Wiley & Sons, New York; 1979; ISBN 0-471-04328-1; 177 pages.

Nandyal, R.; *Making Sense of Software Quality Assurance*; Tata McGraw-Hill Publishing, New Delhi, India; 2007; ISBN 0-07-063378-9; 350 pages.

Radice, R.A.; *High Quality Low Cost Software Inspections*; Paradoxicon Publishing Andover, MA; 2002; ISBN 0-9645913-1-6; 479 pages.

Royce, W.E.; *Software Project Management: A Unified Framework*; Addison-Wesley Longman, Reading, MA; 1998; ISBN 0-201-30958-0.

Wiegers, K.E.; *Peer Reviews in Software: A Practical Guide*; Addison-Wesley Longman, Boston, MA; 2002; ISBN 0-201-73485-0; 232 pages.

Suggested Readings

Abran, A. and Robillard, P.N.; Function point analysis: An empirical study of its measurement processes; *IEEE Transactions on Software Engineering*; 22, 12; 1996; pp. 895–909.

Black, R.; *Managing the Testing Process: Practical Tools and Techniques for Managing Hardware and Software Testing*; Wiley; 2009; ISBN-10 0470404159; 672 pages.

Bogan, C.E. and English, M.J.; *Benchmarking for Best Practices*; McGraw-Hill, New York; 1994; ISBN 0-07-006375-3; 312 pages.

Cohen, L.; *Quality Function Deployment: How to Make QFD Work for You*; Prentice Hall, Upper Saddle River, NJ; 1995; ISBN 10: 0201633302; 368 pages.

Crosby, P.B.; *Quality Is Free*; New American Library, Mentor Books, New York; 1979; 270 pages.

Gack, G.; *Managing the Black Hole: The Executive's Guide to Software Project Risk*; Business Expert Publishing, Thomson, GA; 2010; ISBN10: 1-935602-01-9.

Gilb, T. and Graham, D.; *Software Inspections*; Addison-Wesley, Reading, MA; 1993; ISBN 10: 0201631814.

Humphrey, W.S.; *Managing the Software Process*; Addison-Wesley Longman, Reading, MA; 1989.

Jacobsen, I., Griss, M., and Jonsson, P.; *Software Reuse-Architecture, Process, and Organization for Business Success*; Addison-Wesley Longman, Reading, MA; ISBN 0-201-92476-5; 1997; 500 pages.

Jacobsen, I. et al; *The Essence of Software Engineering; Applying the SEMAT Kernel*; Addison-Wesley Professional, Reading, MA; 2013.

Kan, S.H.; *Metrics and Models in Software Quality Engineering*; 2nd edition; Addison-Wesley Longman, Boston, MA; 2003; ISBN 0-201-72915-6; 528 pages.

Putnam, L.H and Myers, W..; *Industrial Strength Software: Effective Management Using Measurement*; IEEE Press, Los Alamitos, CA; 1997; ISBN 0-8186-7532-2; 320 pages.

Radice, R.A.; *High Quality Low Cost Software Inspections*; Paradoxicon Publishing Andover, MA; 2002; ISBN 0-9645913-1-6; 479 pages.

Royce, W.E.; *Software Project Management: A Unified Framework*; Addison-Wesley Longman, Reading, MA; 1998; ISBN 0-201-30958-0.

Shepperd, M.; A critique of cyclomatic complexity as a software metric; *Software Engineering Journal*; 3; 1988; pp. 30–36.

Strassmann, P.; *The Squandered Computer*; The Information Economics Press, New Canaan, CT; 1997; ISBN 0-9620413-1-9; 426 pages.

Weinberg, Dr. G.; *Quality Software Management: Volume 2 First-Order Measurement*; Dorset House Press, New York; 1993; ISBN 0-932633-24-2; 360 pages.

Wiegers, K.A; *Creating a Software Engineering Culture*; Dorset House Press, New York; 1996; ISBN 0-932633-33-1; 358 pages.

Yourdon, E.; *Death March: The Complete Software Developer's Guide to Surviving "Mission Impossible" Projects*; Prentice Hall PTR, Upper Saddle River, NJ; 1997; ISBN 0-13-748310-4; 218 pages.

Suggested Readings on Software Project Management

Baird, L.M. and Brennan, M.C.; *Software Measurement and Estimation: A Practical Approach*; IEEE Computer Society Press, Los Alamitos, CA; John Wiley & Sons, Hoboken, NJ; 2006; ISBN 0-471-67622-5; 257 pages.

Boehm, B. Dr.; *Software Engineering Economics*; Prentice Hall, Englewood Cliffs, NJ; 1981; 900 pages.

Boehm, B. et al.; *Software Cost Estimating with Cocomo II*; Prentice Hall, Upper Saddle River, NJ; 2000; ISBN-10 0137025769; 544 pages.

Brooks, F.; *The Mythical Man-Month*; Addison-Wesley, Reading, MA; 1974, rev. 1995.

Bundshuh, M. and Dekkers, C.; *The IT Measurement Compendium: Estimating and Benchmarking Success with Functional Size Measurement*; Springer, Berlin, Germany; 2008; ISBN-10 3540681876; 644 pages.

Cohn, M.; *Agile Estimating and Planning;* Prentice Hall PTR, Englewood Cliffs, NJ; 2005; ISBN 0131479415.

Crosby, P.; *Quality Is Free*; New American Library, Mentor Books; New York; 1979, 270 pages.

DeMarco, T.; *Why Does Software Cost So Much*? Dorset House, New York; 1995; ISBN 0-9932633-34-X; 237 pages.

Ebert, C.; Dumke, R.; and Schmeitendorf, A.; *Best Practices in Software Measurement*; Springer, Berlin, Germany; 2004; ISBN-10 3540208674; 344 pages.

Fleming, Q.W. and Koppelman, J.M.; *Earned Value Project Management*; 2nd edition; Project Management Institute, New York; 2000; ISBN 10 1880410273; 212 pages.

Galorath, D.D. and Evans, M.W.; *Software Sizing, Estimation, and Risk Management: When Performance Is Measured, Performance Improves*; Auerbach, Philadelphia, PA; 2006; ISBN 10-0849335930; 576 pages.

Garmus, D. and Herron, D.; *Measuring the Software Process: A Practical Guide to Functional Measurement*; Prentice Hall, Englewood Cliffs, NJ; 1995.

Garmus, D. and Herron, D.; *Function Point Analysis*; Addison-Wesley, Boston, MA; 2001; ISBN 0-201069944-3; 363 pages.

Gilb, T. and Graham, D.; *Software Inspections*; Addison-Wesley, Reading, MA; 1993; ISBN 10: 0201631814.

Glass, R.L.; *Software Runaways: Lessons Learned from Massive Software Project Failures*; Prentice Hall, Englewood Cliffs, NJ; 1998.

Harris, M.; Herron, D.; and Iwanicki, S.; *The Business Value of IT: Managing Risks, Optimizing Performance, and Measuring Results*; CRC Press (Auerbach), Boca Raton, FL; 2008; ISBN 13: 978-1-4200-6474-2; 266 pages.

Humphrey, W.; *Managing the Software Process*; Addison-Wesley, Reading, MA; 1989.

International Function Point Users Group (IFPUG); *IT Measurement: Practical Advice from the Experts*; Addison-Wesley Longman, Boston, MA; 2002; ISBN 0-201-74158-X; 759 pages.

Johnson, J. et al; *The Chaos Report*; The Standish Group, West Yarmouth, MA; 2000.

Jones, C.; *Program Quality and Programmer Productivity*; IBM Technical Report TR 02.764, IBM San Jose, CA; 1977.

Jones, C.; *Patterns of Software System Failure and Success*; International Thomson Computer Press, Boston, MA; 1995; 250 pages; ISBN 1-850-32804-8; 292 pages.

Jones, C.; Sizing up software; *Scientific American* 279, 6; 1998; pp. 104–109.

Jones, C.; *Software Assessments, Benchmarks, and Best Practices*; Addison-Wesley Longman, Boston, MA, 2000; 659 pages.

Jones, C.; Why flawed software projects are not cancelled in time; *Cutter IT Journal*; 10, 12; 2003; pp. 12–17.

Jones, C.; *Conflict and Litigation between Software Clients and Developers*; Version 6; Software Productivity Research, Burlington, MA; 2006; 54 pages.

Jones, C.; Software project management practices: Failure versus success; *Crosstalk*; 19, 6; 2006; pp. 4–8.

Jones, C.; *Estimating Software Costs*; McGraw-Hill, New York; 2007; ISBN 13-978-0-07-148300-1.

Jones, C.; *Applied Software Measurement*; McGraw-Hill; 3rd edition; 2008; ISBN 978-0-07-150244-3; 668 pages.

Jones, C.; *Software Engineering Best Practices*; McGraw-Hill, 2009; ISBN 978-0-07-162161-8660 pages.

Jones, C.; *The Technical and Social History of Software Engineering*; Addison-Wesley, Englewood Cliffs, NJ; 2014.

Jones, C. and Bonsignour, O.; *The Economics of Software Quality*; Addison-Wesley, Englewood Cliffs, NJ; 2012.

Kan, S.H.; *Metrics and Models in Software Quality Engineering;* 2nd edition; Addison-Wesley Longman, Boston, MA; 2003; ISBN 0-201-72915-6; 528 pages.

Kaplan, R.S. and Norton, D.B.; *The Balanced Scorecard*; Harvard University Press, Boston, MA; 2004; ISBN 1591391342.

Love, T.; *Object Lessons: Lessons Learned in Object-Oriented Development Projects*; SIG Books Inc., New York; 1993; ISBN 0-9627477-3-4; 266 pages.

McConnell, S.; *Software Estimation: Demystifying the Black Art*; Microsoft Press, Redmond, WA; 2006; ISBN 10: 0-7356-0535-1.

Parthasarathy, M.A.; *Practical Software Estimation: Function Point Methods for Insourced and Outsourced Projects*; Addison-Wesley, Boston, MA; 2007; ISBN 0-321-43910-4; 388 pages.

Putnam, L.H.; *Measures for Excellence: Reliable Software On-Time within Budget*; Yourdon Press, Prentice Hall, Englewood Cliffs, NJ; 1992; ISBN 0-13-567694-0; 336 pages.
Putnam, L. and Myers, W.; *Industrial Strength Software: Effective Management Using Measurement*; IEEE Press, Los Alamitos CA; 1997; ISBN 0-8186-7532-2; 320 pages.
Strassmann, P.; *The Squandered Computer*; Information Economics Press, Stamford, CT; 1997.
Strassmann, P.; *Governance of Information Management: The Concept of an Information Constitution*; 2nd edition; (eBook); Information Economics Press, Stamford, CT; 2004.
Stutzke, R.D.; *Estimating Software-Intensive Systems: Projects, Products, and Processes*; Addison-Wesley, Boston, MA; 2005; ISBN 0-301-70312-2; 917 pages.
Whitehead, R.; *Leading a Development Team*; Addison-Wesley, Boston, MA; 2001; ISBN 10: 0201675267; 368 pages.
Yourdon, E.: *Death March: The Complete Software Developer's Guide to Surviving "Mission Impossible" Projects,* Prentice Hall, Upper Saddle River, NJ; 1997; ISBN 0-13-748310-4.
Yourdon, E.; *Outsource: Competing in the Global Productivity Race*; Prentice Hall, Upper Saddle River, NJ; 2004; ISBN 0-13-147571-1; 251 pages.

Chapter 5

Measures, Metrics, and Management

Introduction

Project management in every industry is a challenging occupation. But the challenges and hazards of software project management are greater than those of most other industries. This fact is proven by the large number of software project cancellations and the high frequency of software project cost and schedule overruns. Software projects run late and exceed their budgets more than any other modern industry, except for defense projects.

Academic training for software project managers does not seem to be very good in 2017. Some technology companies have recognized the challenges of software project management and created effective in-house training for new software project managers. These companies with effective software project management training include IBM and a number of telecom companies such as AT&T, ITT, Motorola, Siemens, and GTE. Some other technology companies such as Google and Apple also have effective software project management training. (Many of the companies with good software management training build complex physical devices run by software. Most are also more than 75 years old and have had software measurement programs for more than 40 years.)

A few companies even market effective software project management training. One of these is a subsidiary of Computer Aid Inc. called the Information Technology Metrics and Productivity Institute (ITMPI). The nonprofit Project Management Institute (PMI) also offers effective training for software project managers.

Several years ago, a survey by the author on the CEOs of the engineering technology companies (computers, telecom, electronics, medical devices, autos, aircraft) found that the CEOs regarded their software organizations as the least professional of any of the corporate engineering organizations. This was because software projects had higher cancellation rates, longer schedule delays, and higher cost overruns than any of the other engineering organizations.

Lyman Hamilton, a former Chairman of the ITT Corporation, gave an internal speech to ITT executives in which he mentioned that newly hired software engineers just out of college needed about 3 years of internal training before being entrusted with critical projects. Other kinds of engineers such as mechanical and electrical engineers only needed about 12 months of internal training.

Hamilton was troubled by several major software failures of projects that were terminated without being completed. He was also troubled by the dissatisfaction expressed by customers in the quality of the software the corporation produced. He was further dissatisfied by the inability of internal software executives to explain why the problems occurred, and what might be done to eliminate them.

(Harold Geneen, the ITT Chairman prior to Hamilton, and Rand Araskog, the ITT Chairman after Hamilton, also were troubled by software and all three took proactive steps to improve it, including funding a large software research center in Stratford, CT, which had about 150 personnel at its peak before Alcatel acquired the ITT telecom business.)

It is interesting that the failing projects were all large systems in the 10,000-function point size range. Failures in this range are common, and managerial problems are usually a key factor.

Problems, failures, and litigation are directly proportional to the overall size of software applications measured using function point metrics. Table 5.1 shows the approximate distribution of software project results circa 2017.

Table 5.1 Normal Software Results Based on Application Size, circa 2017

Size in Function Points	*Schedule in Calendar Months*	*Total Staffing*	*Productivity in Function Points per Staff Month*	*Cost in U.S. Dollars*	*Odds of Project Failure (%)*	*Odds of Outsource Litigation (%)*
1	0.02	1	50.00	$200	0.10	0.00
10	0.40	1	25.00	$4,000	1.00	0.01
100	3.50	2	14.29	$70,000	2.50	0.25
1,000	15.00	6	11.11	$900,000	11.00	1.20
10,000	35.00	50	5.71	$17,500,000	31.00	7.50
100,000	60.00	575	2.90	$345,000,000	47.50	23.00

Note: Costs are based on $10,000 per month.

As can be seen from Table 5.1, large software projects are distressingly troublesome and have frequent total failures in addition to a high risk of litigation. Poor project management is a key contributing factor.

Some leading companies have recognized the difficulty of successful software project management and taken active steps to improve the situation. Some of these companies include IBM, AT&T, ITT, Motorola, GTE, and Siemens. Google, Apple, and Microsoft have also attempted to improve software management although Microsoft has perhaps been too rigid in some management topics such as employee appraisals.

The companies that are most proactive in software project management tend to build complex engineered products such as computers, medical devices, aircraft controls, and switching systems that depend upon software to operate. The companies also tend to be mature companies founded over 75 years ago and having effective software measurement programs that are more than 40 years old. Most were early adapters of function point metrics and also early adapters of parametric software estimation tools.

The software benchmarks studies carried out by Namcook Analytics LLC often find a significant number of serious software project management problems and issues. Table 5.2 summarizes 40 problems noted in a benchmark study for a Fortune 500 technology corporation.

About 15 of the 40 problems or about 38% were software project management problems. This distribution is not uncommon.

The author of this report is a frequent expert witness in litigation for software projects that either failed without being delivered or operated so poorly after delivery that the clients sued the vendors. It is interesting that project management problems were key factors in every lawsuit. Inaccurate estimation, poor tracking of progress, and poor quality control are endemic problems of the software industry and far too common even in 2017. These problems have been part of every breach of contract case where the author worked as an expert witness.

Improving Software Project Management Tools and Training

From consulting and benchmark studies carried out among top-tier technology corporations, they all have taken effective steps to improve and professionalize software project management. Some of these steps include the 10 steps in Table 5.3:

Let us now consider each of these 10 steps in sequence.

Initial Education for New Project Managers

In many technology companies project managers are often selected from the ranks of technical software engineering personnel. If this is so, they usually had close to

Table 5.2 Corporate Software Risk Factors Found by a Namcook Benchmark Study

1. Project management: no formal training for new managers
2. Project management: no annual benchmark studies
3. Project management: no annual training in state-of-the-art methods
4. Project management: no training in software cost estimation
5. Project management: no training in software quality estimation
6. Project management: no training in software risk analysis
7. Project management: no training in cyber-attack deterrence
8. Project management: no training in function point metrics
9. Project management: no training in schedule planning
10. Project management: lack of accurate productivity measurements
11. Project management: lack of accurate quality metrics
12. Project management: incomplete milestone and progress tracking
13. Project management: historical data leak by over 50%
14. Project management: managers continue to use inaccurate manual estimates
15. Project management: no widespread use of accurate parametric estimation
16. Quality control: no use of requirements models or QFD
17. Quality control: no use of automated proofs for critical features
18. Quality control: no use of cyber-attack inspections
19. Quality control: no use of formal design inspections
20. Quality control: no use of formal code inspections
21. Quality control: no use of static analysis tools
22. Quality control: no use of mathematical test case design (cause–effect graphs)
23. Quality control: no use of test coverage tools
24. Quality control: defect potentials about 4.75 bugs per function point
25. Quality control: DRE below 90%
26. Maintenance: no use of complexity analysis or cyclomatic complexity

(*Continued*)

Table 5.2 (*Continued*) Corporate Software Risk Factors Found by a Namcook Benchmark Study

27. Maintenance: no use of renovation tools or work benches
28. Maintenance: no use of code restructuring tools
29. Maintenance: inconsistent use of defect tracking tools
30. Maintenance: no use of inspections on enhancements
31. No reuse program: requirements
32. No reuse program: design
33. No formal reuse program: source code
34. No reuse program: test materials
35. No reuse program: documentation
36. No reuse program: project plans
37. No formal corporate reuse library
38. No corporate contracts with third-party reuse companies
39. Office space: small open offices, high noise levels, many interruptions
40. Insufficient meeting/breakout space for team meetings; no large meetings

Table 5.3 Ten Steps to Effective Software Project Management

1. Formal internal training for new project managers (10 days)
2. Annual training for project managers and technical staff (5 days)
3. Guest lectures from top software professionals (3 days)
4. Acquisition and use of parametric estimation tools
5. Acquisition and use of effective progress and milestone tracking tools
6. Use of formal project offices for applications >5000 function points
7. Use of and measurement of effective quality control methods
8. Elimination of bad software metrics and adoption of effective metrics
9. Commissioning annual software benchmark studies
10. Formal "best practice" analysis of tools, methods, reuse, and quality

zero management training at the university level. IBM recognized this as a problem back in the 1950s and introduced an effective training program for newly hired or newly appointed project managers.

This training is given to all project managers, but this report only covers software training topics. Since new project management training lasts for 10 days, there were 10 topics covered (Table 5.4).

Eight of the ten topics are technical and deal with actual project issues such as cyber-attacks and risks. Two of the ten topics deal with human resources and appraisals, which are of course a critical part of any manager's job.

(Microsoft has received criticism for their appraisal system, which uses mathematical curves and requires that only a certain percentage of employees can be appraised as "excellent." The problem with this is that technology companies such as Microsoft tend to have more excellent employees than ordinary companies do, so this curve tended to cause voluntary attrition among capable employees who ended up on the wrong side of the excellent barrier.)

Continuing Education for Software Project Managers

A study of software education methods carried out by Namcook Analytics LLC found that in-house education in major companies such as IBM and AT&T was superior to academic or university training for software project managers.

IBM and AT&T both employed more than 100 education personnel. These educators taught in-house courses to software and other personnel, and also taught

Table 5.4 New Software Project Manager Curriculum

	Project Management Courses	*Days*	*Value*
1	Software Milestone tracking	1	10
2	Sizing key software deliverables	1	10
3	Software project planning	1	10
4	Cyber-attack defenses	1	10
5	Software risk management	1	10
6	Software cost estimation: automated	1	10
7	Measurement and metrics of software	1	10
8	Software quality and defect estimation	1	10
9	Human resource policies	1	9
10	Appraisals and employee relations	1	9

customer courses to clients. The education groups operated initially as cost centers, with no charges for in-house or customer training. More recently, these education groups have switched to profit center operations and have started charging for training, at least for some customer trainings.

Quite a few technology companies have at least 10 days of training for new managers and about 1 week of training each year for both managers and technical staff. When Capers Jones was a new manager at IBM, he took this 10-day training series, and later taught some of the IBM project management courses on estimation, measurements, quality control, and software risk analysis.

Table 5.5 shows the rankings of 15 channels of software project management education in order of effectiveness.

A composite software project management curriculum derived from technology companies such as IBM, AT&T, ITT, Microsoft, and Apple is shown in Table 5.6.

Needless to say, this curriculum would be spread over a multiyear period. It is merely a combination of the kinds of software project management courses available in modern technology companies.

Table 5.5 Ranking Software Management Education Channels

1. In-house education in technology companies
2. Commercial education by professional educators
3. University education—graduate
4. University education—undergraduate
5. In-house education in non-technology companies
6. Mentoring by experienced managers
7. On-the-job training
8. Nonprofit education (IEEE, PMI, IFPUG, etc.)
9. Vendor education (management tools)
10. Self-study from work books
11. Self-study from CD-ROMs or DVDs
12. Live conferences with seminars and tutorials
13. Online education via the Internet and the World Wide Web
14. Project management books
15. Project management journals

Table 5.6 Software Project Management Curriculum

	Project Management Courses	*Days*	*Value*
1	Software milestone tracking	1.00	10.00
2	Early sizing before requirements	1.00	10.00
3	Sizing key deliverables	1.00	10.00
4	Controlling creeping requirements	1.00	10.00
5	Software project planning	2.00	10.00
6	Cyber-attack defenses	2.00	10.00
7	Cyber-attack recovery	1.00	10.00
8	Software outsourcing pros and cons	1.00	10.00
9	Optimizing multi-country teams	1.00	10.00
10	Best practices in project management	1.00	10.00
11	Software risk management	1.00	10.00
12	Software cost estimation: automated	2.00	10.00
13	Software high-security architecture	1.00	10.00
14	Benchmark sources: ISBSG, Namcook, etc.	1.00	10.00
15	Measurement and metrics of software	2.00	10.00
16	Software quality and defect estimation	1.00	10.00
17	Software defect tracking	1.00	9.75
18	Software benchmark overview	1.00	9.75
19	Function point analysis: high speed	1.00	9.75
20	Human resource policies	1.00	9.60
21	Software change control	1.00	9.50
22	Principles of software reuse	1.00	9.40
23	Appraisals and employee relations	1.00	9.00
24	Software cost tracking	1.00	9.00
25	Software maintenance and enhancement	1.00	9.00
26	Methodologies: Agile, RUP, TSP, others	1.00	9.00

(*Continued*)

Table 5.6 (*Continued*) Software Project Management Curriculum

	Project Management Courses	*Days*	*Value*
27	The capability maturity model (CMMI)	2.00	9.00
28	Overview of management tools	1.00	9.00
29	Testing for project managers	2.00	8.75
30	Static analysis for project managers	0.50	8.75
31	Inspections for project managers	0.50	8.75
32	PMBOK	1.50	8.70
33	Software metrics for project managers	1.00	8.50
34	Software cost estimation: manual	1.00	8.00
35	Tools: cost accounting	1.00	8.00
36	Tools: project management	1.00	8.00
37	Tools: human resources	1.00	8.00
38	Tools: cost and quality estimation	1.00	8.00
39	Function points for project managers	0.50	8.00
40	ISO standards for functional measures	1.00	8.00
41	Principles of Agile for managers	1.00	7.75
42	Principles of RUP for managers	1.00	7.75
43	Principles of TSP/PSP for managers	1.00	7.75
44	Principles of DevOps for managers	1.00	7.75
45	Principles of containers for managers	1.00	7.75
46	Earned value measurement (EVM)	1.00	6.75
47	Principles of balanced scorecards	1.00	6.50
48	Six Sigma for project managers	2.00	6.00
49	Six Sigma: green belt	3.00	6.00
50	Six Sigma: black belt	3.00	6.00
	Total	60.00	8.82

Guest Lectures from Visiting Experts

Over and above software classroom training, some technology companies have occasional internal seminars for all personnel that feature industry experts and famous software researchers. IBM, AT&T, and ITT had large seminars twice a year. One seminar was open only to employees and discussed some proprietary or confidential information such as new products and market expansion. The second large seminar was intended to demonstrate technical excellence to clients and customers, who were also invited to participate.

Among the well-known experts invited to companies such as AT&T, IBM, Siemens, and ITT were Al Albrecht (inventor of function points), Dr. Barry Boehm (inventor of COCOMO), Dr. Fred Brooks, (author of The Mythical Man-Month), Watts Humphrey (creator of the Software Engineering Institute-Capability Maturity Model Integrated), Dr. Larry Putnam (inventor of SLIM), Dr. Jerry Weinberg (author of The Psychology of Computer Programming), Bill Gates of Microsoft, Donald Knuth (pioneer of computer algorithms), Admiral Grace Hopper of the Navy (inventor of COBOL), Ken Thompson (codeveloper of UNIX), Linus Torvalds (developer of Linux), and many more.

These events usually lasted for about half a day of technical topics, and then had either a lunch or a reception for the guests based on whether it was a morning event or an afternoon event. At the reception or lunch the audience could meet and chat informally with the visiting experts.

Because the guest experts were world famous, these corporate seminars were attended by top software executives as well as software engineers and technical staff. These seminars served as a good means for bringing together all levels of a company to focus on critical software issues.

Having attended a number of these seminars, they were usually very enjoyable and it was good to meet famous software researchers such as Admiral Grace Hopper in person.

Needless to say, this kind of event usually takes place in fairly large companies since they are expensive. However, these seminars were also valuable, and benefitted both the executives and technical staffs of IBM, ITT, AT&T, Microsoft, and other companies that have them.

Acquisition and Use of Software Parametric Estimation Tools

IBM discovered in the early 1970s that manual software cost estimates became increasingly optimistic and inaccurate as application size increased from below 100 function points to more than 10,000 function points. Since applications grew rapidly in this era, IBM commissioned Capers Jones and Dr. Charles Turk to build its first parametric estimation tool in 1973. ITT did the same in 1979, and AT&T commissioned Capers Jones and colleagues to build a custom parametric estimation tool for its electronic switching systems in 1983.

The technology and telecom sectors have been pioneers in the development and usage of parametric estimation tools.

Software has achieved a bad reputation as a troubling technology. Large software projects have tended to have a very high frequency of schedule overruns, cost overruns, quality problems, and outright cancellations of large systems. While this bad reputation is often deserved, it is important to note that some large software projects are finished on time, stay within their budgets, and operate successfully when deployed.

The successful software projects differ in many respects from the failures and disasters. One important difference is how the successful projects arrived at their schedule, cost, resource, and quality estimates in the first place.

It often happens that projects exceeding their planned schedules or cost estimates did not use state-of-the-art methods for determining either their schedules or their costs. Although the project is cited for overruns, the root problem is inadequate planning and estimation.

From large-scale studies first published in the author's books *Software Engineering Best Practices* (Jones, 2010) and *The Economics of Software Quality* (Jones and Bonsignour, 2011), usage of automated parametric estimating tools, automated project scheduling tools, and automated defect and quality estimating tools [all of these are combined in some tools such as Software Risk Master (SRM)] are strongly correlated with successful outcomes.

Conversely, software project failures tended to use casual and manual methods of arriving at initial estimates. Indeed, for many software failures there was no formal estimation at all. In the 15 lawsuits for failure, delays, and poor quality where the author has been an expert witness, all the projects were larger than 10,000 function points and all used manual estimating methods.

The author's book *Estimating Software Costs* (Jones, 2007) discusses the history of the software cost estimating business. His newer book on *The Technical and Social History of Software Engineering* (Jones, 2014) also covered the history and arrival of commercial estimation tools.

The first software parametric cost estimation tools were created by researchers who were employed by large enterprises that built large and complex software systems: IBM, Hughes, RCA, TRW, and the U.S. Air Force were the organizations whose research led to the development of commercial parametric cost estimating tools.

Some of the estimating pioneers who developed the first parametric estimating tools include, in alphabetical order: Dr. Barry Boehm (TRW), Frank Freiman (RCA), Dan Galorath (SEER), Capers Jones (IBM), Dr. Larry Putnam (Air Force), and Dr. Howard Rubin (academic-SUNY).

In 1973, the author and his colleague Dr. Charles Turk at IBM, San Jose, built IBM's first automated parametric estimation tool for systems software. This tool was called the "Interactive Productivity and Quality (IPQ)" tool. This internal IBM tool was proprietary and not put on the commercial market since it gave IBM competitive advantages. This tool was developed at IBM's San Jose complex and soon had over 200 IBM users at over 20 locations around the world.

At present, in 2017, most technology and telecom companies use parametric estimation tools. These are normally used by project offices or by special estimating teams that provide estimates as a service to specific projects.

Some of the major parametric estimation tools currently, in 2017, include, in alphabetical order, the following:

1. Constructive Cost Model (COCOMO)
2. CostXpert
3. ExcelerPlan
4. KnowledgePlan
5. SEER
6. SLIM
7. SRM
8. True Price

These commercial parametric estimation tools are widely used by technology companies, defense contractors, and other large corporations. They are fairly expensive at costs of over $5000 per seat for most.

However, there is also the COCOMO developed by Dr. Barry Boehm. This estimating tool is free. Because COCOMO is free, it is widely used by universities and small companies that cannot afford the more expensive commercial estimation tools, but COCOMO does not have as wide usage among U.S. technology companies that need the more detailed estimates provided by the commercial parametric tools.

For example, the major features of the author's SRM commercial software estimation tool include the following:

- Sizing logic for specifications, source code, and test cases
- Sizing logic for function points and lines of code (LOC)
- Sizing logic for story points and use case points
- Sizing logic for requirements creep during development and post release
- Activity=level estimation for requirements, design, code, testing, etc.
- Sophisticated quality estimates that predict defects and defect removal efficiency (DRE)
- Support for 60 development methods such as agile, containers, DevOps, TSP, spiral, etc.
- Support for development, user costs, and 3 years of maintenance
- Support for International Function Point Users Group (IFPUG) function point metrics
- Support for the new SNAP point metric for nonfunctional requirements

- Support for other function point metrics such as COSMIC, NESMA, FISMA, etc.
- Support for older LOC metrics (both physical and logical)
- Support for modern topics such as cyber-defenses and cyber-attack recovery
- Support for proprietary metrics such as feature points and object points
- Support for software reusability of various artifacts
- Support for 84 modern languages such as Java, Ruby, mySQL, and others
- Support for systems applications such as operating systems and telecom
- Support for IT applications such as finance and insurance
- Support for Web-based applications
- Support for cloud applications
- Support for ERP deployment and customization
- Support for including commercial off-the-shelf (COTS) software
- Support for software portfolio analysis

SRM also supports many advanced functions:

- Quality and reliability estimation
- Numbers of test cases needed and test coverage
- Litigation costs for breach of contract
- Cyber-attack costs and recovery costs
- ERP deployment and modification
- Portfolio sizing and annual portfolio maintenance
- Risk and value analysis
- Measurement modes for collecting historical data
- Cost and time to complete estimates mixing historical data with projected data
- Support for software process assessments
- Support for results from the five levels of the SEI CMMI
- Special estimates such as the odds and costs of outsource litigation
- Special estimates such as venture funding for software startups

The other commercial software parametric estimation tools have similar features. The U.S. Air Force used to perform annual trials of commercial parametric estimation tools. All the major parametric tools were within about 10% of one another, and all were significantly more accurate than manual estimates for large projects above 1000 function points in size. Manual estimates were often optimistic on cost and schedules by more than 50% above 1000 function points. Parametric estimation tools are almost never optimistic and usually come within 10% of actual results.

Of course, another industry problem is that most companies do not have accurate results. Among the author's clients, the average accuracy of software project historical data is only 37%. The most common "leaks" from project historical data include business analysts, project managers, software quality assurance, technical writers, project office personnel, configuration control, and integration specialists. Measuring only developers and test personnel such as "design, code, and unit test" or DCUT is a common but fundamentally inaccurate and inadequate for economic or quality analysis.

Acquisition and Use of Progress and Milestone Tracking Tools

In addition to parametric estimation tools, there are also many commercial project management tools. The phrase "project management tools" has been applied to a large family of tools whose primary purpose is sophisticated scheduling for projects with hundreds or even thousands of overlapping and partially interdependent tasks and large teams in the hundreds. These tools are capable of dropping down to very detailed task levels and can even handle the schedules of individual workers. Microsoft Project and Artemis Views are two samples of project management tools. The new automated project office (APO) tool of Computer Aid is a modern project management tool only recently put on the market. There are also open-source project management and tracking tools such as JIRA.

However, the family of project management tools are general purpose in nature, and do not include specialized software sizing and estimating capabilities as do the software–cost estimating tools. Neither do these general project management tools deal with quality issues such as DRE. Project management tools are useful, but software requires additional capabilities to be under full management control.

Project management tools are an automated form of several management aids developed by the Navy for controlling large and complex weapons systems: the "program evaluation and review technique" (PERT), critical path analysis, resource leveling, and the classic Gantt charts. Project management tools used for defense projects also support earned value analysis (EVA) although this is seldom used in the civilian sectors.

Project management tools have no built-in expertise regarding software, as do software cost estimating tools. For example, if you wish to explore the quality and cost impact of an object-oriented programming language such as Objective C, a standard project management tool is not the right choice.

By contrast, many software cost estimating tools have built-in tables of programming languages and will automatically adjust the estimate based on which language is selected for the application. Project management tools and software cost estimating tools provide different but complementary functions. Most software cost estimating tools interface directly with common project management tools such as Microsoft Project.

Project Management Tools	*Software Cost Estimating Tools*
Work breakdown structures	Sizing logic for function points, code, etc.
Activity-based cost analysis	Quality estimates
Earned-value calculations	Integral risk estimates
Effort by staff members	Staffing predictions (testers, programmers, etc.)
Cost accumulation	Cost estimation

Both kinds of tools are useful for large and complex software projects and are generally used concurrently by project office personnel. An average software project using both parametric estimation tools and project management tools would be a significant project of over 1000 function points in size. Small projects below 100 function points are often estimated informally and use only normal corporate cost accounting tools rather than sophisticated project management tools.

The Use of Formal Project Offices for Applications >1000 Function Points

When software was created most applications were small and below 100 function points in size. During the late 1960s and early 1970s, software applications such as operating systems and telephone switching systems grew above 10,000 function points.

These larger applications had development teams that often topped 100 workers and they had over a dozen managers, including some second and third line managers.

Due to the poor training of most software managers in topics such as sizing, planning, estimating, and measurement, it soon became obvious that expert help was needed for these critical tasks.

The project management office (PMO) concept is much older than software and appeared in the late 1800s for manufacturing and even for agriculture and commodity training. Many industries were known to use PMOs before software became important in the 1950s. IBM was a pioneer in the creation and effective use of software project offices in the late 1960s.

There are a number of levels and kinds of project offices, but for this report the main emphasis is on specific software projects that are fairly large and complex such as 10,000 function points in size.

For these large systems, the main roles played by the software project office will be the following:

1. Identifying important standards for the project such as International Organization for Standardization (ISO) or Object Management Group (OMG) standards

2. Identifying important corporate standards such as the IBM standards for software quality
3. Identifying and monitoring government mandates such as Federal Aviation Administration (FAA), Food and Drug Administration (FDA) or Sarbanes–Oxley
4. Early sizing of project deliverables using one or more parametric estimation tools
5. Early prediction of schedules, costs, and staffing using one or more parametric estimation tools
6. Early prediction of requirements creep
7. Early prediction of defect potentials and DRE
8. Establishing project cost and milestone tracking guidelines for all personnel
9. Continuous monitoring of progress, accumulated costs, and schedule adherence
10. Continuous monitoring of planned vs. actual DRE

Project offices are a useful and valuable software structure. They usually contain from 3 to more than 10 people based on the size of the project being controlled. Among the kinds of personnel employed in software project offices are estimating specialists, metric specialists such as certified function point counters, quality control specialists, and standards specialists.

Project offices usually have several parametric estimation tools available and also both general and specialized project tracking tools. Shown below are samples of some of the kinds of tools and information that a project office might utilize in a technology company for a major software application (Table 5.7).

As is evident, software project offices add knowledge and rigor to topics where ordinary project managers may not be fully trained or highly experienced.

Use and Measurement of Effective Quality Control Methods

The #1 cost driver for the software industry for more than 50 years has been "the cost of finding and fixing bugs." Since bug repairs are the top cost drivers, it is impossible to have an accurate cost estimate without including quality costs. It is also impossible to have an accurate cost estimate, or to finish a project on time, unless it uses state-of-the-art quality control methods.

The #1 reason for software schedule delays and cost overruns for more than 50 years has been excessive defects present when testing starts, which stretches out the planned test duration by over 100% and also raises planned development costs. Without excellent quality control, software projects will always run late, exceed their budgets, and be at risk of cancellation if excessive costs make the return on investment (ROI) a negative value.

It is alarming that in several lawsuits where Capers Jones has been an expert witness, depositions showed that project managers deliberately cut back on pretest removal such as inspections and truncated testing early "in order to meet

Table 5.7 Tools for Software Projects

	Tasks	*Tools Utilized*	
1	Architecture	QEMU	
2	Automated test	HP QuickTest Professional	
3	Benchmarks	ISBSG, Namcook SRM, Davids, Q/P Management	
4	Coding	Eclipse, Slickedit	
5	Configuration	Perforce	
6	Cost estimate	SRM, SLIM, SEER, COCOMO	
7	Cost tracking	APO, Microsoft Project	
8	Cyclomatic	BattleMap	
9	Debugging	GHS probe	
10	Defect tracking	Bugzilla	
11	Design	Projects Unlimited, Visio	
12	Earned value	DelTek Cobra	
13	ERP	Microsoft Dynamics	
14	Function points 1	SRM	
15	Function points 2	Function point workbench	
16	Function points 3	CAST automated function points	
17	Graphics design	Visio	
18	Inspections	SlickEdit	
19	Integration	Apache Camel	
20	ISO tools	ISOXpress	
21	Maintenance	Mpulse	
22	Manual test	DevTest	
23	Milestone track	KIDASA Software Milestone Professional	
24	Progress track	Jira, APO	

(Continued)

Table 5.7 (*Continued*) Tools for Software Projects

	Tasks	*Tools Utilized*	
25	Project management	APO	
26	Quality estimate	SRM	
27	Requirements	Rational Doors	
28	Risk analysis	SRM	
29	Source code size 1	SRM	
30	Source code size 2	Unified code counter (UCC)	
31	Software Quality Assurance	NASA Goddard ARM tool	
32	Static analysis	OptimMyth Kiuwin, Coverity, Klocwork	
33	Support	Zendesk	
34	Test coverage	Software verify suite	
35	Test library	DevTest	
36	Value analysis	Excel and value stream tracking	
ISO and Other Standards Used for Project			
	IEEE 610.12-1990	Software engineering terminology	
	IEEE 730-1999	Software assurance	
	IEEE 12207	Software process tree	
	ISO/IEC 9001	Software quality	
	ISO/IEC 9003	Software quality	
	ISO/IEC 12207	Software engineering	
	ISO/IEC 25010	Software quality	
	ISO/IEC 29119	Software testing	
	ISO/IEC 27034	Software security	
	ISO/IEC 20926	Function point counting	
	OMG Corba	Common object request broker architecture	

(*Continued*)

Table 5.7 (*Continued*) Tools for Software Projects

	Tasks	*Tools Utilized*	
	OMG Models	Meta models for software	
	OMG funct. pts.	Automated function points (legacy applications)	
	UNICODE	Globalization and internationalization	
Professional Certifications Used on Project			
	Certification used for project=1		
	Certification not used for project=0		
	Note: Some team members have multiple certifications		
	Certification—Apple		0
	Certification—Computer Aid Inc.		0
	Certification—Computer Associates		0
	Certification—FAA		0
	Certification—FDA		0
	Certification—Hewlett Packard		0
	Certification—IBM		1
	Certification—Microsoft		1
	Certification—Oracle		0
	Certification—PMI		1
	Certification— Quality Assurance Institute (QAI)		0
	Certification—Red Hat		0
	Certification— Rex Black Consulting Services (RBCS)		0
	Certification—SAP		0
	Certification—Sarbanes–Oxley		0
	Certification— Software Engineering Institute (SEI)		0
	Certification—Sun		0

(*Continued*)

Table 5.7 (*Continued*) Tools for Software Projects

	Tasks	*Tools Utilized*	
	Certification—Symantec		0
	Certification—TickIT		0
	Certification of computing professionals		0
	Certified configuration management specialist		1
	Certified function point analyst		1
	Certified project managers		1
	Certified requirements engineers		0
	Certified scrum master		1
	Certified secure software lifecycle professional		0
	Certified security engineer		0
	Certified SEI appraiser		1
	Certified software architect		0
	Certified software business analyst		1
	Certified software development professional		0
	Certified software engineers		0
	Certified software quality assurance		1
	Certified test managers		0
	Certified testers		1
	Certified webmaster		1
	Certified software auditor		1
	Total		13

schedules." From the fact that the projects were late and over budget, cutting back on quality control raised costs and lengthened schedules, but the project managers did not know that this would happen.

Table 5.8 shows the 15 major cost drivers for software projects in 2017.

Table 5.8 U.S. Software Costs in Rank Order

1. The cost of finding and fixing bugs
2. The cost of canceled projects
3. The cost of producing English words
4. The cost of programming or code development
5. The cost of requirements changes
6. The cost of successful cyber-attacks
7. The cost of customer support
8. The cost of meetings and communication
9. The cost of project management
10. The cost of renovation and migration
11. The cost of innovation and new kinds of software
12. The cost of litigation for failures and disasters
13. The cost of training and learning
14. The cost of avoiding security flaws
15. The cost of assembling reusable components

Table 5.8 illustrates an important but poorly understood economic fact about the software industry. Four of the 15 major cost drivers can be attributed specifically to poor quality. The poor quality of software is a professional embarrassment and a major drag on the economy of the software industry and for that matter a drag on the entire U.S. and global economies.

Poor quality is also a key reason for cost driver #2. A common reason for canceled software projects is because quality is so bad that schedule slippage and cost overruns turned the project's ROI from positive to negative.

Note the alarming location of successful cyber-attacks in sixth place (and rising) on the cost driver list. Since security flaws are another form of poor quality, it is obvious that high quality is needed to deter successful cyber-attacks.

Poor quality is also a key factor in cost driver #12 or litigation for breach of contract. (The author has worked as an expert witness in 15 lawsuits. Poor software quality is an endemic problem with breach of contract litigation. In one case against a major ERP company, the litigation was filed by the company's own shareholders who asserted that the ERP package quality was so bad that it was lowering stock values!)

If you can't measure a problem, then you can't fix the problem either. Software quality has been essentially unmeasured and therefore unfixed for 50 years. A very useful quality metric developed by IBM around 1970 is that of "defect potentials."

Software defect potentials are the sum total of bugs found in requirements, architecture, design, code, and other sources of error. The approximate U.S. average for defect potentials is shown in Table 5.9 using IFPUG function points version 4.3.

Note that the phrase bad fix refers to new bugs accidentally introduced in bug repairs for older bugs. The current U.S. average for bad-fix injections is about 7%; i.e., 7% of all bug repairs contain new bugs. For modules that are high in cyclomatic complexity and for "error-prone modules," bad-fix injections can top 75%. For applications with low cyclomatic complexity, bad fixes can drop below 0.5%.

Defect potentials are necessarily measured using function point metrics. The older LOC metric cannot show requirements, architecture, and design defects nor any other defect outside the code itself. (As of 2017, function points are the most widely used software metric in the world. There are more benchmarks using function point metrics than all other metrics put together.)

Successful and effective software quality control requires these 15 technical factors:

1. Quality estimation before starting project using parametric estimation tools
2. Accurate defect tracking from requirements through post-release maintenance
3. Effective defect prevention such as requirements models and automated proofs
4. Effective pretest defect removal such as inspections and static analysis

Table 5.9 Average Software Defect Potentials circa 2017 for the United States

Requirements	0.70 defects per function point
Architecture	0.10 defects per function point
Design	0.95 defects per function point
Code	1.15 defects per function point
Security code flaws	0.25 defects per function point
Documents	0.45 defects per function point
Bad fixes	0.65 defects per function point
Totals	4.25 defects per function point

5. Effective mathematical test case design using cause–effect graphs or design of experiments
6. Effective cyber-attack prevention methods such as security inspections
7. Cyclomatic complexity analysis of all code modules in application
8. Keeping cyclomatic complexity <10 for critical software modules
9. Automated test coverage analysis for all forms of testing
10. Achieving defect potentials below 3.00 per function point
11. Achieving >95% test coverage
12. Achieving >97% DRE for all applications
13. Achieving >99% DRE for critical applications
14. Achieving <1% bad-fix injection (bad fixes are bugs in bug repairs)
15. Reuse of certified materials that approach zero-defect status

The bottom line is that poor software quality is the main weakness of the software industry. But poor software quality can be eliminated by better education for project managers and technical staff, and by using quality methods of proven effectiveness. The good news is that high quality is faster and cheaper than poor quality.

Elimination of Bad Metrics and Adoption of Effective Software Metrics

The software industry has the worst metrics and worst measurement practices of any industry in human history. It is one of the very few industries that cannot measure its own quality and its own productivity. This is professionally embarrassing.

Some of the troubling and inaccurate metrics used by the software industry are the following:

Cost per defect metrics penalize quality and make the buggiest software look cheapest. There are no ISO or other standards for calculating cost per defect. Cost per defect does not measure the economic value of software quality. The urban legend that it costs 100 times as much to fix post-release defects as early defects is not true and is based on ignoring fixed costs. Due to fixed costs of writing and running test cases, cost per defect rises steadily because fewer and fewer defects are found. This is caused by a standard rule of manufacturing economics: "If a process has a high percentage of fixed costs and there is a reduction in the units produced, the cost per unit will go up." This explains why cost per defects seems to go up over time even though actual defect repair costs are flat and do not change very much. There are of course very troubling defects that are expensive and time-consuming, but these are comparatively rare.

Defect density metrics measure the number of bugs released to clients. There are no ISO or other standards for calculating defect density. One method counts only code defects released. A more complete method used by the author includes bugs originating in requirements, architecture, design, and documents as well as code defects. The author's method also includes bad fixes or bugs in defect repairs themselves. About 7% of bug repairs contain new bugs. There is more than a 500% variation between counting only released code bugs and counting bugs from all sources. For example, requirements defects comprise about 20% of released software problem reports.

LOC metrics penalize high-level languages and make low-level languages look better than they are. LOC metrics also make requirements and design invisible. There are no ISO or other standards for counting LOC metrics. About half of the papers and journal articles use physical LOC and half use logical LOC. The difference between counts of physical and logical LOC can top 500%. The overall variability of LOC metrics has reached an astounding 2200% as measured by Joe Schofield, the former president of IFPUG! LOC metrics make requirements and design invisible and also ignore requirements and design defects, which outnumber code defects. Although there are benchmarks based on LOC, the intrinsic errors of LOC metrics make them unreliable. Due to lack of standards for counting LOC, benchmarks from different vendors for the same applications can contain widely different results.

SNAP point metrics are a new variation on function points introduced by IFPUG in 2012. The term SNAP is an awkward acronym for "software nonfunctional assessment process." The basic idea is that software requirements have two flavors: (1) functional requirements needed by users; (2) nonfunctional requirements due to laws, mandates, or physical factors such as storage limits or performance criteria. The SNAP committee view is that these nonfunctional requirements should be sized, estimated, and measured separately from function point metrics. Thus, SNAP and function point metrics are not additive, although they could have been. Having two separate metrics for economic studies is awkward at best and inconsistent with other industries. For that matter, it seems inconsistent with standard economic analysis in every industry. Almost every industry has a single normalizing metric such as "cost per square foot" for home construction or "cost per gallon" for gasoline and diesel oil. As of 2017, none of the parametric estimation tools have fully integrated SNAP and it may be that they won't be doing so since the costs of adding SNAP are painfully expensive. As a rule of thumb, nonfunctional requirements are about equal to 15% of functional requirements, although the range is very wide.

Story point metrics are widely used for agile projects with "user stories." Story points have no ISO standard for counting or any other standard. They are highly ambiguous and vary by as much as 400% from company to company and project to project. There are few if any useful benchmarks using story points. Obviously, story points can't be used for projects that don't utilize user stories, so they are worthless for comparisons against other design methods.

Technical debt is a new metric and is rapidly spreading. It is a brilliant metaphor developed by Ward Cunningham. The concept of "technical debt" is that topics deferred during development in the interest of schedule speed will cost more after release than they would have cost initially. However, there are no ISO standards for technical debt and the concept is highly ambiguous. It can vary by over 500% from company to company and project to project. Worse, technical debt does not include all the costs associated with poor quality and development short cuts. Technical debt omits canceled projects, consequential damages or harm to users, and the costs of litigation for poor quality.

Use case points are used by projects with designs based on "use cases" that often utilize IBM's Rational Unified Process (RUP). There are no ISO standards for use cases. Use cases are ambiguous and vary by over 200% from company to company and project to project. Obviously, use cases are worthless for measuring projects that don't utilize use cases, so they have very little benchmark data. This is yet another attempt to imitate the virtues of function point metrics, only with somewhat less rigor and with imperfect counting rules as of 2015.

Velocity is an agile metric that is used for prediction of sprint and project outcomes. It uses historical data on completion of past work units combined with the assumption that future work units will be about the same. Of course, it is necessary to know future work units for the method to operate. The concept of velocity is basically similar to the concept of using historical benchmarks for estimating future results. However, as of 2017, velocity has no ISO standards and no certification.

There are no standard work units for velocity and these can be story points or other metrics such as function points or use case points, or even synthetic concepts such as "days per task." If agile projects used function points then they could gain access to large volumes of historical data using activity-based costs; i.e., requirements effort, design effort, code effort, test effort, integration effort, documentation effort, etc. So long as agile continues to use quirky and unstandardized metrics without any certification exams, then agile productivity and quality will continue to be a mystery to clients who will no doubt be dismayed to find as many schedule delays and cost overruns as they had with waterfall.

As already stated in other reports, there are 11 primary metrics and 10 supplementary metrics that allow software projects to be measured with about 1% precision:

The *only* software metrics that allow quality and productivity to be measured with 1% precision are these 16 primary software metrics topics and 10 supplemental metrics topics:

Primary Software Metrics for High Precision

1. Application size in function points including requirements creep
2. SNAP points for nonfunctional requirements
3. Size of reusable materials (design, code, documents, etc.)
4. Activity-based costs using function points for normalization
5. Staff work hours per month including paid and unpaid overtime
6. Work hours per function point by activity
7. Function points per month by activity
8. Defect potentials using function points (requirements, design, code, document, and bad-fix defect categories)
9. DRE or the percentage of defects removed before release
10. Delivered defects per function point
11. High-severity defects per function point
12. Security flaw defects per function point
13. Cost of quality (COQ)
14. Cyber-attack defense and recovery costs
15. At least 3 years of maintenance and enhancements costs
16. Total cost of ownership (TCO)

Supplemental Software Metrics for High Precision

1. Software project taxonomy (nature, scope, class type)
2. Occupation groups (business analysts, programmers, testers, managers, Quality Assurance (QA), etc.)
3. Team experience levels (expert to novice)
4. CMMI level (1–5)
5. Development methodology used on application
6. Programming language(s) used on application
7. Complexity levels (problem, data, code complexity)
8. User effort for internal applications
9. Documents produced (type, pages, words, illustrations, etc.)
10. Meeting and communication costs

What is important are the technical features of the metrics themselves and not the numbers of metric users. Even if 50,000 companies use LOC, it is still a bad metric that distorts reality and should be viewed as professional malpractice. Following are the characteristics of a sample of current software metrics.

	Function Points	*LOC*	*Story Points*	*Use Case Points*
Software Metric Attributes				
ISO standard?	Yes	No	No	No
OMG standard?	Yes	No	No	No
Professional associations?	Yes	No	No	No
Formal training?	Yes	No	No	No
Certification exam?	Yes	No	No	No
Automated counting?	Yes	Yes	No	No
Required by governments?	Yes	No	No	No
Good for productivity?	Yes	No	Yes	No
Good for quality?	Yes	No	No	No
Good for estimates?	Yes	No	Yes	No
Published conversion rules?	Yes	No	No	No
Accepted by benchmark groups?	Yes	Yes	No	No
Used for IT projects?	Yes	Yes	Yes	Yes
Used for Web projects?	Yes	Yes	Yes	Yes
Used for cloud projects?	Yes	Yes	Yes	No
Used for embedded projects?	Yes	Yes	No	No
Used for systems software?	Yes	Yes	No	No
Used for telecom software?	Yes	Yes	No	No
Used for defense software?	Yes	Yes	No	No
Productivity Measures				
Activity-based costs?	Yes	No	No	No
Requirements productivity?	Yes	No	No	No
Design productivity?	Yes	No	No	No
Coding productivity?	Yes	Yes	No	No
Testing productivity?	Yes	Yes	No	No

	Function Points	*LOC*	*Story Points*	*Use Case Points*
Quality assurance productivity?	Yes	No	No	No
Technical writer productivity?	Yes	No	No	No
Project management productivity?	Yes	No	No	No
Net productivity of projects	Yes	Yes	Yes	Yes
Quality Measures				
Requirements defects?	Yes	No	No	No
Architecture defects?	Yes	No	No	No
Design defects?	Yes	No	No	No
Document defects?	Yes	No	No	No
Coding defects?	Yes	Yes	No	No
Bad-fix defects?	Yes	Yes	No	No
Net quality of projects?	Yes	Yes	Yes	Yes

As can be seen, function point metrics are the only metrics that can be used for all software activities and for both quality and productivity analysis. This paper uses IFPUG function points version 4.3 but other function point variations such as COSMIC, FISMA, and NESMA would produce similar but slightly different results.

Commissioning Annual Software Benchmark Studies

Software benchmarks are collections of data on software costs, schedules, staffing, quality, and technology usage that allow companies to compare results against similar projects in other companies. Usually, the benchmarks are "sanitized" and do not reveal the names of the companies or projects themselves.

Major corporations should commission software benchmark studies about once a year in order to judge progress. Some companies such as IBM can produce their own internal benchmarks with high accuracy, but still commission external benchmarks to compare results against other companies.

However, most companies are incompetent in collecting historical data. Their data leak and the average is only about 37% complete. This is why self-reported benchmark data often have higher productivity than benchmarks collected by professional benchmark consultants. Self-reported data leak and are usually incomplete.

A more fundamental problem is that most enterprises simply do not record data for anything but a small subset of the activities actually performed. In carrying out interviews with project managers and project teams to validate and correct historical data, the author has observed the following patterns of incomplete and missing data, using the 25 activities of a standard chart of accounts as the reference model (Table 5.10).

When the author and his colleagues collect benchmark data, we ask the managers and personnel to try and reconstruct any missing cost elements. Reconstruction of data from memory is plainly inaccurate, but it is better than omitting the missing data entirely.

Unfortunately, the bulk of the software literature and many historical studies only report information to the level of complete projects, rather than to the level of specific activities. Such gross "bottom line" data cannot readily be validated and is almost useless for serious economic purposes.

As of 2017, there are about 40 companies and nonprofit organizations that perform software benchmarks of various kinds. Some of the many forms of available software benchmarks include the following:

1. Compensation studies for software occupation groups
2. Voluntary and involuntary attrition of software personnel
3. Customer satisfaction for quality, reliability, etc.
4. Cyber-attack statistics for hacking, denial of service, data theft, etc.
5. Software productivity using function points per month and work hours per function point
6. Software quality using defect potentials in function points and DRE

Usually, software benchmarks are commissioned by individual business units rather than at the corporate level. Some companies spend over $5,000,000 per year on various kinds of benchmark studies, but may not realize this because the costs are scattered across various business and operating units. Table 5.11 shows some 40 software benchmark organizations circa 2017. The great majority of these are located in the United States, South America, or Europe. Asia is sparse on software benchmarks except for Japan, South Korea, and Malaysia. South America has large benchmark organizations in Brazil, and other more local benchmark groups in Mexico and Peru.

Benchmarks are hard to do with accuracy, but useful when done well. When done poorly, they add to the confusion about software productivity and quality that has blinded the software industry for more than 50 years.

Formal Best Practice Analysis of Software Tools, Methods, and Quality

The software literature has many articles and some books on software best practices. However, these usually lack quantitative data. To the author, a "best practice"

Table 5.10 Gaps and Omissions Observed in Data for a Software Chart of Accounts

Activities Performed	*Completeness of Historical Data*
1. Requirements	Missing or incomplete
2. Prototyping	Missing or incomplete
3. Architecture	Missing or incomplete
4. Project planning	Missing or incomplete
5. Initial analysis and design	Missing or incomplete
6. Detail design	Incomplete
7. Design reviews	Missing or incomplete
8. Coding	Complete
9. Reusable code acquisition	Missing or incomplete
10. Purchased package acquisition	Missing or incomplete
11. Code inspections	Missing or incomplete
12. Independent verification and validation	Complete
13. Configuration management	Missing or incomplete
14. Integration	Missing or incomplete
15. User documentation	Missing or incomplete
16. Unit testing	Incomplete
17. Function testing	Incomplete
18. Integration testing	Incomplete
19. System testing	Incomplete
20. Field testing	Missing or incomplete
21. Acceptance testing	Missing or incomplete
22. Independent testing	Complete
23. Quality assurance	Missing or incomplete
24. Installation and training	Missing or incomplete
25. Project management	Missing or incomplete
26. Total project resources, costs	Incomplete

Table 5.11 Software Benchmark Providers 2017

1. 4SUM Partners
2. Bureau of Labor Statistics, Department of Commerce
3. Capers Jones (Namcook Analytics LLC)
4. CAST Software
5. Congressional Cyber Security Caucus
6. Construx
7. COSMIC function points
8. Cyber Security and Information Systems
9. David Consulting Group
10. Economic Research Center (Japan)
11. Forrester Research
12. Galorath Incorporated
13. Gartner Group
14. German Computer Society
15. Hoovers Guides to Business
16. IDC
17. IFPUG
18. ISBSG Limited
19. ITMPI
20. Jerry Luftman (Stevens Institute)
21. Level 4 Ventures
22. Metri Group, Amsterdam
23. Namcook Analytics LLC
24. Price Systems
25. Process Fusion
26. QuantiMetrics

(*Continued*)

Table 5.11 (*Continued*) Software Benchmark Providers 2017

27. Quantitative Software Management (QSM)
28. Q/P Management Group
29. RBCS, Inc.
30. Reifer Consultants LLC
31. Howard Rubin
32. SANS Institute
33. Software Benchmarking Organization (SBO)
34. SEI
35. Software Improvement Group (SIG)
36. Software Productivity Research
37. Standish Group
38. Strassmann, Paul
39. System Verification Associates LLC
40. Test Maturity Model Integrated

should improve quality or productivity by at least 10% compared to industry averages. A "worst practice" might degrade productivity and quality by 10%.

With over 26,000 projects to examine, the author has published a number of quantitative tables and reports on software best (and worst) practices. Although we have evaluated about 335 methods and practices, the list is too big for convenience. A subset of 115 methods and practices shows best practices at the top and worst practices at the bottom. Namcook recommends using as many as possible from the top and avoiding the bottom (Table 5.12).

Because new practices come out at frequent intervals, companies need to have a formal method and practice evaluation group. At ITT, the Applied Technology Group evaluated existing and commercial tools, methods, and practices. The ITT Advanced Technology Group developed new tools and methods beyond the state of the art.

It is of minor historical interest that the Objective C programming language selected by Steve Jobs for all Apple software was actually developed by Dr. Tom Love and Dr. Brad Cox at the ITT Advanced Technology Group. When Alcatel acquired the ITT telecom research labs, the ownership of Objective C was transferred to Dr. Love at his new company.

Table 5.12 Software Technology Stack Scoring

	Methods and Practices in Technology Stack	*Value Scores*
1	Benchmarks (validated historical data from similar projects)	10.00
2	Defect potential <2.5	10.00
3	DRE >99%	10.00
4	Estimates: activity-based cost estimates	10.00
5	Estimates: parametric estimation tools	10.00
6	Estimates: TCO cost estimates	10.00
7	Formal and early quality predictions	10.00
8	Formal and early risk abatement	10.00
9	Inspection of all critical deliverables	10.00
10	Methods: patterns and >85% reuse of key deliverables	10.00
11	Metrics: defect potential measures	10.00
12	Metrics: DRE measures	10.00
13	Metrics: IFPUG function points	10.00
14	Metrics: SRM pattern matching sizing	10.00
15	Pre-requirements risk analysis	10.00
16	Static analysis of all source codes	10.00
17	APO	9.75
18	Metrics: bad-fix injections	9.70
19	Accurate cost tracking	9.50
20	Accurate defect tracking	9.50
21	Accurate status tracking	9.50
22	Estimates: COQ estimates	9.25
23	Metrics: COSMIC function points	9.25
24	Metrics: FISMA function points	9.25
25	Metrics: NESMA function points	9.25
26	Metrics: COQ measures	9.00

(*Continued*)

Table 5.12 (*Continued*) Software Technology Stack Scoring

	Methods and Practices in Technology Stack	*Value Scores*
27	Metrics: DDE measures	9.00
28	Reusable test materials	9.00
29	SEMAT usage on project	9.00
30	Test coverage >96%	9.00
31	DRE >95%	8.75
32	Methods: DAD	8.65
33	Mathematical test case design	8.60
34	CMMI 5	8.50
35	Methods: TSP/PSP	8.50
36	Test coverage tools used	8.50
37	Metrics: requirements growth before and after release	8.50
38	Metrics: deferred features	8.50
39	Methods: containers	8.40
40	Methods: DevOps	8.40
41	Methods: Hybrid: (agile/TSP)	8.25
42	Automated requirements modeling	8.15
43	Methods: Git	8.10
44	Methods: Mashups	8.10
45	Methods: RUP	8.00
46	Methods: Evolutionary Development (EVO)	8.00
47	Metrics: Automated function points	8.00
48	Reusable requirements	8.00
49	Reusable source code	8.00
50	Methods: (hybrid/agile/waterfall)	7.80
51	Static analysis of text requirements	7.80

(*Continued*)

Table 5.12 (*Continued*) Software Technology Stack Scoring

	Methods and Practices in Technology Stack	*Value Scores*
52	Methods: Kanban/Kaizen	7.70
53	Methods: Iterative development	7.60
54	CMMI 4	7.50
55	Methods: service-oriented models	7.50
56	Metrics: cyclomatic complexity tools	7.50
57	Requirements change tracking	7.50
58	Reusable designs	7.50
59	Automated proofs of correctness	7.50
60	Methods: continuous development	7.40
61	Methods: QFD	7.35
62	CMMI 3	7.00
63	Methods: JAD	7.00
64	Methods: spiral development	7.00
65	Requirements change control board	7.00
66	Reusable architecture	7.00
67	Reusable user documents	7.00
68	Methods: Extreme Programming	6.90
69	Metrics: FOG/Flesch readability scores	6.85
70	DRE >90%	6.50
71	Methods: Agile <1000 function points	6.50
72	Methods: Correctness proofs—automated	6.25
73	Automated testing	6.00
74	Certified quality assurance personnel	6.00
75	Certified test personnel	6.00
76	Defect potential 2.5–4.9	6.00

(*Continued*)

Table 5.12 (*Continued*) Software Technology Stack Scoring

	Methods and Practices in Technology Stack	*Value Scores*
77	Maintenance: data mining	6.00
78	Metrics: EVA	6.00
79	Six-Sigma for software	5.50
80	ISO risk standards	5.00
81	Metrics: Unadjusted function points	5.00
82	ISO quality standards	4.75
83	Maintenance: Information Technology Infrastructure Library (ITIL)	4.75
84	Metrics: Mark II function points	4.00
85	Requirements modeling—manual	3.00
86	Metrics: SNAP nonfunctional metrics	2.50
87	CMMI 2	2.00
88	Estimates: phase-based cost estimates	2.00
89	Metrics: story point metrics	2.00
90	Metrics: technical debt measures	2.00
91	Metrics: use case metrics	1.00
92	CMMI 0 (not used)	0.00
93	CMMI 1	–1.00
94	Methods: correctness proofs—manual	–1.00
95	Test coverage <90%	–1.00
96	Benchmarks (unvalidated self-reported benchmarks)	–1.50
97	Testing by untrained developers	–2.00
98	Methods: waterfall development	–3.00
99	Methods: Agile >5000 function points	–4.00
100	Cyclomatic complexity >20	–6.00

(Continued)

Table 5.12 (*Continued*) Software Technology Stack Scoring

	Methods and Practices in Technology Stack	*Value Scores*
101	Metrics: no productivity measures	–7.00
102	Methods: pair programming	–7.50
103	Methods: Cowboy development	–8.00
104	No static analysis of source code	–8.00
105	Test coverage not used	–8.00
106	Estimates: manual estimation >250 function points	–9.00
107	Inaccurate defect tracking	–9.00
108	Metrics: cost per defect metrics	–9.00
109	Inaccurate status tracking	–9.50
110	Defect potential >5	–10.00
111	DRE <85%	–10.00
112	Inaccurate cost tracking	–10.00
113	Metrics: LOC for economic study	–10.00
114	Metrics: no function point measures	–10.00
115	Metrics: no quality measures	–10.00

Summary and Conclusions on Software Project Management

Software is viewed by a majority of corporate CEOs as the most troublesome engineering technology of the modern era. It is true that software has very high rates of canceled projects and also of cost and schedule overruns. It is also true that poor project management practices are implicated in these problems.

However, some companies have been able to improve software project management and thereby improve software results. These improvements need better estimates, better metrics and measures, and better quality control.

Since academic training in software project management is marginal, the best source of project management training is in-house education in large companies followed by professional education companies such as the ITMPI, and then by nonprofit associations such as IFPUG, PMI, American Society for Quality (ASQ), etc.

Metrics and Measures for Achieving Software Excellence

As of the year 2017, software applications are the main operational component of every major business and government organization in the world. But software quality is still not good for a majority of these applications. Software schedules and costs are both frequently much larger than planned. Cyber-attacks have become more frequent and more serious.

This portion of Chapter 5 discusses the proven methods and results for achieving software excellence. It also provides quantification of what the term "excellence" means for both quality and productivity. Formal sizing and estimation using parametric estimation tools, excellent progress and quality tracking also using special tools, and a comprehensive software quality program can lead to shorter schedules, lower costs, and higher quality at the same time.

Software is the main operating tool of business and government in 2016 to 2017. But software quality remains marginal; software schedules and costs remained much larger than desirable or planned. Canceled projects are about 35% in the 10,000-function point size range and about 5% of software outsource agreements end up in court in litigation. Cyber-attacks are increasing in numbers and severity. This short study identifies the major methods for bringing software under control and achieving excellent results.

The first topic of importance is to show the quantitative differences between excellent, average, and poor software projects in quantified form. Table 5.13 shows the essential differences between software excellence, average, and unacceptably poor results for a mid-sized project of 1,000 function points or about 53,000 Java statements.

The data comes from benchmarks performed by Namcook Analytics LLC. These were covered by nondisclosure agreements and hence specific companies are not shown. However, the "excellent" column came from technology and medical device companies; the average from insurance and manufacturing; and the poor column from state and local governments.

As stated, the data in Table 5.13 comes from the author's clients, which consist of about 750 companies of whom 150 are Fortune 500 companies. About 40 government and military organizations are also clients, but the good and average columns in Table 5.13 are based on corporate results rather than government results. State and local governments provided data for the poor-quality column.

(Federal Government and defense software tend to have large overhead costs and extensive status reporting that are not found in the civilian sector. Some big defense projects have produced so much paperwork that there were over 1400 English words for every Ada statement, and the words cost more than the source code.)

(Note that the data in this report was produced using the Namcook Analytics Software Risk Master (SRM) tool. SRM can operate as an estimating tool prior to requirements or as a benchmark measurement tool after deployment.)

At this point it is useful to discuss and explain the main differences between the best, average, and poor results.

Table 5.13 Comparisons of Excellent, Average, and Poor Software Results

Topics	*Excellent*	*Average*	*Poor*
Monthly Costs			
(Salary+overhead)	$10,000	$10,000	$10,000
Size at Delivery			
Size in function points	1,000	1,000	1,000
Programming language	Java	Java	Java
Language levels	6.25	6.00	5.75
Source statements per function point	51.20	53.33	55.65
Size in logical code statements	51,200	53,333	55,652
Size in KLOC	51.20	53.33	55.65
Certified reuse percentage	20.00%	10.00%	5.00%
Quality			
Defect potentials	2,818	3,467	4,266
Defects per function point	2.82	3.47	4.27
Defects per KLOC	55.05	65.01	76.65
DRE	99.00%	90.00%	83.00%
Delivered defects	28	347	725
High-severity defects	4	59	145
Security vulnerabilities	2	31	88
Delivered per function point	0.03	0.35	0.73
Delivered per KLOC	0.55	6.50	13.03
Key Quality Control Methods			
Formal estimates of defects	Yes	No	No
Formal inspections of deliverables	Yes	No	No
Static analysis of all code	Yes	Yes	No

(Continued)

Table 5.13 (*Continued*) Comparisons of Excellent, Average, and Poor Software Results

Topics	*Excellent*	*Average*	*Poor*
Formal test case design	Yes	Yes	No
Testing by certified test personnel	Yes	No	No
Mathematical test case design	Yes	No	No
Project Parameter Results			
Schedule in calendar months	12.02	13.80	18.20
Technical staff+management	6.25	6.67	7.69
Effort in staff months	75.14	92.03	139.98
Effort in staff hours	9,919	12,147	18,477
Costs in dollars	$751,415	$920,256	$1,399,770
Cost per function point	$751.42	$920.26	$1,399.77
Cost per KLOC	$14,676	$17,255	$25,152
Productivity Rates			
Function points per staff month	13.31	10.87	7.14
Work hours per function point	9.92	12.15	18.48
LOC per staff month	681	580	398
Cost Drivers			
Bug repairs	25.00%	40.00%	45.00%
Paper documents	20.00%	17.00%	20.00%
Code development	35.00%	18.00%	13.00%
Meetings	8.00%	13.00%	10.00%
Management	12.00%	12.00%	12.00%
Total	100.00%	100.00%	100.00%
Methods, Tools, Practices			
Development methods	TSP/PSP	Agile	Waterfall
Requirements methods	JAD	Embedded	Interview

(*Continued*)

Table 5.13 (*Continued*) Comparisons of Excellent, Average, and Poor Software Results

Topics	*Excellent*	*Average*	*Poor*
CMMI levels	5	3	1
Work hours per month	132	132	132
Unpaid overtime	0	0	0
Team experience	Experienced	Average	Inexperienced
Formal risk analysis	Yes	Yes	No
Formal quality analysis	Yes	No	No
Formal change control	Yes	Yes	No
Formal sizing of project	Yes	Yes	No
Formal reuse analysis	Yes	No	No
Parametric estimation tools	Yes	No	No
Inspections of key materials	Yes	No	No
Static analysis of all code	Yes	Yes	No
Formal test case design	Yes	No	No
Certified test personnel	Yes	No	No
Accurate status reporting	Yes	Yes	No
Accurate defect tracking	Yes	No	No
More than 15% certified reuse	Yes	Maybe	No
Low cyclomatic complexity	Yes	Maybe	No
Test coverage >95%	Yes	Maybe	No

Software Sizing, Estimation, and Project Tracking Differences

High-quality projects with excellent results all use formal parametric estimating tools, perform formal sizing before starting, and have accurate status and cost tracking during development.

A comparative study by the author of accuracy differences between manual estimates and parametric estimates showed that the manual estimates averaged about 34% optimistic for schedules and costs.

Worse, manual estimating errors increased with application size. Below 250 function points, manual and parametric estimates were both within 5%. Above

10,000 function points, manual estimates were optimistic by almost 40% while parametric estimates were often within 10%. Overall, parametric estimates usually differed by <10% from actual results for schedules and costs, sometimes <5%, and were almost never optimistic.

The parametric estimation tools included COCOMO, Excelerator, KnowledgePlan, SEER, SLIM, SRM, and TruePrice. All these parametric tools were more accurate than manual cost and schedule estimates for all size ranges and application types.

High-quality projects also track results with high accuracy for progress, schedules, defects, and cost accumulation. Some excellent projects use specialized tracking tools such as Computer Aid's APO, which was built to track software projects. Others use general tools such as Microsoft Project, which supports many kinds of projects in addition to software.

Average projects with average results sometimes used parametric estimates but more often use manual estimates. However, some of the average projects did utilize estimating specialists, who are more accurate than untrained project managers.

Project tracking for average projects tends to be informal and uses general-purpose tools such as Excel rather than specialized software tracking tools such as APO, Jira, Asana, and others. Average tracking also leaks and tends to omit topics such as unpaid overtime and project management.

Poor-quality projects almost always use manual estimates. Tracking of progress is so bad that problems are sometimes concealed rather than revealed. Poor-quality cost tracking has major gaps and omits over 50% of total project costs. The most common omissions are unpaid overtime, project managers, and the work of part-time specialists such as business analysts, technical writers, and software quality assurance.

Quality tracking is embarrassingly bad and omits all bugs found before testing via static analysis or reviews, and usually omits bugs found during unit testing. Some poor-quality companies and government organizations don't track quality at all. Many others don't track until late testing or deployment.

Software Quality Differences for Best, Average, and Poor Projects

Software quality is the major point of differentiation between excellent results, average results, and poor results.

While software executives demand high productivity and short schedules, the vast majority do not understand how to achieve them. Bypassing quality control does not speed projects up: it slows them down.

The number one reason for enormous schedule slips noted in breach of contract litigation where the author has been an expert witness is starting testing with so many bugs that test schedules are at least double their planned duration.

The major point of this article is: *High quality using a synergistic combination of defect prevention, pretest inspections, and static analysis combined with formal testing is fast and cheap.*

Poor quality is expensive, slow, and unfortunately far too common. Because most companies do not know how to achieve high quality, poor quality is the norm and at least twice as common as high quality.

High quality does not come from testing alone. It requires defect prevention such as Joint Application Design (JAD), quality function deployment (QFD), or embedded users; pretest inspections and static analysis; and of course, formal test case development combined with certified test personnel. New methods of test case development based on cause–effect graphs and design of experiments are quite a step forward.

The defect potential information in the following table includes defects from five origins: requirements defects, design defects, code defects, document defects, and bad fixes or new defects accidentally included in defect repairs. The approximate distribution among these five sources is:

1. Requirements defects	15%
2. Design defects	30%
3. Code defects	40%
4. Document defects	8%
5. Bad fixes	7%
6. Total defects	100%

Note that a bad fix is a bug in a bug repair. These can sometimes top 25% of bug repairs for modules with high cyclomatic complexity.

However, the distribution of defect origins varies widely based on the novelty of the application, the experience of the clients and the development team, the methodologies used, and the programming languages. Certified reusable material also has an impact on software defect volumes and origins.

Table 5.14 shows approximate U.S. ranges for defect potentials based on a sample of 1500 software projects that include systems software, Web projects, embedded software, and information technology projects that range from 100 to 100,000 function points.

It is unfortunate that buggy software projects outnumber low-defect projects by a considerable margin.

Because the costs of finding and fixing bugs have been the #1 cost driver for the entire software industry for more than 50 years, the most important difference between excellent and mediocre results are in the areas of defect prevention, pretest defect removal, and testing.

Table 5.14 Defect Potentials for 1000 Projects

Defect Potentials	*Projects*	*Percentage*
<1.00	5	0.50
2 to 1	35	3.50
3 to 2	120	12.00
4 to 3	425	42.50
5 to 4	350	35.00
>5.00	65	6.50
Totals	1000	100.00

All three examples are assumed to use the same set of test stages, including the following:

1. Unit test
2. Function test
3. Regression test
4. Component test
5. Performance test
6. System test
7. Acceptance test

The overall DRE levels of these seven test stages range from below 80% for the worst case up to about 95% for the best case.

Note that these seven test stages are generic and are used on a majority of software applications. Additional forms of testing may also be used, and can be added to SRM for specific clients and specific projects:

1. Independent testing (mainly government and military software)
2. Usability testing (mainly software with complex user controls)
3. Performance testing (mainly real-time software)
4. Security testing
5. Limits testing
6. Supply chain testing
7. Nationalization testing (for international projects)

Testing alone is not sufficient to top 95% in DRE. Pretest inspections and static analysis are needed to approach or exceed the 99% range of the best case. Also, requirements models and "quality strong" development methods such as team software process (TSP) need to be part of the quality equation.

Excellent Quality Control

Excellent projects have rigorous quality control methods that include formal estimation of quality before starting, full defect measurement and tracking during development, and a full suite of defect prevention, pretest removal and test stages. The combination of low-defect potentials and high DRE is what software excellence is all about.

The most common companies that are excellent in quality control are usually the companies that build complex physical devices such as computers, aircraft, embedded engine components, medical devices, and telephone switching systems. Without excellence in quality these physical devices will not operate successfully. Worse, failure can lead to litigation and even criminal charges. Therefore, all companies that use software to control complex physical machinery tend to be excellent in software quality.

Examples of organizations noted as excellent software quality in alphabetical order include Advanced Bionics, Apple, AT&T, Boeing, Ford for engine controls, General Electric for jet engines, Hewlett Packard for embedded software, IBM for systems software, Motorola for electronics, NASA for space controls, the Navy for surface weapons, Raytheon, and Siemens.

Companies and projects with excellent quality control tend to have low levels of code cyclomatic complexity and high test coverage; i.e., test cases cover >95% of paths and risk areas.

These companies also measure quality well and all know their DRE levels. (Any company that does not measure and know their DRE is probably below 85% in DRE.)

Excellent quality control has DRE levels between about 97% for large systems in the 10,000-function point size range and about 99.6% for small projects <1000 function points in size.

A DRE of 100% is theoretically possible but is extremely rare. The author has only noted DRE of 100% in 10 projects out of a total of about 25,000 projects examined. As it happens, the projects with 100% DRE were all compilers and assemblers built by IBM and using >85% certified reusable materials. The teams were all experts in compilation technology and of course a full suite of pretest defect removal and test stages were used as well.

Average Quality Control

In today's world, agile is the new average. Agile development has proven to be effective for smaller applications below 1000 function points in size. Agile does not scale up well and is not a top method for quality. Agile is weak in quality measurements and does not normally use inspections, which have the highest DRE of any known form of defect removal. Disciplined Agile Development (DAD) can be used successfully on large systems where vanilla agile/scrum is not effective. Inspections top 85% in DRE and also raise testing DRE levels. Among the author's clients that use Agile, the average value for DRE is about 92%–94%. This is certainly better than the 85%–90% industry average for waterfall projects, but not up to the 99% actually needed to achieve optimal results.

Some but not all agile projects use "pair programming" in which two programmers share an office and a work station and take turns coding while the other watches and "navigates." Pair programming is very expensive but only benefits quality by about 15% compared to single programmers. Pair programming is much less effective in finding bugs than formal inspections, which usually bring three to five personnel together to seek out bugs using formal methods.

Agile is a definite improvement for quality compared to waterfall development, but is not as effective as the quality-strong methods of TSP and the RUP for larger applications >1000 function points. An average agile project among the author's clients is about 275 function points. DAD is a good choice for larger information software applications.

Average projects usually do not know defects by origin, and do not measure DRE until testing starts; i.e., requirements and design defects are underreported and sometimes invisible.

A recent advance in software quality control now frequently used by average as well as advanced organizations is that of static analysis. Static analysis tools can find about 55% of code defects, which is much higher than most forms of testing.

Many test stages such as unit test, function test, regression test, etc. are only about 35% efficient in finding code bugs, or in finding one bug out of three. This explains why 6–10 separate kinds of testing are needed.

The kinds of companies and projects that are "average" would include internal software built by hundreds of banks, insurance companies, retail and wholesale companies, and many government agencies at federal, state, and municipal levels.

Average quality control has DRE levels from about 85% for large systems up to 97% for small and simple projects.

Poor Quality Control

Poor quality control is characterized by weak defect prevention and almost a total omission of pretest defect removal methods such as static analysis and formal inspections. Poor quality control is also characterized by inept and inaccurate quality measures that ignore front-end defects in requirements and design. There are also gaps in measuring code defects. For example, most companies with poor quality control have no idea how many test cases might be needed or how efficient various kinds of test stages are.

Companies or government groups with poor quality control also fail to perform any kind of up-front quality predictions and hence they jump into development without a clue as to how many bugs are likely to occur and what are the best methods for preventing or removing these bugs.

One of the main reasons for the long schedules and high costs associated with poor quality is the fact that so many bugs are found when testing starts that the test interval stretches out to two or three times longer than planned.

Some of the kinds of software that are noted for poor quality control include the Obamacare website, municipal software for property tax assessments, and software for programmed stock trading, which has caused several massive stock crashes.

Poor quality control is often below 85% in DRE levels. In fact, for canceled projects or those that end up in litigation for poor quality, the DRE levels may drop below 80%, which is low enough to be considered professional malpractice. In litigation where the author has been an expert witness, DRE levels in the low 80% range have been the unfortunate norm.

Table 5.15 shows the ranges in DRE noted from a sample of 1000 software projects. The sample included systems and embedded software, Web projects, cloud projects, information technology projects, and also defense and commercial packages.

As can be seen, high DRE does not occur often. This is unfortunate because projects that are above 95% in DRE have shorter schedules and lower costs than projects below 85% in DRE. The software industry does not measure either quality or productivity well enough to know this.

However, the most important economic fact about high quality is: *Projects >97% in DRE have shorter schedules and lower costs than projects <90% in DRE.* This is because projects that are low in DRE have test schedules that are at least twice as long as projects with high DRE due to omission of pretest inspections and static analysis!

Table 5.15 Distribution of DRE for 1000 Projects

DRE (%)	*Projects*	*Percentage (%)*
>99.00	10	1.00
95–99	120	12.00
90–94	250	25.00
85–89	475	47.50
80–85	125	12.50
<80.00	20	2.00
Totals	1000	100.00

Reuse of Certified Materials for Software Projects

So long as software applications are custom designed and coded by hand, software will remain a labor-intensive craft rather than a modern professional activity. Manual software development even with excellent methodologies cannot be much more than 15% better than average development due to the intrinsic limits in human performance and legal limits in the number of hours that can be worked without fatigue.

The best long-term strategy for achieving consistent excellence at high speed would be to eliminate manual design and coding in favor of construction from certified reusable components.

It is important to realize that software reuse encompasses many deliverables and not just source code. A full suite of reusable software components would include at least the following 10 items, as shown in Table 5.16.

These materials need to be certified to near zero-defect levels of quality before reuse becomes safe and economically viable. Reusing buggy materials is harmful and expensive. This is why excellent quality control is the first stage in a successful reuse program.

The need for being close to zero defects and formal certification adds about 20% to the costs of constructing reusable artifacts, and about 30% to the schedules for construction. However, using certified reusable materials subtracts over 80% from the costs of construction and can shorten schedules by more than 60%. The more frequently materials are reused, the greater their cumulative economic value.

Table 5.16 Reusable Software Artifacts, circa 2016

1. Reusable requirements
2. Reusable architecture
3. Reusable design
4. Reusable code
5. Reusable project plans and estimates
6. Reusable test plans
7. Reusable test scripts
8. Reusable test cases
9. Reusable user manuals
10. Reusable training materials

One caution to readers: Reusable artifacts may be treated as taxable assets by the Internal Revenue Service. It is important to check this topic out with a tax attorney to be sure that formal corporate reuse programs will not encounter unpleasant tax consequences.

The three samples in the following table showed only moderate reuse typical for the start of 2017:

Excellent project	>25% certified reuse
Average project	±10% certified reuse
Poor projects	<5% certified reuse

In future it is technically possible to make large increases in the volumes of reusable materials. By around 2025 we should be able to construct software applications with perhaps 85% certified reusable materials. In fact, some "mashup" projects already achieve 85% reuse, but the reused materials are not certified and some may contain significant bugs and security flaws.

Table 5.17 shows the productivity impact of increasing volumes of certified reusable materials. The table uses whole numbers and generic values to simplify the calculations.

Table 5.17 Productivity Gains from Software Reuse

Reuse Percentage (%)	*Months of Staff Effort*	*Function Points per Month*	*Work Hours per Function Point*	*LOC per Month*	*Project Costs*
0.00	100	10.00	13.20	533	$1,000,000
10.00	90	11.11	11.88	592	$900,000
20.00	80	12.50	10.56	666	$800,000
30.00	70	14.29	9.24	761	$700,000
40.00	60	16.67	7.92	888	$600,000
50.00	50	20.00	6.60	1,066	$500,000
60.00	40	25.00	5.28	1,333	$400,000
70.00	30	33.33	3.96	1,777	$300,000
80.00	20	50.00	2.64	2,665	$200,000
90.00	10	100.00	1.32	5,330	$100,000
100.00	1	1,000.00	0.13	53,300	$10,000

Note: Assumes 1,000 function points and 53,300 LOC.

Software reuse from certified components instead of custom design and hand coding is the only known technique that can achieve order-of-magnitude improvements in software productivity. True excellence in software engineering must derive from replacing costly and error-prone manual work with construction from certified reusable components.

Because finding and fixing bugs is the major software cost driver, increasing volumes of high-quality certified materials can convert software from an error-prone manual craft into a very professional high-technology profession. Table 5.18 shows probable quality gains from increasing volumes of software reuse.

Since the current maximum for software reuse from certified components is only in the range of 15% or a bit higher, it is evident that there is a large potential for future improvement.

Note that uncertified reuse in the form of mashups or extracting materials from legacy applications may top 50%. However, uncertified reusable materials often have latent bugs, security flaws, and even error-prone modules (EPM), so this is not a very safe practice. In several cases the reused material was so buggy that it had to be discarded and replaced by custom development.

Several emerging development methodologies such as mashups are pushing reuse values up above 90%. However, the numbers and kinds of applications built

Table 5.18 Quality Gains from Software Reuse

Reuse Percentage (%)	*Defects per Function Point*	*Defect Potential*	*DRE (%)*	*Delivered Defects*
0.00	5.00	1000	90.00	100
10.00	4.50	900	91.00	81
20.00	4.00	800	92.00	64
30.00	3.50	700	93.00	49
40.00	3.00	600	94.00	36
50.00	2.50	500	95.00	25
60.00	2.00	400	96.00	16
70.00	1.50	300	97.00	9
80.00	1.00	200	98.00	4
90.00	0.50	100	99.00	1
100.00	—	1	99.99	0

Note: Assumes 1,000 function points and 53,300 LOC.

from these emerging methods are small. Reuse needs to become generally available with catalogs of standard reusable components organized by industries, i.e., banking, insurance, telecommunications, firmware, etc.

Software Methodologies

Unfortunately, selecting a methodology is more like joining a cult than making an informed technical decision. Most companies don't actually perform any kind of due diligence on methodologies and merely select the one that is most popular.

In the present-day world, agile is definitely most popular. Fortunately, agile is also a pretty good methodology and much superior to the older waterfall method. However, there are some caveats about methodologies.

Agile has been successful primarily for smaller applications <1000 function points in size. It has also been successful for internal applications where users can participate or be "embedded" with the development team to work out requirements issues.

Agile has not scaled up well to large systems >10,000 function points. Agile has also not been visibly successful for commercial or embedded applications where there are millions of users and none of them work for the company building the software, so their requirements have to be collected using focus groups or special marketing studies.

A variant of agile that uses pair programming or two programmers working in the same cubical with one coding and the other "navigating" has become popular. However, it is very expensive since two people are being paid to do the work of one person. There are claims that quality is improved, but formal inspections combined with static analysis achieve much higher quality for much lower costs.

Another agile variation, extreme programming, in which test cases are created before the code itself is written has proven to be fairly successful for both quality and productivity, compared to traditional waterfall methods. However, both TSP and RUP are just as good and even better for large systems. Another successful variation on agile is DAD, which expands the agile concept up above 5000 function points.

There are more than 80 available methodologies circa 2016 and many are good; some are better than agile for large systems; some older methods such as waterfall and cowboy development are at the bottom of the effectiveness list and should be avoided on modern applications.

For major applications in the 10,000-function point size range and above, the TSP and the RUP have the best track records for successful projects and fewest failures. Table 5.19 ranks 50 current software development methodologies. The rankings show their effectiveness for small projects below 1000 function points and for large systems above 10,000 function points. Table 5.19 is based on data from around 600 companies and 26,000 project results.

Table 5.19 Methodology Rankings for Small and Large Software Projects

	Small Projects <1,000 Function Points	*Large Systems >10,000 Function Points*
1	Agile scrum	TSP/PSP
2	Crystal	Reuse-oriented
3	DSDM	Pattern-based
4	Feature driven development (FDD)	IntegraNova
5	Hybrid	Product line engineering
6	IntegraNova	Model-driven
7	Lean	DevOps
8	Mashup	Service-oriented
9	Microsoft solutions	Specifications by Example
10	Model-driven	Mashup
11	Object-oriented	Object-oriented
12	Pattern-based	Information engineering (IE)
13	Product line engineering	Feature-driven development (FDD)
14	PSP	Microsoft solutions
15	Reuse-oriented	Structured development
16	Service-oriented modeling	Spiral development
17	Specifications by Example	T-VEC
18	Structured development	Kaizen
19	Test-driven development (TDD)	RUP
20	CASE	Crystal
21	Clean room	DSDM
22	Continuous development	Hybrid
23	DevOps	CASE
24	EVO	Global 24 h

(*Continued*)

Table 5.19 (*Continued*) Methodology Rankings for Small and Large Software Projects

	Small Projects <1,000 Function Points	*Large Systems >10,000 Function Points*
25	Information engineering (IE)	Continuous development
26	Legacy redevelopment	Legacy redevelopment
27	Legacy renovation	Legacy renovation
28	Merise	Merise
29	Open-source	Iterative
30	Spiral development	Legacy data mining
31	T-VEC	Custom by client
32	Kaizen	CMMI 3
33	Pair programming	Agile scrum
34	Reengineering	Lean
35	Reverse engineering	EVO
36	XP	Open-source
37	Iterative	Reengineering
38	Legacy data mining	V-Model
39	Prototypes-evolutionary	Clean room
40	RAD	Reverse engineering
41	RUP	Prototypes-evolutionary
42	TSP/PSP	RAD
43	V-Model	Prince 2
44	Cowboy	Prototypes-disposable
45	Prince 2	Test-driven development (TDD)
46	Waterfall	Waterfall
47	Global 24 h	Pair programming
48	CMMI 3	XP
49	Prototypes-disposable	Cowboy
50	Anti-patterns	Anti-patterns

The boldface type of the top 20 entries highlights the methods with the most successful project outcomes. In general, the large-system methods are quality strong methodologies that support inspections and rigorous quality control. Some of these are a bit "heavy" for small projects although quality results are good. However, the overhead of some rigorous methods tends to slow down small projects.

Starting in 2014 and expanding fairly rapidly is the new "software engineering methods and theory" or SEMAT approach. This is not a "methodology" per se but new way of analyzing software engineering projects and applications themselves.

SEMAT had little or no empirical data as this chapter was being written, but the approach seems to have merit. The probable impact, although this is not yet proven, will be a reduction in software defect potentials and perhaps an increase in certified reusable components.

Unfortunately, SEMAT seems to be aimed at custom designs and manual development of software, both of which are intrinsically expensive and error-prone. SEMAT would be better used for increasing the supply of certified reusable components. As SEMAT usage expands, it will be interesting to measure actual results, which to date are purely theoretical.

Quantifying Software Excellence

Because the software industry has a poor track record for measurement, it is useful to show what excellence means in quantified terms.

Excellence in software quality combines defect potentials of no more than 2.50 bugs per function point combined with DRE of 99.00%. This means that delivered defects will not exceed 0.025 defects per function point.

By contrast, current average values circa 2016 are about 3.00–5.00 bugs per function point for defect potentials and only 90%–94% DRE, leading to as many as 0.50 bugs per function point at delivery. There are projects that top 99.00% but the distribution is <5% of U.S. projects.

Poor projects that are likely to fail and end up in court for poor quality or breach of contract often have defect potentials of >6.00 per function point combined with DRE levels <85%. Some poor projects deliver >0.75 bugs per function point and also excessive security flaws.

Excellence in software productivity and development schedules are not fixed values but vary with the size of the applications. Table 5.20 shows two "flavors" of productivity excellence: (1) the best that can be accomplished with 10% reuse and (2) the best that can be accomplished with 50% reuse.

From Table 5.20 it can be seen that software reuse is the most important technology for improving software productivity and quality by really significant amounts. Methods, tools, CMMI levels, SEMAT, and other minor factors are certainly beneficial. However, so long as software applications are custom designed and hand coded, software will remain an expensive craft and not a true professional occupation.

Table 5.20 Excellent Productivity with Varying Quantities of Certified Reuse

	Schedule Months	*Staffing*	*Effort Months*	*Function Point per Month*
With <10% Certified Reuse				
100 function points	4.79	1.25	5.98	16.71
1,000 function points	13.80	6.25	86.27	11.59
10,000 function points	33.11	57.14	1,892.18	5.28
100,000 function points	70.79	540.54	38,267.34	2.61
With 50% Certified Reuse				
100 function points	3.98	1.00	3.98	25.12
1,000 function points	8.51	5.88	50.07	19.97
10,000 function points	20.89	51.28	1,071.43	9.33
100,000 function points	44.67	487.80	21,789.44	4.59

The Metaphor of Technical Debt

Ward Cunningham's interesting metaphor of technical debt has become a popular topic in the software industry. The concept of technical debt is that in order to get software released in a hurry, short cuts and omissions occur that will need to be repaired after release, for much greater cost; i.e., like interest builds up on a loan.

Although the metaphor has merit, it is not yet standardized and therefore can vary widely. In fact, a common question at conferences is "What do you include in technical debt?"

Technical debt is not a part of standard costs of quality. There are some other topics that are excluded also. The most important and also the least studied are "consequential damages" or actual financial harm to clients of buggy software. These show up in lawsuits against vendors and are known to attorneys and expert witnesses, but otherwise are not widely published.

A major omission from technical debt circa 2016 is the cost of cyber-attacks and recovery from cyber-attacks. In cases where valuable data are stolen, cyber-attack costs can be more expensive than total development costs for the attacked application.

Another omission from both COQ and technical debt are the costs of litigation and damage awards when software vendors or outsourcers are sued for poor quality. The final table in this chapter puts all these costs together to show the full set of costs that might occur for excellent quality, average quality, and poor quality. Note that Table 5.21 uses "defects per function point" for the quality results.

Table 5.21 Technical Debt and Software Quality for 1000 Function Points

	High Quality	*Average Quality*	*Poor Quality*
Defect potential	2	4	6
Removal efficiency	99.00%	92.00%	80.00%
Delivered defects	0.02	0.32	1.2
Post-release defect repair $	$5,000	$60,000	$185,000
Technical debt problems	1	25	75
Technical debt costs	$1,000	$62,500	$375,000
Excluded from Technical Debt			
Consequential damages	$0.00	$281,250	$2,437,500
Cyber-attack costs	$0.00	$250,000	$5,000,000
Litigation costs	$0.00	$2,500,000	$3,500,000
Total COQ	$6,000	$3,153,750	$11,497,500

As of early 2016, almost 85% of the true costs of poor quality software are invisible and not covered by either technical debt or standard COQ. No one has yet done a solid study of the damages of poor quality to clients and users but these costs are much greater than internal costs.

(This is a topic that should be addressed by both the CMMI and the SEMAT approach, although neither has studied consequential damages.)

No data has yet been published on the high costs of litigation for poor quality and project failures, or even the frequency of such litigation.

(The author has been an expert witness in 15 cases for project failure or poor quality, and therefore has better data than most on litigation frequencies and costs. Also, the author's SRM tool has a standard feature that predicts probable litigation costs for both the plaintiff and the defendant in breach of contract litigation.)

Table 5.21 illustrates two important but poorly understood facts about software quality economics:

1. High-quality software is faster and cheaper to build than poor-quality software; maintenance costs are many times cheaper; and technical debt is many times cheaper.
2. Poor-quality software is slower and more expensive to build than high-quality software; maintenance costs are many times more expensive; and technical debt is many times more expensive.

Companies that skimp on quality because they need to deliver software in a hurry don't realize that they are slowing down software schedules; not speeding them up.

High quality also causes little or no consequential damages to clients, and the odds of being sued are below 1%, as opposed to about 15% for poor quality software built by outsource vendors. Incidentally, state governments seem to have more litigation for failing projects and poor quality than any other industry sector.

High-quality projects are also less likely to experience cyber-attacks because many of these attacks are due to latent security flaws in deployed software. These flaws might have been eliminated prior to deployment if security inspections and security testing plus static analysis had been used.

For software projects, high quality is more than free; it is one of the best investments companies can make. High quality has a large and positive ROI. Poor quality software projects have huge risks of failure, delayed schedules, major cost overruns, and more than double the cost per function point compared to high quality.

Stages in Achieving Software Excellence

Readers are probably curious about the sequence of steps needed to move from average to excellent in software quality. They are also curious about the costs and schedules needed to achieve excellence. Following are short discussions of the sequence and costs needed for a company with about 1000 software personnel to move from average to excellent results.

Stage 1: Quantify Your Current Software Results

In order to plan improvements rationally, all companies should know their current status using effective quantified data points. This means that every company should measure and know these topics:

1. Defect potentials
2. Defect severity levels
3. Defects per function point
4. Defect detection efficiency (DDE)
5. DRE
6. Cyclomatic complexity of all applications
7. EPM in deployed software
8. Test coverage of all applications
9. Test cases and test scripts per function point
10. Duplicate or incorrect test cases in test libraries
11. Bad-fix injection rates (bugs in defect repairs)

12. The existence or absence of EPM in operational software
13. Customer satisfaction with existing software
14. Defect repair turnaround
15. Technical debt for deployed software
16. COQ
17. Security flaws found before release and then after deployment
18. Current set of defect prevention, pretest, and test quality methods in use
19. The set of software development methodologies in use for all projects
20. Amount of reusable materials utilized for software projects

For a company with 1000 software personnel and a portfolio of perhaps 3000 software applications, this first stage can take from two to three calendar months. The effort would probably be in the range of 15–25 internal staff months, plus the use of external quality consultants during the fact-finding stage.

The most likely results will be the discovery that defect potentials top 3.5 per function point and DRE is below 92%. Other likely findings will include <80% test coverage and cyclomatic complexity that might be >50 for key modules. Probably a dozen or more EPM will be discovered. Quantitative goals for every software company should be to have defect potentials <2.5 per function point combined with DRE levels >97% for every software project and >99% for mission critical software projects. Software reuse will probably be <15% and mainly be code modules that are picked up informally from other applications.

The analogy for this stage would be like going to a medical clinic for a thorough annual medical checkup. The checkup does not cure any medical problems by itself, but it identifies the problems that physicians will need to cure, if any exist.

Once the current quality results have been measured and quantified, it is then possible to plan rational improvement strategies that will reduce defect potentials and raise DRE to approximately 99% levels.

Stage 2: Begin to Adopt State-of-the-Art Quality Tools and Methods

Software excellence requires more than just adopting a new method such as agile and assuming everything will get better. Software excellence is the result of a web of related methods and tools that are synergistic.

The second stage, which occurs as the first stage is ending, and perhaps overlaps the last month, is to acquire and start to use proven methods for defect prevention, pretest defect removal, and formal testing.

This stage can vary by the nature and size of the software produced. Real-time and embedded applications will use different tools and methods compared to Web and information technology applications. Large systems will use different methods than small applications. However, a nucleus of common techniques is used for all software. These include the following:

Formal Sizing, Estimation, and Tracking

1. Use parametric estimation tools on projects >250 function points
2. Carry out formal risk analysis before starting
3. Use formal tracking of progress, quality, and costs

Defect Prevention

1. JAD
2. QFD
3. Requirements models
4. Formal reuse programs
5. Formal defect measurements
6. Data mining of legacy applications for lost requirements
7. Training and certification of quality personnel
8. Acquisition of defect measurements tools and methods
9. Formal methodology analysis and selection for key projects
10. Formal quality and defect estimation before projects start

Pre-Test Defect Removal

1. Static analysis of all legacy applications
2. Static analysis of all new applications
3. Static analysis of all changes to applications
4. Inspections of key deliverables for key projects (requirements, design, code, etc.)
5. Automated proofs of correctness for critical features

Test Defect Removal

1. Formal test case design, often using design of experiments or cause–effect graphs
2. Acquisition of test coverage tools
3. Acquisition of cyclomatic complexity tools
4. Review of test libraries for duplicate or defective test cases
5. Formal training of test personnel
6. Certification of test personnel
7. Planning optimal test sequences for every key project
8. Measuring test coverage for all projects
9. Measuring cyclomatic complexity for all code
10. Formal test and quality measures of all projects

This second stage normally lasts about a year and includes formal training of managers, development personnel, quality assurance personnel, test personnel, and other software occupation groups.

Because there is a natural tendency to resist changes, the best way of moving forward is to treat the new tools and methods as experiments. In other words, instead of directing that certain methods such as inspections be used, treat them as experiments and make it clear that if the inspections don't seem useful after trying them out, the teams will not be forced to continue with them. This is how IBM introduced inspections in the 1970s, and the results were so useful that inspections became a standard method without any management directives.

This second stage will take about a year for a company with 1000 software personnel, and less or more time for larger or smaller organizations. Probably all technical personnel will receive at least a week of training, and so will project managers.

Probably the costs during this phase due to training and learning curves can top $1000 per staff member. Some costs will be training; others will be acquisitions of tools. It is difficult to establish a precise cost for tools due to the availability of a large number of open-source tools that have no costs.

Improvements in quality will start to occur immediately during stage 2. However, due to learning curves, productivity will drop down slightly for the first 4 months due to having formal training for key personnel. But by the end of a year, productivity may be 15% higher than when the year started. Defect potentials will probably drop by 20% and DRE should go up by >7% from the starting point, and top 95% for every project.

Stage 3: Continuous Improvements Forever

Because stages 1 and 2 introduce major improvements, some interesting sociological phenomena tend to occur. One thing that may occur is that the technical and management leaders of stages 1 and 2 are very likely to get job offers from competitive companies or from other divisions in large corporations.

It sometimes happens that if the stage 1 and 2 leaders are promoted or change jobs, their replacements may not recognize the value of the new tools and methods. For example, many companies that use inspections and static analysis find that defects are much reduced compared to previous years.

When quality improves significantly, unwise managers may say, "Why keep using inspections and static analysis when they are not finding many bugs?" Of course, if the inspections and static analysis stop, the bug counts will soon start to climb back up to previous levels and DRE will drop down to previous levels.

In order to keep moving ahead and staying at the top, formal training and formal measurements are both needed. Annual training is needed, and also formal training of new personnel and new managers. Companies that provide five or more days of training for software personnel have higher annual productivity than companies with 0 days of training.

When the ITT Corporation began a successful 4-year improvement program, one of the things that was part of their success was an annual report for corporate executives. This report was produced on the same schedule as the annual corporate financial report to shareholders; i.e., in the first quarter of the next fiscal year.

The ITT annual reports showed accomplishments for the prior year; comparisons to earlier years; and projected accomplishments for the following year. Some of the contents of the annual reports included the following:

1. Software personnel by division
2. Software personnel by occupation groups
3. Year-by-year COQ
4. Total costs of software ownership
5. Changes in software personnel by year for 3 years
6. Average and ranges of defect potentials
7. Average and ranges of DRE
8. Three-year running averages of defect potentials and DRE
9. Customer satisfaction year by year
10. Plans for the next fiscal year for staffing, costs, quality, etc.

ITT was a large corporation with over 10,000 software personnel located in a number of countries and more than 25 software development labs. As a result, the overall corporate software report was a fairly large document of about 50 pages in size.

For a smaller company with a staffing of about 1000 personnel, the annual report would probably be in the 20-page size range.

Once software is up to speed and combines high quality and high productivity, that opens up interesting business questions about the best use of the savings. For example, ITT software personnel had been growing at more than 5% per year for many years. Once quality and productivity improved, it was clear that personnel growth was no longer needed. In fact, the quality and productivity were so good after a few years that perhaps 9000 personnel instead of 10,000 personnel could build and maintain all needed software.

Some of the topics that need to be considered when quality and productivity improve are related to what the best use of resources no longer devoted to fixing bugs is. Some of the possible uses include:

- Reduce corporate backlogs to zero by tackling more projects per year.
- Move into new kinds of applications using newly available personnel no longer locked into bug repairs.
- Allow natural attrition to lower overall staffing down to match future needs.

For commercial software companies, expanding into new kinds of software and tackling more projects per year are the best use of available personnel that will be freed up when quality improves.

For government software or for companies that are not expanding their businesses, probably allowing natural attrition to reduce staffing might be considered. For large organizations, transfers to other business units might occur.

One thing that could be a sociological disaster would be to have layoffs due to the use of improved technologies that reduced staffing needs. In this case, resistance to changes and improvements would become a stone wall and progress would stop cold.

Since most companies have large backlogs of applications that are awaiting development, and since most leading companies have needs to expand software into new areas, the best overall result would be to use the available personnel for expansion.

Stage 3 will run for many years. The overall costs per function point should be about 30% lower than before the improvement program started. Overall schedules should be about 25% shorter than before the improvement program started.

Defect potentials will be about 35% lower than when the improvement program started and corporate DRE should top 97% for all projects and 99% for mission critical projects.

Going beyond Stage 3 into Formal Reuse Programs

As mentioned previously, custom designs and manual coding are intrinsically expensive and error-prone no matter what methodologies and programming languages are used.

For companies that need peak performance, moving into a full and formal software reuse program can achieve results even better than Stage 3.

Summary and Conclusions

Because software is the driving force of both industry and government operations, it needs to be improved in terms of both quality and productivity. The most powerful technology for making really large improvements in both quality and productivity will be from eliminating costly custom designs and labor-intensive hand coding, and moving toward manufacturing software applications from libraries of well-formed standard reusable components that approach zero-defect quality levels.

The best combinations of methods, tools, and programming languages today are certainly superior to waterfall or cowboy development using unstructured methods and low-level languages. But even the best current methods continue to involve error-prone custom designs and labor-intensive manual coding.

Suggested Readings on Software Project Management

Abran, A. and Robillard, P.N.; Function point analysis: An empirical study of its measurement processes; *IEEE Transactions on Software Engineering*; 22, 12; 1996; pp. 895–909.

Abrain, A.; *Software Maintenance Management: Evolution and Continuous Improvement*; Wiley-IEEE Computer Society; Los Alamitos, CA; 2008.

Abrain, A.; *Software Metrics and Metrology*; Wiley-IEEE Computer Society; Los Alamitos, CA; 2010.

Abrain, A.; *Software Estimating Models*; Wiley-IEEE Computer Society; Los Alamitos, CA; 2015.

Baird, L.M. and Brennan, M.C.; *Software Measurement and Estimation: A Practical Approach*; IEEE Computer Society Press, Los Alamitos, CA; John Wiley & Sons, Hoboken NJ; 2006; ISBN 0-471-67622-5; 257 pages.

Boehm, B. Dr.; *Software Engineering Economics*; Prentice Hall, Englewood Cliffs, NJ; 1981; 900 pages.

Boehm, B. et al.; *Software Cost Estimating with Cocomo II*; Prentice Hall, Upper Saddle River, NJ; 2000; ISBN-10 0137025769; 544 pages.

Brooks, F.: *The Mythical Man-Month,* Addison-Wesley, Reading, MA; 1974, rev. 1995.

Bundshuh, M. and Dekkers, C.; *The IT Measurement Compendium; Estimating and Benchmarking Success with Functional Size Measurement*; Springer; Berlin, Germany; 2008; ISBN-10 3540681876; 644 pages.

Cohn, M.; *Agile Estimating and Planning*; Prentice Hall, Englewood Cliffs, NJ; 2005; ISBN 0131479415.

Crosby, P.; *Quality Is Free*; New American Library, Mentor Books; New York; 1979, 270 pages.

DeMarco, T.; *Why Does Software Cost So Much*? Dorset House, New York; 1995; ISBN 0-9932633-34-X; 237 pages.

Ebert, C.; Dumke, R. and Schmeitendorf, A.; *Best Practices in Software Measurement*; Springer; Berlin, Germany; 2004; ISBN-10 3540208674; 344 pages.

Fleming, Q.W. and Koppelman, J.M.; *Earned Value Project Management*; 2nd edition; Project Management Institute, New York; 2000; ISBN 10 1880410273; 212 pages.

Galorath, D.D. and Evans, M.W.; *Software Sizing, Estimation, and Risk Management: When Performance Is Measured, Performance Improves*; Auerbach, Philadelphia, PA; 2006; ISBN 10-0849335930; 576 pages.

Gack, G.; *Managing the Black Hole: The Executive's Guide to Project Risk*; Business Expert Publishing; New York; 2010.

Garmus, D. and Herron, D.; *Function Point Analysis*; Addison-Wesley, Boston, MA; 2001; ISBN 0-201069944-3; 363 pages.

Garmus, D.; Russac J. and Edwards, R.; *Certified Function Point Counters Examination Guide*; CRC Press; Boca Raton, FL; 2010.

Garmus, D. and Herron, D.; *Measuring the Software Process: A Practical Guide to Functional Measurement*; Prentice Hall, Englewood Cliffs, NJ; 1995.

Gilb, T. and Graham, D.; *Software Inspections*; Addison-Wesley, Reading, MA; 1993; ISBN 10: 0201631814.

Glass, R.L.; *Software Runaways: Lessons Learned from Massive Software Project Failures*; Prentice Hall, Englewood Cliffs, NJ; 1998.

Harris, M.; Herron, D. and Iwanicki, S.; *The Business Value of IT: Managing Risks, Optimizing Performance, and Measuring Results*; CRC Press (Auerbach), Boca Raton, FL; 2008; ISBN 13: 978-1-4200-6474-2; 266 pages.

Hill, P.; Jones C. and Reifer, D.; *The Impact of Software Size on Productivity*; International Software Standards Benchmark Group (ISBSG), Melbourne, Australia; 2013.

Humphrey, W.; *Managing the Software Process*; Addison-Wesley, Reading, MA; 1989.

International Function Point Users Group (IFPUG); *IT Measurement: Practical Advice from the Experts*; Addison-Wesley Longman, Boston, MA; 2002; ISBN 0-201-74158-X; 759 pages.

Jacobsen, I., Griss, M. and Jonsson, P.; *Software Reuse-Architecture, Process, and Organization for Business Success*; Addison-Wesley Longman, Reading, MA; 1997; ISBN 0-201-92476-5; 500 pages.

Jacobsen, I. et al; *The Essence of Software Engineering; Applying the SEMAT Kernel*; Addison-Wesley Professional, Reading, MA; 2013.

Johnson, J. et al.; *The Chaos Report*; The Standish Group, West Yarmouth, MA; 2000.

Jones, C.; *Program Quality and Programmer Productivity*; IBM Technical Report TR 02.764, IBM San Jose, CA; 1977.

Jones, C.; *Patterns of Software System Failure and Success*; International Thomson Computer Press, Boston, MA; 1995; 250 pages; ISBN 1-850-32804-8; 292 pages.

Jones, C.; Sizing up software; *Scientific American* 279, 6; 1998; pp. 104–109.

Jones, C.; *Software Assessments, Benchmarks, and Best Practices*; Addison-Wesley Longman, Boston, MA, 2000; 659 pages.

Jones, C.; Why flawed software projects are not cancelled in time; *Cutter IT Journal*; 10, 12; 2003; pp. 12–17.

Jones, C.; *Conflict and Litigation between Software Clients and Developers*; Version 6; Software Productivity Research, Burlington, MA; 2006; 54 pages.

Jones, C.; Software project management practices: Failure versus success; *Crosstalk*; 19, 6; 2006; pp. 4–8.

Jones, C.; *Estimating Software Costs*; McGraw-Hill, New York; 2007; ISBN 13-978-0-07-148300-1.

Jones, C.; *Applied Software Measurement*; 3rd edition; McGraw-Hill; New York; 2008; ISBN 978-0-07-150244-3; 668 pages.

Jones, C.; *Software Engineering Best Practices*; McGraw-Hill, New York; 2009; ISBN 978-0-07-162161-8; 660 pages.

Jones, C.; *The Technical and Social History of Software Engineering;* Addison-Wesley; Reading, MA; 2014.

Jones, C. and Bonsignour, O.; *The Economics of Software Quality*; Addison-Wesley; Reading, MA; 2011.

Kan, S.H.; *Metrics and Models in Software Quality Engineering,* 2nd edition; Addison-Wesley Longman, Boston, MA; 2003; ISBN 0-201-72915-6; 528 pages.

Kaplan, R.S. and Norton, D. B.; *The Balanced Scorecard*; Harvard University Press, Boston, MA; 2004; ISBN 1591391342.

Love, T.; *Object Lessons: Lessons Learned in Object-Oriented Development Projects*; SIG Books Inc., New York; 1993; ISBN 0-9627477-3-4; 266 pages.

McConnell, S.; *Software Estimation: Demystifying the Black Art*; Microsoft Press, Redmond, WA; 2006; ISBN 10: 0-7356-0535-1.

Parthasarathy, M.A.; *Practical Software Estimation: Function Point Methods for Insourced and Outsourced Projects*; Addison-Wesley, Boston, MA; 2007; ISBN 0-321-43910-4; 388 pages.

Paulk M. et al.; *The Capability Maturity Model; Guidelines for Improving the Software Process*; Addison-Wesley, Reading, MA; 1995; ISBN 0-201-54664-7; 439 pages.

Pressman, R.; *Software Engineering: A Practitioner's Approach*; McGraw-Hill, New York; 1982.

Putnam, L.H.; *Measures for Excellence: Reliable Software On-Time within Budget*; Yourdon Press, Prentice Hall, Englewood Cliffs, NJ; 1992; ISBN 0-13-567694-0; 336 pages.

Putnam, L. and Myers, W.; *Industrial Strength Software: Effective Management Using Measurement*; IEEE Press, Los Alamitos, CA; 1997; ISBN 0-8186-7532-2; 320 pages.

Robertson, S. and Robertson, J.; *Requirements-Led Project Management*; Addison-Wesley, Boston, MA; 2005; ISBN 0-321-18062-3.

Roetzheim, W.H. and Beasley, R.A.; *Best Practices in Software Cost and Schedule Estimation*; Prentice Hall, Saddle River, NJ; 1998.

Royce, W.; *Software Project Management: A Unified Framework*; Addison-Wesley; 1998.
Strassmann, P.; *The Squandered Computer*; Information Economics Press, Stamford, CT; 1997.
Strassmann, P.; *Governance of Information Management: The Concept of an Information Constitution*; 2nd edition; (eBook); Information Economics Press, Stamford, CT; 2004.
Stutzke, R.D.; *Estimating Software-Intensive Systems: Projects, Products, and Processes*; Addison-Wesley, Boston, MA; 2005; ISBN 0-301-70312-2; 917 pages.
Weinberg, Dr. G.; *Quality Software Management: Volume 2 First-Order Measurement*; Dorset House Press, New York; 1993; ISBN 0-932633-24-2; 360 pages.
Wiegers, K.A; *Creating a Software Engineering Culture*; Dorset House Press, New York; 1996; ISBN 0-932633-33-1; 358 pages.
Wiegers, K.E.; *Peer Reviews in Software: A Practical Guide*; Addison-Wesley Longman, Boston, MA; 2002; ISBN 0-201-73485-0; 232 pages.
Whitehead, R.; *Leading a Development Team*; Addison-Wesley, Boston, MA; 2001; ISBN 10: 0201675267; 368 pages.
Yourdon, E.; *Death March: The Complete Software Developer's Guide to Surviving "Mission Impossible" Projects*, Prentice Hall, Upper Saddle River, NJ; 1997; ISBN 0-13-748310-4.
Yourdon, E.; *Outsource: Competing in the Global Productivity Race*; Prentice Hall, Upper Saddle River, NJ; 2004; ISBN 0-13-147571-1; 251 pages.

Suggested Websites

http://www.IASAhome.org: This is the website for the nonprofit International Association of Software Architects (IASA). Software architecture is the backbone of all large applications. Good architecture can lead to applications whose useful life expectancy is 20 years or more. Questionable architecture can lead to applications whose useful life expectancy is <10 years, coupled with increasing complex maintenance tasks and high defect levels. The IASA is working hard to improve both the concepts of architecture and the training of software architects via a modern and extensive curriculum.

http://www.IIBA.org: This is the website for the nonprofit International Institute of Business Analysis (IIBA). This institute deals with the important linkage between business knowledge and software that supports business operations. Among the topics of concern are the Business Analysis Body of Knowledge (BABOK), training of business analysts, and certification to achieve professional skills.

http://www.IFPUG.org: This is the website for the nonprofit International Function Point Users Group (IFPUG). IFPUG is the largest software metrics association in the world, and the oldest association of function point users. This website contains information about IFPUG function points themselves, and also citations to the literature dealing with function points. IFPUG also offers training in function point analysis and administers. IFPUG also administers a certification program for analysts who wish to become function point counters.

http://www.ISBSG.org: This is the website for the nonprofit International Software Benchmarking Standards Group (ISBSG). ISBSG, located in Australia, collects benchmark data on software projects throughout the world. The data is self-reported by companies using a standard questionnaire. About 4000 projects comprise the ISBSG collection as of 2017, and the collection has been growing at a rate of about 500 projects per year. Most of the data is expressed in terms of IFPUG function point metrics, but some of the data is also expressed in terms of COSMIC function points, NESMA function points, Mark II function points, and several other function point variants. Fortunately, the data in variant metrics is identified. It would be statistically invalid to include attempts to average IFPUG and COSMIC data, or to mix up any of the function point variations.

http://www.iso.org: This is the website for the International Organization for Standardization (ISO). The ISO is a nonprofit organization that sponsors and publishes a variety of international standards. As of 2007, the ISO published about 1,000 standards a year, and the total published to date is ~17,000. Many of the published standards affect software. These include the ISO 9000–9004 quality standards and the ISO standards for functional size measurement.

http://www.namcook.com: This website contains a variety of quantitative reports on software quality and risk factors. It also contains a patented high-speed sizing tool that can size applications of any size in 90 s or less. It also contains a catalog of software benchmark providers that currently lists 20 organizations that provide quantitative data about software schedules, costs, quality, and risks.

http://www.PMI.org: This is the website for the Project Management Institute (PMI). PMI is the largest association of managers in the world. PMI performs research and collects data on topics of interest to managers in every discipline: software, engineering, construction, and so forth. This data is assembled into the well-known Project Management Body of Knowledge or PMBOK.

http://www.ITMPI.org: This is the website for the Information Technology Metrics and Productivity Institute. ITMPI is a wholly-owned subsidiary of Computer Aid Inc. The ITMPI website is a useful portal with a broad range of measurement, management, and software engineering information. The ITMPI website also provides useful links to many other websites that contain topics of interest on software issues.

http://www.sei.cmu.edu: This is the website for the Software Engineering Institute (SEI). The SEI is a federally sponsored nonprofit organization located on the campus of Carnegie Mellon University in Pittsburgh, PA. The SEI carries out a number of research programs dealing with software maturity and capability levels, with quality, risks, measurement and metrics, and other topics of interest to the software community.

www.Crosstalkonline.org This is the website of both the Air Force Software Technology Support Center (STSC) and also the CrossTalk journal, which is published by the STSC. The STSC gathers data and performs research into a wide variety of software engineering and software management issues. The CrossTalk journal is one of the few technical journals that publish full-length technical articles of 4000 words or more. Although the Air Force is the sponsor of STSC and CrossTalk, many topics are also relevant to the civilian community. Issues such as quality control, estimation, maintenance, measurement, and metrics have universal relevance.

Chapter 6

50 Years of Global Software Benchmark Results

Introduction

As this chapter is written, it is calendar year 2017. The author's first benchmark collection of software data took place while working at IBM in the year 1967, some 50 years ago. Quite a lot has changed in software since 1967:

- Thousands of new programming languages have been created.
- Dozens of new software development methodologies have been created.
- Global software personnel have increased from thousands to millions.
- Software applications have increased from hundreds to millions.
- Software has become a critical technology for all companies and governments.
- Cyber-attacks and cyber-crime are new threats to citizens, companies, and governments.

Unfortunately for the software industry, quite a few topics have not changed, or at least not changed far enough to have significant benefits for software projects:

- Canceled projects still top 35% for large applications >5000 function points.
- Schedule delays occur for more than 70% of large applications >5000 function points.
- Cost overruns occur for more than 50% of large applications >5000 function points.

- Average productivity hovers around 8 function points per staff month.
- Defect removal efficiency (DRE) remains below 92% globally.
- Software metrics and measurements remain incompetent and unprofessional.
- Manual estimates remain optimistic, although parametric estimates are accurate.

This chapter summarizes the author's benchmark data collection from more than 26,000 projects derived from more than 400 companies and 50 government organizations in about 27 countries gathered between 1967 and 2017. Some data are also used from other sources, such as other benchmark consulting groups who also publish data using function point metrics.

During peak data collection years, the author had a dozen consultants bringing in benchmark data from more than 75 projects per month. The author also had the good fortune to have access to thousands of software projects recorded by several leading companies such as IBM and ITT while working for them. Both companies had large volumes of accurate data. ITT was one of the very few companies to commission an international portfolio benchmark study of all major software projects in all software locations.

When collecting benchmarks for other Fortune 500 companies, the author also had access to many thousands of their internal software benchmark data collections. However, some of this data was of questionable accuracy.

The purpose of this chapter is to show the readers software productivity and quality results with consistent and accurate metrics such as International Function Point Users Group (IFPUG) function points, defect potentials, and DRE.

Measuring U.S. Software Productivity and Quality

The author and his colleagues have collected on-site benchmark data from 27 countries, and remote data via Skype or webinar tools from 19 other countries—46 countries in total. However, about 70% of the author's total data comes from U.S. corporations and government agencies.

The majority of the author's clients have been Fortune 500 companies employing between 5,000 and more than 150,000 software personnel. Our data are somewhat sparse for small companies with less than 100 software personnel.

Typical large U.S. clients include AT&T, Boeing, IBM, Microsoft, GTE, Aetna Insurance, ITT, Raytheon, Motorola, Hartford Insurance, Bank of America, Chase, American Express, General Motors, Ford, Lockheed Martin, McKesson, Sears, J.C. Penney, COSTCO, etc. We have had about a dozen telecommunication companies and a dozen banks as clients and quite a few technology companies such as CISCO and Intel.

(International clients include Barclays Bank, Bell Northern Research, British Airways, Alcatel, Ericsson, Huawei, Kozo Keikaku, Nippon Telephone, Nokia,

Siemens, Scandinavian Airlines, Sony, the governments of Malaysia, South Korea, and Thailand, and the government of Quebec.)

The majority of our U.S. government clients are large: the Department of Defense, the U.S. Air Force, the U.S. Navy, the Internal Revenue Service, NASA, and the Department of Health, Education, and Welfare, and over a dozen state governments such as California, New York, South Carolina, Kansas, and Rhode Island.

We have nondisclosure agreements with our clients, so we can't divulge specific results. This is one of the reasons why our benchmarks are usually aggregated from dozens of projects.

Unfortunately, small companies and small government groups can't afford to hire professional benchmark services and tend to use self-reported data, which is less accurate. We do acquire and consider self-reported benchmark data from groups such as the International Software Benchmark Standards Group (ISBSG), but of course we prefer our own benchmark data since it is more complete.

Because our clients are large companies and government groups that build large software applications, we have many good data points for large systems between 10,000 and 250,000 function points in size.

We are seldom hired to collect benchmark data below a size of about 1000 function points. Most companies don't have any serious problems with small applications, so they concentrate on the high end where problems are numerous and severe.

Out of about a total of 26,000 software projects examined, the average size is about 3,500 function points in size. The smallest projects are about 100 function points in size, and these are mainly done for colleagues who made special requests. We have examined several very large civilian and military projects that top 300,000 function points. These are definitely outliers and are usually in serious trouble.

As examples of large applications, defense software such as Star Wars antimissile software tops 350,000 function points and so does the Worldwide Military Command and Control System. Enterprise resource planning (ERP) packages such as SAP and Oracle top 250,000 function points for all features. Operating systems such as Windows and IBM's mainframe operating system top 100,000 function points. But such large applications comprise <3% of all software.

By contrast, the average size of all U.S. software applications when small companies and large companies are both in the mix is probably around 450 function points. Small projects far outnumber large applications in the United States and indeed in every country. This is probably a good thing because large software projects are very risky and seldom successful.

Even though we don't measure many small projects ourselves, our clients do and they provide us with their own internal benchmark data. For example, among our clients, an average agile project is about 275 function points in size, and disciplined agile delivery projects are about 1200 function points in size. Agile seldom goes as high as 5000 function points and is rarely successful when it does.

Also among client-collected data are Team Software Process (TSP) projects that average about 2500 function points in size and run up above 20,000 function points. DevOps projects are about 600 function points in size. Container projects are about 500 function points in size. Waterfall projects average about 1000 function points in size. Of course, there are big ranges for all kinds of software.

Our overall benchmark data is very strong for large systems and large applications above 1,000 function points and up through 100,000 function points. Our benchmark data is less complete below 500 function points, but we have still examined over 800 small projects below 500 function points in size.

Table 6.1 shows the total number of software projects examined by the author and his team between 1967 and 2017 sorted by application size. This table is global and includes data from all countries.

As can be seen from the table, the author and his colleagues are usually commissioned for applications more than 1000 function points in size.

Readers may also be interested in the chronological distribution since the author's data run from 1967 through 2017 (Table 6.2).

As can be seen from the table, the peak data collection period was 1987–1996. However, there are still more than 10,000 newer projects to be examined out of the total of more than 26,000 projects.

Life Expectancy of Software Benchmark Data

Because the author has old data as well as new data, it is interesting to consider the life expectancy of historical software data. Once benchmark data is collected and entered into a database, how long will it be useful? In other words, is data collected in 1997 still relevant in 2017?

Table 6.1 Application Benchmark Size Distribution (1967 through 2017)

Function Points	*Projects Examined*	*Percentage of Projects (%)*
<10	500	1.87
10–100	3,500	13.06
100–1,000	10,000	37.31
1,000–10,000	12,000	44.78
10,000–100,000	725	2.80
>100,000	75	0.19
Total	26,800	100.00

Table 6.2 Application Benchmark Time Distribution (1967 through 2017)

	Decade	*Percentage (%)*
1967–1976	3,000	11.19
1977–1986	5,000	18.66
1987–1996	8,000	29.85
1997–2006	7,000	26.12
2007–2016	3,800	14.18
Total	26,800	100.00

As it happens, the value of historical benchmark data is just about as good as the value of historical medical records. Historical data allows longitudinal studies that can show improvements over long time spans. This is why medical data can show life expectancies of human beings from the 1700s through modern times: medicine has been keeping good data for hundreds of years.

It is interesting that waterfall projects of 1000 function points developed in 1985, 1990, 1995, 2000, 2005, and 2010 are remarkably similar in their productivity and quality rates and resemble waterfall projects delivered in early 2017. On the other hand, projects using Agile, Rational Unified Process (RUP), and TSP all show higher productivity rates and better quality levels than waterfall projects at any year.

Thus, historical data can reveal interesting trends about the effectiveness of new methods by allowing them to be compared against historical records for similar classes and types of applications from earlier decades.

When comparing modern projects developed in 2017, it is possible to limit the years for comparative projects and have a cutoff point of perhaps 2010. However, calendar time is less significant than methodology. If you want to compare a new Agile project against others, you only need to search on "Agile" because there are no old Agile projects as yet. The database itself would contain no Agile entries prior to 2000 because the methodology did not exist.

Once benchmark data is collected, it will always have value. But benchmark data collections need to be refreshed with new projects on a monthly basis. To keep track of new tools, languages, and methods, monthly benchmark data for the United States should increase by about 250 projects per month in round numbers.

For specialized benchmarks such as those that deal with special topics—embedded software or Agile development, benchmark volumes should grow at about 25–75 projects per month in round numbers.

The author would like to see cooperation and data sharing among the major software benchmark collection groups such as Quantitative Software Management

(QSM), ISBSG, Q/P Management Group, David Consulting Group, and the author's own Namcook Analytics. There are a few cooperative studies and reports, but unfortunately the benchmark groups are also competitors and therefore not totally open with data other than to their own paying clients.

It might be of interest to show readers what software productivity has looked like over the past 50 years. Table 6.3 shows aggregate results For every decade from 1960 through 2015, although the author has included data through early 2017. It uses the metric of function points per staff month and is set for an average of 132 work hours per calendar month.

It is obvious that many U.S. companies and many software engineers work for more than 132 h per month, but for national data including thousands of companies and millions of software personnel, a constant value needs to be used.

Function point counting rules have changed over the years, so Tables 6.3 and 6.4 are based on IFPUG 4.3. The Software Nonfunctional Assessment Process (SNAP) metric for nonfunctional requirements is not included.

Table 6.4 shows the same information as Table 6.3, but is expressed in terms of "work hours per function point" using IFPUG function point metrics.

The two metrics have similar purposes, but differ based on circumstances. Work hours per function point can be used in any country, for any project, and even for any activity within a project. It provides a stable base metric for benchmark data.

Function points per month is mathematically related to work hours per function point, but has a major difference in possible results. Suppose Programmer John works on 10 function points for one calendar month and works 150 h. Productivity using work hours per function point is of course 15. Productivity using function points per month would be 10.

On the other hand, suppose Programmer Alice works on a similar 10-function point project and also uses 150 h. She puts in lots of overtime and finishes the project in two calendar weeks. Like Programmer John, Alice has a productivity rate of 15 work hours per function point. But her productivity using function points per month would be 20 because she finishes in 2 weeks due to unpaid overtime.

Because work hours vary from country to country, industry to industry, and company to company, the two metrics are often quite different. Work hours per function point is more stable for global comparisons. But function points per staff month is also useful because it highlights speed differences in software development.

Apologies to readers for having to use 9-point type. Unfortunately, there is no other good solution for showing this data on a printed page in a portrait format.

Note that some forms of software, such as Web applications, did not appear until the 1990s. The oldest forms of software are management information systems (MIS) applications, systems and embedded software, and military software, all of which began in the late 1950s and the early 1960s. Commercial software barely

Table 6.3 U.S. Productivity 1960–2015 Using IFPUG Function Points per Staff Month

Decade	*MIS Projects*	*Web Projects*	*Domestic Outsourced Projects*	*Systems and Embedded Projects*	*Commercial Projects*	*Civilian Government Projects*	*Military Projects*	*Average*
1960–1969	4.50	0.00	0.00	5.20	4.55	3.90	3.25	3.06
1970–1979	6.30	0.00	0.00	6.50	6.50	5.20	4.23	4.10
1980–1989	7.40	0.00	0.00	7.15	7.80	5.85	4.88	4.73
1990–1999	8.40	10.40	8.95	8.40	9.43	6.75	5.75	8.30
2000–2010	9.05	13.50	9.10	8.70	10.50	7.25	6.30	9.20
2011–2015	9.60	14.30	9.85	8.95	11.40	7.30	6.45	9.69
Average	7.54	12.73	9.30	7.48	8.36	6.04	5.14	8.09

Table 6.4 U.S. Productivity 1960–2015 Using Work Hours per IFPUG Function Point

Decade	*MIS Projects*	*Web Projects*	*Domestic Outsourced Projects*	*Systems and Embedded Projects*	*Commercial Projects*	*Civilian Government Projects*	*Military Projects*	*Average*
1960–1969	29.33	0.00	0.00	25.38	29.01	33.85	40.62	43.18
1970–1979	20.95	0.00	0.00	20.31	20.31	25.38	31.24	32.17
1980–1989	17.84	0.00	0.00	18.46	16.92	22.56	27.08	27.94
1990–1999	15.71	12.69	14.75	15.71	14.01	19.56	22.96	15.91
2000–2010	14.59	9.78	14.51	15.17	12.57	18.21	20.95	14.35
2011–2015	13.75	9.23	13.40	14.75	11.58	18.08	20.47	13.62
Average	17.50	10.37	14.19	17.64	15.78	21.85	25.67	17.57

existed until after the IBM antitrust suit and did not explode until the 1980s. Outsource software barely existed in the 1960s but has been expanding rapidly since the 1990s.

Productivity improved between 1960 and 2017 due to a synergistic combination of factors that include better programming languages, maturing experience of software personnel, and more reusable software components, especially for Web applications.

Software quality has also changed over the decades from 1960 through 2017 although it still is far worse than it should be.

The software industry does change and therefore benchmarks need to be identified by the year of project completion. Table 6.5 illustrates approximate software quality averages from the 1960s through today, with projections out to 2030.

As can be seen from the table, there are continuous changes in software results over time. However, the software industry resembles a drunkard's walk, with about as many companies going backward and getting worse as there are companies improving and getting better.

In fact, if every company got better every year, assessments and benchmarks would probably not be needed. It is the combination of regressions and advances that make benchmarks valuable for companies and government agencies.

Currently, in 2017, computers and software are the main engines that power both government operations and all businesses. Software is also produced in more than 200 countries. More than 50 countries produce software under outsource contracts.

Table 6.5 Software Quality Levels by Decade

Era	*Languages*	*Methods*	*Defect Potential*	*Defect Removal (%)*	*Delivered Defects*
1960s	Assembly	Cowboy	6.00	83	1.02
1970s	COBOL/ FORTRAN	Waterfall	5.50	85	0.83
1980s	C, Ada	Structured	5.00	85	0.75
1990s	COBOL/ FORTRAN	Object Oriented	4.50	87	0.59
2000s	Java, C#	Agile/RUP/TSP	4.25	90	0.43
2010s	PHP/mySQL/ JavaScript	Agile/RUP/TSP/ Mashups	4.00	93	0.28
2020s	Unknown	Mashups/75% reuse	3.50	96	0.14
2030s	Unknown	SEMAT/90% reuse	2.50	99	0.03

U.S. Software Benchmark Results

The tables in this chapter show aggregate U.S. software development productivity rates. The topics include schedules, staffing, effort, and productivity in terms of both work hours per function point and function points per month. They are based on about 20,000 domestic projects out of our total of 26,800 (Table 6.6).

Now we look at software personnel staffing for the same sample. Our staffing data include many occupation groups including business analysts, architects, software engineers, test specialists, technical writers, quality assurance, project managers and project office staff, etc. Benchmarks for software should include all personnel and not just programmers, designers, and testers.

Next, we show project effort in total staff months for all occupation groups (Tables 6.7 and 6.8).

Now that schedules, staffing, and effort data have been presented, we can show productivity using function points per staff month and work hours per function point. Function points per staff month is the most common and hence it is shown first (Table 6.9).

Work hours per function point is the most stable metric available, and can be used for projects developed anywhere in the world. However, it tends to conceal unpaid overtime unless that is broken out and displayed separately, which seldom occurs (Table 6.10).

Our benchmarks for clients always include both function points per month and work hours per function point. The former metric of function points per month shows the impact of unpaid overtime and working long days. The latter metric of work hours per function point shows the actual amount of effort needed to construct software.

Assume that two teams are both building 10–function point components. Team A works 8 h per day for 12 days or 96 h. Team B works 12 h per day for 8 days or 96 h, with 4 h per day being unpaid overtime. Both teams clearly spent the same amount of actual effort but Team B's schedule was shorter by one-third.

Work hours per function point for both teams was 9.6 h. When the view changes to function points per month, Team A's rate was 13.75 function points per month while that of Team B was 17.50 function points per month.

Assume that the burdened compensation of both teams was $100 per hour. The cost for the component by Team A was $9600 or $960 per function point. The second team put in 4 h per day of unpaid overtime, so the Team B cost was only $6400 or $640 per function point.

The same issues could occur with the older lines of code (LOC) metric. Assume both versions were 500 logical Java statements in size. The cost per LOC for Team A would be $19.20, while the cost per LOC for Team B would be only $12.80.

As can be seen to understand software economics, both metrics are useful. Because software is the most labor-intensive of all engineering fields, the metric of work hours per month is an important metric. But since software development

Table 6.6 Average Software Development Schedules in Calendar Months by Type and Size of Software Project

	MIS Projects	*Web Projects*	*Domestic Outsourced Projects*	*Systems and Embedded Projects*	*Commercial Projects*	*Civilian Government Projects*	*Military Projects*	*Average*
Size in FP								
1	0.03	0.02	0.03	0.03	0.03	0.04	0.04	0.03
10	0.35	0.28	0.33	0.60	0.65	0.66	0.70	0.51
100	4.00	3.25	3.50	4.50	4.00	4.75	5.00	4.14
1,000	13.00	10.00	11.00	14.00	13.00	14.50	16.50	13.14
10,000	44.00	37.00	42.00	41.00	42.00	46.00	48.00	42.86
100,000	62.00	56.00	58.00	57.00	58.00	63.00	66.00	60.00
Average	20.56	17.76	19.14	19.52	19.61	21.49	22.71	20.11

Note: Schedules run from start of requirements until delivery to initial customers.

Table 6.7 Average Staff Headcount by Type and Size of Software Projects

	MIS Projects	*Web Projects*	*Domestic Outsourced Projects*	*Systems and Embedded Projects*	*Commercial Projects*	*Civilian Government Projects*	*Military Projects*	*Average*
Size in FP								
1	1.00	1.00	1.00	1.00	1.00	1.00	1.00	1.00
10	1.00	1.00	1.00	1.00	1.00	1.05	1.20	1.04
100	1.33	1.54	1.33	1.43	1.41	1.60	1.75	1.48
1,000	5.88	5.75	6.06	6.10	6.33	6.70	7.04	6.27
10,000	61.73	59.52	60.98	62.50	64.10	68.00	72.00	64.12
100,000	632.91	625.00	617.28	641.03	628.93	649.35	675.68	638.60
Average	117.31	115.63	114.61	118.84	117.13	121.28	126.44	118.75

Note: All technical personnel and project managers have been assumed.

Table 6.8 Average Effort in Person Months by Type and Size of Software Projects

	MIS Projects	*Web Projects*	*Domestic Outsource Projects*	*Systems and Embedded Projects*	*Commercial Projects*	*Civilian Government Projects*	*Military Projects*	*Average*
Size in FP								
1	0.03	0.02	0.03	0.03	0.03	0.04	0.04	0.03
10	0.35	0.28	0.33	0.60	0.70	0.69	0.84	0.54
100	5.33	5.00	4.67	6.43	5.63	7.60	8.75	6.20
1,000	76.47	57.47	66.67	85.37	82.28	97.15	116.20	83.09
10,000	2,716.05	2,202.38	2,560.98	2,562.50	2,692.31	3,128.00	3,456.00	2,759.74
100,000	39,240.51	35,000.00	35,802.47	36,538.46	36,477.99	40,909.09	44,594.59	38,366.16
Average	7,006.46	6,210.86	6,405.86	6,532.23	6,543.16	7,357.09	8,029.40	6,869.29

Note: All technical personnel and project managers have been assumed.

Table 6.9 Average Productivity Rates in Function Points per Staff Month by Type and Size of Software Project

	MIS Projects	*Web Projects*	*Domestic Outsourced Projects*	*Systems and Embedded Projects*	*Commercial Projects*	*Civilian Government Projects*	*Military Projects*	*Average*
Size in FP								
1	33.33	50.00	33.33	33.33	33.33	28.57	25.00	*33.84*
10	28.57	36.36	30.30	16.67	14.29	14.43	11.90	*21.79*
100	18.75	21.67	21.43	15.56	17.75	13.16	8.14	*16.64*
1,000	14.17	17.40	15.00	11.71	12.15	10.33	7.89	*12.67*
10,000	3.68	4.54	3.90	3.90	3.71	3.37	3.13	*3.75*
100,000	2.55	2.86	2.79	2.74	2.74	2.44	2.24	*2.62*
Average	*16.84*	*22.14*	*17.79*	*13.98*	*14.00*	*12.05*	*9.72*	*15.22*

Note: Complete life cycle from the start of requirements to delivery has been assumed. Technical workers and managers have been assumed.

Table 6.10 Average Productivity Rates in Work Hours per Function Point by Type and Size of Software Project

	MIS Projects	*Web Projects*	*Domestic Outsource Projects*	*Systems and Embedded Projects*	*Commercial Projects*	*Civilian Government Projects*	*Military Projects*	*Average*
Size in FP								
1	3.96	2.64	3.96	3.96	3.96	4.62	5.28	4.05
10	4.62	3.63	4.36	7.92	9.24	9.15	11.09	7.14
100	7.04	6.09	6.16	8.49	7.44	10.03	16.21	8.78
1,000	9.32	7.59	8.80	11.27	10.86	12.77	16.73	11.05
10,000	35.85	29.07	33.80	33.83	35.54	39.17	42.24	35.64
100,000	51.80	46.20	47.26	48.23	48.15	54.00	58.86	50.64
1,000,000	62.62	55.07	58.48	57.91	59.53	67.74	71.35	61.81
Average	25.03	21.47	23.26	24.51	24.96	28.21	31.68	25.59

Note: Complete life cycle from start of requirements to delivery is assumed. Technical workers and managers are assumed.

costs and schedules are important too, function points per month are also an important metric.

It is an unfortunate fact of life that the majority of U.S. companies do not record unpaid overtime. As a result, benchmarks with large volumes of invisible unpaid overtime seem to have higher productivity than the same-size applications where the team worked normal hours with zero unpaid overtime. The fact that unpaid overtime is both common and usually not recorded makes software benchmarks difficult. This is why we always check with development teams about unpaid overtime and ask them to reconstruct overtime patterns if their tracking systems did not show it.

Because costs are also important for software economic analysis, the final table in this section shows cost per function point.

Note an important aspect of Table 6.11. The compensation levels and overhead percentages vary widely among the seven types of software. The lowest burdened monthly cost is $9,750 and the highest is $13,725. Here too benchmarks can be tricky and need to include cost structures.

Software Cost Drivers

The phrase "software cost drivers" was first introduced by Dr. Barry Boehm in the 1980s. Software cost drivers are synthetic aggregate metrics that accumulate costs into a small number of "cost buckets."

Cost drivers are usually listed in rank order with the most expensive factors at the top of the list. The author of this book uses cost drivers to show C-level executives the most essential cost factors of software systems.

Cost drivers should not be confused with "activity-based costs." The cost drivers are artificial collections of similar costs. For example, the cost driver of "producing paper documents" can actually include the costs of over 50 kinds of documents: requirements, architecture, designs, user manuals, training manuals, slide shows, plans, etc.

Although cost drivers are small in numbers (the author uses 10), they should not be confused with "phases," which was an early attempt to show software costs chronologically: (1) requirements phase, (2) design phase, (3) code phase, (4) test phase, and (5) deployment phase.

Phases are much less accurate and effective than either cost drivers or activity-based costs. There are many activities such as project management and quality assurance that span multiple phases. There are other activities such as business analysis and user documentation that are embedded with phases and are not really visible without activity-based costs or at least cost drivers.

The author uses a set of 10 cost drivers for a variety of educational purposes. Cost drivers are simple and easy to understand, but they can convey useful information.

Table 6.11 Average $ per Function Point by Type and Size of Software Projects

	MIS Projects	*Web Projects*	*Domestic Outsourced Projects*	*Systems and Embedded Projects*	*Commercial Projects*	*Civilian Government Projects*	*Military Projects*	*Average*
Compensation	$6,500	$9,200	$5,750	$9,000	$6,300	$6,200	$6,100	$7,007
Burden rate (%)	50	55	70	50	60	75	125	69
Burdened cost	$9,750	$14,260	$9,775	$13,500	$10,080	$10,850	$13,725	$11,706
Size in FP								
1	$292.50	$285.20	$293.25	$405.00	$302.40	$379.75	$549.00	$358.16
10	$341.25	$392.15	$322.58	$810.00	$705.60	$751.91	$1,152.90	$639.48
100	$520.00	$713.00	$456.17	$867.86	$567.89	$824.60	$1,200.94	$735.78
1,000	$745.59	$819.54	$651.67	$1,152.44	$829.37	$1,054.08	$1,594.81	$978.21
10,000	$2,648.15	$3,140.60	$2,503.35	$3,459.38	$2,713.85	$3,393.88	$4,743.36	$3,228.94
100,000	$3,825.95	$4,991.00	$3,499.69	$4,932.69	$3,676.98	$4,438.64	$6,120.61	$4,497.94
Average	$1,396.34	$1,723.58	$1,287.78	$1,937.89	$1,466.01	$1,807.14	$2,560.27	$1,739.75

Note: Burden or overhead rates and average compensation vary by more than 50% in either direction.

Table 6.12 illustrates how cost drivers vary as application size grows from 100 function points to 100,000 function points. Coding costs are the #1 cost driver for small applications, but "finding and fixing bugs" is the #1 cost driver above 1000 function points. One of the advantages of cost drivers is that the sequence displays useful information about relative costs.

As can be observed from the simplicity of cost drivers, application costs go up rapidly as application size increases. But note the more subtle changes in costs with size: the costs of finding and fixing bugs, the costs of deferred features, the costs of requirements changes, and the costs of cyber-attacks rise more steeply than other costs.

Note that although the size of nonfunctional costs might be measured via the new SNAP metric, the costs of these nonfunctional requirements should be added to the overall cost of the application and can be measured by "cost per function point."

There would be no nonfunctional requirements without user features, so they are subordinate costs. In other words, the metric cost per function point can serve as a neutral metric for all costs, just as "cost per square foot" can serve as a neutral metric for all home construction costs such as framing, roofing, plumbing, etc.

The key value of the cost driver concept is that it can show useful information in a simple format that makes the data easier to understand than conventional activity-based costs or earned-value costs.

Phase-Based Costs versus Activity-Based Costs

The first attempt to move beyond single-point benchmarks was to use software phases. The idea of software phases dates to the early 1960s but has always been somewhat ambiguous because of the lack of a standard definition of software phases. One common six-layer phase definition includes:

1. Requirements
2. Design
3. Development
4. Testing
5. Deployment
6. Maintenance

Not all phase definitions include maintenance. Few other common phase definitions noted on the Web cover only development and include the following:

1. Analysis
2. Design
3. Implementation
4. Testing
5. Release

Table 6.12 Software Cost Drivers by Application Size

100 Function Point Cost Drivers	*Percentage*	*Work Hours per FP*	*FP per Month*	*Project Costs*	*Cost per FP*
1) The cost of programming or coding	37.00	4.44	29.73	$33,636	$336.36
2) The cost of finding and fixing bugs	17.00	2.04	64.71	$15,455	$154.55
4) The cost of project management	11.00	1.32	100.00	$10,000	$100.00
5) The cost of producing paper documents	10.00	1.20	110.00	$9,091	$90.91
3) The cost of meetings and communications	9.00	1.08	122.22	$8,182	$81.82
6) The costs of training and learning	6.00	0.72	183.33	$5,455	$54.55
7) The costs of requirements changes	4.00	0.48	275.00	$3,636	$36.36
8) The cost of nonfunctional requirements	4.00	0.48	275.00	$3,636	$36.36
9) The costs of deferred features	1.00	0.12	1,100.00	$909	$9.09
10) Cyber-attack prevention and recovery	1.00	0.12	1,100.00	$909	$9.09
Total	100.00	12.00	11.00	$90,909	$909.09
1,000 Function Point Cost Drivers	*Percentage*	*Work Hours per FP*	*FP per Month*	*Project Costs*	*Cost per FP*
1) The cost of programming or coding	28.00	4.48	29.46	$339,394	$339.39
2) The cost of finding and fixing bugs	19.00	3.04	43.42	$230,303	$230.30
3) The cost of producing paper documents	12.00	1.92	68.75	$145,455	$145.45
4) The cost of project management	12.00	1.92	68.75	$145,455	$145.45
5) The cost of meetings and communications	8.00	1.28	103.13	$96,970	$96.97

(*Continued*)

Table 6.12 (*Continued*) Software Cost Drivers by Application Size

1,000 Function Point Cost Drivers	*Percentage*	*Work Hours per FP*	*FP per Month*	*Project Costs*	*Cost per FP*
6) The cost of requirements changes	7.00	1.12	117.86	$84,848	$84.85
7) The cost of nonfunctional requirements	5.00	0.80	165.00	$60,606	$60.61
8) The costs of training and learning	4.00	0.64	206.25	$48,485	$48.48
9) The costs of deferred features	3.00	0.48	275.00	$36,364	$36.36
10) Cyber-attack prevention and recovery	2.00	0.32	412.50	$24,242	$24.24
Total	100.00	16.00	8.25	$1,212,121	$1,212.12
10,000 Function Point Cost Drivers	*Percentage*	*Work Hours per FP*	*FP per Month*	*Project Costs*	*Cost per FP*
1) The cost of finding and fixing bugs	22.00	5.28	25.00	$4,000,000	$400.00
2) The cost of programming or coding	20.00	4.80	27.50	$3,636,364	$363.64
3) The cost of producing paper documents	12.00	2.88	45.83	$2,181,818	$218.18
4) The cost of project management	12.00	2.88	45.83	$2,181,818	$218.18
5) The cost of requirements changes	8.00	1.92	68.75	$1,454,545	$145.45
9) The costs of deferred features.	8.00	1.92	68.75	$1,454,545	$145.45
7) The cost of meetings and communications	6.00	1.44	91.67	$1,090,909	$109.09
8) The cost of nonfunctional requirements	6.00	1.44	91.67	$1,090,909	$109.09
9) Cyber-attack prevention and recovery	4.00	0.96	137.50	$727,273	$72.73

(*Continued*)

Table 6.12 (*Continued*) Software Cost Drivers by Application Size

10,000 Function Point Cost Drivers	*Percentage*	*Work Hours per FP*	*FP per Month*	*Project Costs*	*Cost per FP*
10) The cost of training and learning	2.00	0.48	275.00	$363,636	$36.36
Total	100.00	24.00	5.50	$18,181,818	$1,818.18
100,000–Function Point Cost Drivers	*Percentage*	*Work Hours per FP*	*FP per Month*	*Project Costs*	*Cost per FP*
1) The cost of finding and fixing bugs	23.00	7.82	16.88	$59,242,424	$592.42
2) The cost of programming or coding	15.00	5.10	25.88	$38,636,364	$386.36
3) The cost of producing paper documents	12.00	4.08	32.35	$30,909,091	$309.09
4) The cost of project management	12.00	4.08	32.35	$30,909,091	$309.09
9) The costs of deferred features	9.00	3.06	43.14	$23,181,818	$231.82
6) The cost of requirements changes	8.00	2.72	48.53	$20,606,061	$206.06
7) The cost of nonfunctional requirements	7.00	2.38	55.46	$18,030,303	$180.30
8) The cost of meetings and communications	6.00	2.04	64.71	$15,454,545	$154.55
9) Cyber-attack prevention and recovery	6.00	2.04	64.71	$15,454,545	$154.55
10) The cost of training and learning	2.00	0.68	194.12	$5,151,515	$51.52
Total	100.00	34.00	3.88	$257,575,758	$2,575.76

The general problems with all phase definitions are that they are simplistic and don't include all major cost elements. For example, some activities such as quality assurance and project management span multiple phases and hence are invisible using phase-based costs. Other activities such as technical user guides are usually in one phase, but not broken out or costed separately from code development.

Instead of phase-based cost analysis, it is far more accurate to use activity-based cost analysis. The number of actual software development activities goes up with application size. Small projects may have <10 activities, whereas large systems may top 50 activities. Activity-based costs need to be matched to the specific activity patterns actually deployed.

For large applications, the author uses a generic 40-activities chart of accounts for benchmark data collection, as shown in Table 6.13, for a sample application of 1000 function points coded in Java.

As can be seen, the granularity of activity-based cost structures far exceeds simplistic phase structures. These details are necessary to really understand software engineering economics. Of course, small applications below 500 function points have much simpler activity patterns.

As stated, these 40 activities are generic. For example, "independent verification and validation (IV&V)" is shown in the table but only occurs for military software projects; civilian projects don't use IV&V. This is why activity patterns need to be matched to specific projects.

The Strange Mystery of Why Software Has 3000 Programming Languages

One of the strangest sociological mysteries of any industry in human history is why software has developed over 3000 programming languages, and new ones are being released at approximately monthly intervals. Adding to the mystery, there is little or no solid research on the value of any known programming language for either software quality or software productivity.

Lots of programming languages are probably developed just because they are fairly easy to develop. But why they are developed does not have a good sociological answer in 2017.

In the early days of programming, circa 1950, it was easy to see why software engineers wanted a better language than machine language, which was both arcane and prolix. Basic assembly language was the initial solution, and it was not bad for the time. In fact, the author started programming with basic assembly in the mid-1960s.

Soon after, other "high-level" languages such as COBOL, FORTRAN, APL, PL/I, ALGOL, ADA, and quite a few more were developed. The first batch of programming languages were in theory general purpose languages. Later, more specialized languages were developed such as Job Control Language (JCL) and

Table 6.13 Example of Software Activity-Based Costs

	Development Activities	*Monthly Function Points per Month*	*Work Hours per Function Point*	*Burdened Cost per Function Point*	*Project Cost at $10,000 per Month*
1	Business analysis	50,000	0.00	$0.21	$2,100.00
2	Risk analysis/sizing	100,000	0.00	$0.11	$1,050
3	Risk solution planning	50,000	0.00	$0.21	$2,100
4	Requirements	450	0.29	$23.33	$233,333
5	Requirement inspection	550	0.24	$19.09	$190,909
6	Prototyping	500	0.26	$21.00	$210,000
7	Architecture	2,500	0.05	$4.20	$42,000
8	Architecture inspection	3,500	0.04	$3.00	$30,000
9	Project plans/estimates	3,500	0.04	$3.00	$30,000
10	Initial design	300	0.44	$35.00	$350,000
11	Detail design	225	0.59	$46.67	$466,667
12	Design inspections	250	0.53	$42.00	$420,000
13	Coding	25	5.28	$420.00	$4,200,000
14	Code inspections	75	1.76	$140.00	$1,400,000
15	Reuse acquisition	75,000	0.00	$0.14	$1,400
16	Static analysis	15,000	0.01	$0.70	$7,000
17	COTS package purchase	25,000	0.01	$0.42	$4,200
18	Open-source acquisition	50,000	0.00	$0.21	$2,100
19	Code security audit	2,000	0.07	$5.25	$52,500
20	IV&V	10,000	0.01	$1.05	$10,500
21	Configuration control	5,000	0.03	$2.10	$21,000

(Continued)

Table 6.13 (*Continued*) Example of Software Activity-Based Costs

	Development Activities	*Monthly Function Points per Month*	*Work Hours per Function Point*	*Burdened Cost per Function Point*	*Project Cost at $10,000 per Month*
22	Integration	6,000	0.02	$1.75	$17,500
23	User documentation	500	0.26	$21.00	$210,000
24	Unit testing	175	0.75	$60.00	$600,000
25	Function testing	150	0.88	$70.00	$700,000
26	Regression testing	150	0.88	$70.00	$700,000
27	Integration testing	250	0.53	$42.00	$420,000
28	Performance testing	500	0.26	$21.00	$210,000
29	Security testing	350	0.38	$30.00	$300,000
30	Usability testing	600	0.22	$17.50	$175,000
31	System testing	175	0.75	$60.00	$600,000
32	Cloud testing	2,400	0.06	$4.38	$43,750
33	Field (Beta) testing	4,000	0.03	$2.63	$26,250
34	Acceptance testing	5,000	0.03	$2.10	$21,000
35	Independent testing	6,000	0.02	$1.75	$17,500
36	Quality assurance	750	0.18	$14.00	$140,000
37	Installation/training	4,000	0.03	$2.63	$26,250
38	Project measurement	9,500	0.01	$1.11	$11,053
39	Project office	750	0.18	$14.00	$140,000
40	Project management	100	1.32	$105.00	$1,050,000
	Cumulative results	8.03	16.44	$1,307.99	$13,079,912

database languages such as DBL and SQL arrived. Soon new languages became a torrent instead of a trickle, and the torrent has never slowed down.

Out of the sum total of about 3000 programming languages, only about 50 are in widespread use at any given decade. Today, we have Ruby, Python, Java, JavaScript, MySQL, HTML, and quite a few others. In the earlier decades, we had COBOL, FORTRAN, ADA, PL/I, BASIC, and some others.

The result of this deluge of programming languages is that thousands of legacy applications were coded in old languages such as MUMPS, CHILL, and CORAL, which have few modern programmers and sometimes not even working compilers or interpreters. All these aging languages in aging legacy software make software maintenance very expensive.

The author recommends that a major university or large corporation such as IBM or Microsoft create a "museum" of programming languages that would have working compilers and interpreters for all languages with existing applications that are still running. There should also be tutorial materials so modern programmers can get up to speed in obscure languages if they are tasked with maintaining legacy applications coded in dead languages.

When high-level languages first came out there was no exact definition of what "high level" meant. In the early 1970s, IBM was the first company to assign a mathematical level score to various programming languages.

The IBM scoring system made the reasonable assumption that the level of a language could be defined by comparing the number of basic assembly language statements needed to create the functionality of one statement in the target language. Using this rule, COBOL and FORTRAN were "level 3" languages. PL/I was a "level 4" language. This was because programmers would need to code three assembly statements to create the functionality of one COBOL statement. They would need to code four assembly statements to create the functionality of one PL/I statement.

Later, when function points were developed in the 1970s, the definition of language level was expanded to include the number of code statements needed to implement one function point. This method was developed at the same time as function points, and in fact Al Albrecht and the original IBM function point team were the first to explore the ratio of code statements to function points.

It was soon discovered by examining applications measured with both LOC and function points that basic assembly, the original "level 1" language, needed about 320 code statements to create the features contained in one function point. A level 2.5 language such as C took about 128 statements per function point. A level 3 language such as COBOL took about 106.7 code statements per function point. A level 6 language such as Java took about 53 statements per function point.

These values were statistical averages, but of course there are wide ranges for all programming languages based on individual programming skills and programming styles. For example, COBOL may average 106.7 code statements per function point but the range is from <60 statements per function point to over 200 code statements per function point. We provide our estimating tool users with default values but they are free to change language levels if they wish to.

All estimating tools need to support modern languages with high usage, and also older languages with large numbers of legacy applications. Our Software Risk Master (SRM) tool currently supports 84 programming languages and we add new ones every year. Table 6.14 shows our current list of supported languages sorted in ascending levels.

Table 6.14 Common Programming Languages and Levels circa 2017

Language Levels	*Languages*	*Logical code statements per function point*	
0.10	English text	3,200.00	
0.50	Machine language	640.00	
1.00	Basic assembly	320.00	
1.45	JCL	220.69	
1.50	Macro assembly	213.33	
2.00	HTML	160.00	
2.50	C	128.00	
2.50	XML	128.00	
3.00	Algol	106.67	
3.00	Bliss	106.67	
3.00	Chill	106.67	
3.00	COBOL	106.67	
3.00	Coral	106.67	
3.00	Fortran	106.67	2014
3.00	Jovial	106.67	
3.25	GW Basic	98.46	
3.50	Pascal	91.43	
3.50	PL/S	91.43	2016
4.00	ABAP	80.00	
4.00	Modula	80.00	2014
4.00	PL/I	80.00	
4.50	ESPL/I	71.11	
4.50	JavaScript	71.11	
5.00	Basic (interpreted)	64.00	

(Continued)

Table 6.14 (*Continued*) Common Programming Languages and Levels circa 2017

Language Levels	*Languages*	*Logical code statements per function point*	
5.00	Forth	64.00	2014
5.00	Haxe	64.00	
5.00	Lisp	64.00	
5.00	Prolog	64.00	
5.00	SH (shell scripts)	64.00	
5.25	Quick Basic	60.95	2014
5.50	Zimbu	58.18	2014
6.00	C++	53.33	
6.00	Go	53.33	
6.00	Java	53.33	
6.00	PHP	53.33	
6.00	Python	53.33	2014
6.25	C#	51.20	2015
6.25	X10	51.20	
6.50	Ada 95	49.23	
6.50	Ceylon	49.23	2014
6.50	Fantom	49.23	
6.75	Dart	47.41	
6.75	RPG III	47.41	
7.00	CICS	45.71	
7.00	DTABL	45.71	
7.00	F#	45.71	
7.00	Groovy	45.71	
7.00	Ruby	45.71	2014

(*Continued*)

Table 6.14 (*Continued*) Common Programming Languages and Levels circa 2017

Language Levels	*Languages*	*Logical code statements per function point*	
7.00	Simula	45.71	
7.50	ColdFusion	42.67	2014
7.50	Erlang	42.67	
8.00	DB2	40.00	
8.00	LiveScript	40.00	
8.00	Oracle	40.00	2014
8.00	R	40.00	2014
8.50	Elixir	37.65	2015
8.50	Haskell	37.65	
8.50	Mixed Languages	37.65	
8.50	Ruby on Rails	37.65	
9.00	Julia	35.56	
9.00	M	35.56	2014
9.00	OPA	35.56	
9.00	Perl	35.56	
10.00	APL	32.00	
11.00	Delphi	29.09	
12.00	MATLAB	26.67	
12.00	Objective C	26.67	
12.00	Swift	26.67	
12.00	Visual Basic	26.67	
13.00	ASP NET	24.62	
13.00	Visual J++	24.62	
14.00	Eiffel	22.86	2015

(*Continued*)

Table 6.14 (*Continued*) Common Programming Languages and Levels circa 2017

Language Levels	*Languages*	*Logical code statements per function point*	
14.00	WebDNA	22.86	
15.00	Smalltalk	21.33	
16.00	IBM ADF	20.00	2016
17.00	MUMPS	18.82	2014
18.00	Forte	17.78	
19.00	APS	16.84	
20.00	TELON	16.00	2015
25.00	Mathematica 9	12.80	
25.00	QBE	12.80	2014
25.00	SQL	12.80	
25.00	Transcript-SQL	12.80	
25.00	X	12.80	2016
35.00	Mathematica 10	9.14	2016
45.00	BPM	7.11	
45.00	Generators	7.11	2014
50.00	Excel	6.40	2014
60.00	IntegraNova	5.33	2014
10.42	Average	99.10	

Table 6.15 shows default values we provide to our clients, but clients can change the language levels if they feel their local teams might create less (or more) code than average.

Another curious issue of programming languages that is partly technical and partly sociological is the fact that most applications need and use multiple programming languages such as Java and HTML or Ruby and MySQL. Indeed, some applications have over a dozen programming languages. To the author, the need for multiple languages is a proof that no single programming language is truly a general-purpose language.

Table 6.15 Productivity Rates for 10 Versions of the Same Software Project (A Private Branch Exchange Switching System of 1500 Function Points in Size)

Language	*Effort (Months)*	*Function Point per Staff Month*	*Work Hours per Function Point*	*LOC per Staff Month*	*LOC per Staff Hour*
Assembly	781.91	1.92	68.81	480	3.38
C	460.69	3.26	40.54	414	3.13
CHILL	392.69	3.82	34.56	401	3.04
PASCAL	357.53	4.20	31.46	382	2.89
PL/I	329.91	4.55	29.03	364	2.76
Ada83	304.13	4.93	26.76	350	2.65
C++	293.91	5.10	25.86	281	2.13
Ada95	269.81	5.56	23.74	272	2.06
Objective C	216.12	6.94	19.02	201	1.52
Smalltalk	194.64	7.71	17.13	162	1.23
Average	360.13	4.17	31.69	366	2.77

Our SRM tool has a built-in calculator for handling the effective level of up to three programming languages. Suppose an application comprised 70% C# (level 6.25) 20% SQL (level 25), and 10% HTML (level 2). Using our calculator, the combined level of these three would be 9.58, which is equivalent to 33.42 code statements per function point.

It might be asked why bother with predicting and calculating code size when function points are the metric used for estimation. The two main reasons are (1) to predict code defect volumes that are inversely proportional to language levels, (2) to predict bad-fix injections that are inversely proportional to language levels. In other words, low-level programming languages tend to create a buggier code that is harder to fix safely without accidentally putting in new bugs.

A third reason helps solve a sociological problem. Many companies and software engineers still try and measure productivity using LOC per month. This unfortunate metric distorts reality, penalizes high-level programming languages, and conceals the impact of the work of requirements and design. The author regards LOC metrics as professional malpractice for software economic analysis.

Table 6.15 shows the results of a study done for a European telecommunications company. We compared productivity for 10 languages using both LOC per month and function points per month.

As can be seen, the LOC column and the function point column move in opposite directions. Function point metrics correctly show improved productivity for higher level languages; LOC shows reduced productivity for higher level languages. Using a metric such as LOC for 60 years that distorts reality and conceals true economic productivity is a professional disgrace and should be considered professional malpractice, and is one of the reasons why software is not yet a true profession.

Readers can buy tables of conversion factors between logical code statements and function points from several vendors including Gartner Group, QSM, and David Consulting Group. Conversion ratios are also published. The author's book *Applied Software Measurement* has ratios for about 600 programming languages in an appendix.

The LOC metric originated in the 1950s when machine language and basic assembly were the only languages in use. In those early days coding was over 95% of the total effort, so the fixed costs of noncode work barely mattered. It was only after high-level programming languages began to reduce coding effort and requirements and design became progressively larger components that the LOC problems occurred. Table 6.16 shows the coding and noncoding percentages by language with the caveat that the noncode work is artificially held constant at 3000 h.

As is evident, the problems of LOC metrics are minor for very low-level languages. But as language levels increase, a higher percentage of effort goes to noncode

Table 6.16 Percentages of Coding and Noncoding Tasks (Percentage of Work Hours for Code and Noncode)

	Languages	*Work Hours for Noncode (%)*	*Work Hours for Code (%)*
1	Machine language	2.51	97.49
2	Basic Assembly	4.90	95.10
3	JCL	6.96	93.04
4	Macro Assembly	7.18	92.82
5	HTML	9.35	90.65
6	C	11.42	88.58
7	XML	11.42	88.58
8	Algol	13.40	86.60
9	Bliss	13.40	86.60

(Continued)

Table 6.16 (*Continued*) Percentages of Coding and Noncoding Tasks (Percentage of Work Hours for Code and Noncode)

	Languages	*Work Hours for Noncode (%)*	*Work Hours for Code (%)*
10	Chill	13.40	86.60
11	COBOL	13.40	86.60
12	Coral	13.40	86.60
13	Fortran	13.40	86.60
14	Jovial	13.40	86.60
15	GW Basic	14.35	85.65
16	Pascal	15.29	84.71
17	PL/S	15.29	84.71
18	ABAP	17.10	82.90
19	Modula	17.10	82.90
20	PL/I	17.10	82.90
21	ESPL/I	18.83	81.17
22	JavaScript	18.83	81.17
23	Basic (interpreted)	20.50	79.50
24	Forth	20.50	79.50
25	Haxe	20.50	79.50
26	Lisp	20.50	79.50
27	Prolog	20.50	79.50
28	SH (shell scripts)	20.50	79.50
29	Quick Basic	21.30	78.70
30	Zimbu	22.09	77.91
31	C++	23.63	76.37
32	Go	23.63	76.37
33	Java	23.63	76.37

(Continued)

Table 6.16 (*Continued*) Percentages of Coding and Noncoding Tasks (Percentage of Work Hours for Code and Noncode)

	Languages	*Work Hours for Noncode (%)*	*Work Hours for Code (%)*
34	PHP	23.63	76.37
35	Python	23.63	76.37
36	C#	24.37	75.63
37	X10	24.37	75.63
38	Ada 95	25.10	74.90
39	Ceylon	25.10	74.90
40	Fantom	25.10	74.90
41	Dart	25.82	74.18
42	RPG III	25.82	74.18
43	CICS	26.52	73.48
44	DTABL	26.52	73.48
45	F#	26.52	73.48
46	Ruby	26.52	73.48
47	Simula	26.52	73.48
48	Erlang	27.89	72.11
49	DB2	29.20	70.80
50	LiveScript	29.20	70.80
51	Oracle	29.20	70.80
52	Elixir	30.47	69.53
53	Haskell	30.47	69.53
54	Mixed Languages	30.47	69.53
55	Julia	31.70	68.30
56	M	31.70	68.30
57	OPA	31.70	68.30

(*Continued*)

Table 6.16 (*Continued*) Percentages of Coding and Noncoding Tasks (Percentage of Work Hours for Code and Noncode)

	Languages	*Work Hours for Noncode (%)*	*Work Hours for Code (%)*
58	Perl	31.70	68.30
59	APL	34.02	65.98
60	Delphi	36.19	63.81
61	Objective C	38.22	61.78
62	Visual basic	38.22	61.78
63	ASP NET	40.13	59.87
64	Eiffel	41.92	58.08
65	Smalltalk	43.61	56.39
66	IBM ADF	45.21	54.79
67	MUMPS	46.71	53.29
68	Forte	48.14	51.86
69	APS	49.49	50.51
70	TELON	50.77	49.23
71	Mathematica9	56.31	43.69
72	Transcript-SQL	56.31	43.69
73	QBE	56.31	43.69
74	X	56.31	43.69
75	Mathematica10	64.35	35.65
76	BPM	69.88	30.12
77	Generators	69.88	30.12
78	Excel	72.05	27.95
79	IntegraNova	75.57	24.43
	Average	29.08	70.92

work while coding effort progressively gets smaller. Thus, LOC metrics are invalid and hazardous for high-level languages.

U.S. Industry Work Hour Variations

An importation topic for benchmark accuracy is how many work hours per month the development teams put in. Work hours vary from country to country, industry to industry, company to company, and project to project. We provide clients with default values based on historical data, but we suggest that clients substitute their own local values for work hours (assuming they know them.)

Table 6.17 shows our default work hours for a variety of U.S. industry sectors. Having worked in several start-up technology companies myself, I know from first-hand experience that professional staff in new technology companies put in A LOT of unpaid overtime. In fact, this is such a serious sociological issue that it probably elevates the divorce rates among technology entrepreneurs.

As can be observed, it is important to record work-hour patterns in order to have accurate software benchmark data.

This brings up an endemic business problem. A majority of the author's clients, and probably a majority of U.S. companies, do not record unpaid overtime. The kinds of companies that do record unpaid overtime are contractors whose contracts mandate recording all hours.

This failure to record unpaid overtime leads to some anomalies in software benchmarks. For example, large systems above 10,000 function points average more than 16 h per month of unpaid overtime that is usually invisible because it is not recorded. The overtime is due to schedule pressure from management.

Small applications below 500 function points use little or no unpaid overtime because they are usually on time and less troublesome than large systems. This can lead to the curious anomaly that large applications may seem to have higher productivity than small applications because of the intensive but unmeasured unpaid overtime that is a normal part of large software system development.

The bottom line is that unless benchmarks record unpaid overtime they don't actually measure productivity with enough accuracy to be useful. When we collect benchmark data from clients, we ask these questions:

1. Did you use unpaid overtime?
2. Does your tracking system record unpaid overtime?

If the project had overtime but it was not recorded, we ask the development team to reconstruct overtime patterns from memory or from informal records. This is not as good as accurate time recording but it is better than omitting unpaid overtime because it was invisible and not recorded.

Table 6.17 U.S. Ranges on Work Hours per Month

	U.S. Industry Segments	*Software Work Hours per Month*	*Software Unpaid Overtime per Month*	*Software Total Hours per Month*	*Average Work Hours (%)*
1	Startup technology companies	191.67	16.00	207.67	150.48
2	Technology companies	175.00	14.00	189.00	136.96
3	Computer games	165.00	8.00	173.00	125.36
4	Open source	160.42	8.00	168.42	122.04
5	Web/cloud	150.00	8.00	158.00	114.49
6	Bioengineering/medicine	147.50	10.00	157.50	114.13
7	Fixed-price contracts	138.28	12.00	150.28	108.90
8	Management consulting	142.00	8.00	150.00	108.70
9	Outsource contractors	140.13	8.00	148.13	107.34
10	Manufacturing	136.44	6.00	142.44	103.22
11	Finance/insurance	134.59	6.00	140.59	101.88
12	Telecom	134.59	6.00	140.59	101.88
13	Entertainment	132.75	6.00	138.75	100.54
14	U.S. average	132.00	6.00	138.00	100.00
15	Wholesale/retail	131.00	6.00	137.00	99.28
16	Health care	130.00	4.00	134.00	97.10
17	Avionics	129.06	4.00	133.06	96.42
18	Energy	127.29	4.00	131.29	95.14
19	Profit-center projects	125.38	4.00	129.38	93.75
20	Time and material contracts	129.06	0.00	129.06	93.52
21	Education	123.53	2.00	125.53	90.96
22	Federal government	123.53	0.00	123.53	89.52
23	Cost-center projects	119.84	2.00	121.84	88.29
24	Defense	121.69	0.00	121.69	88.18
25	State/local government	117.26	0.00	117.26	84.97
	Average	142.61	7.20	149.81	108.56

U.S. Industry Productivity and Quality Results circa 2017

The United States has several hundred industries but the author does not work in all of them. Table 6.18 shows approximate productivity for 2017 for 75 major industry segments where we have either collected benchmarks ourselves or access to data collected by other researchers. All these industries create large volumes of software.

As is evident, technology companies such as computer games and smartphones are on top, and government is on the bottom. Table 6.18 is sorted in terms of work hours per function point.

The productivity data show a wide range of productivity across various industries. The industries at the top of the list tend to be new, dynamic, fast moving, and have quite a bit of unpaid overtime. The industries at the bottom of the list tend to be mature, slow moving, and seldom have much in the way of unpaid overtime.

When the view switches from productivity to quality, the sequence of industries changes significantly. Medical devices are #1 for software quality, as indeed they should be. Computer games have dropped down to #30.

Software quality is actually more important to clients than software productivity because poor quality causes "consequential damages" or financial harm to clients. A number of author's clients have lost millions of dollars due to bugs in software applications such as financial packages. One client had to restate their prior year financial statements and tax returns due to a software bug, which cost them a favorable credit rating and a preferred banking relationship. Needless to say, the client sued the vendor.

In the State of Rhode Island, as this chapter was being written, hundreds of families on welfare had not been able to receive their assistance checks on time, and sometimes the delays were several months, due to bugs in a new state welfare system.

It is interesting that productivity and quality are somewhat coupled, in that high levels of DRE tend to have high productivity rates as well (Table 6.19).

The quality metrics of defect potentials, DRE, and defects per function point were all developed by IBM in the 1970s. This combination of metrics is currently the best available for measuring software quality.

Cost per defect is invalid and penalizes quality. LOC penalize high-level languages and make requirements and design bugs invisible. Technical debt is not a standard metric and varies widely from company to company. Defect density using LOC ignores requirements and design bugs, which often outnumber code bugs. Story points and use case points have no International Organization for Standardization (ISO) standards and vary by over 400% from company to company.

Software metrics are a professional embarrassment. Except for function points that were validated by IBM prior to release, most software metrics are just pushed out to the world with no validation and no proof of accuracy.

Table 6.18 U.S. Software Industry Productivity circa 2017

	Industry	*Function Points*	*Work Hours per Function Point*
1	Games—computer	15.75	8.38
2	Smartphone/tablet applications	15.25	8.66
3	Software (commercial)	15.00	8.80
4	Social networks	14.90	8.86
5	Software (outsourcing)	14.00	9.43
6	Open source development	13.75	9.60
7	Entertainment—films	13.00	10.15
8	Consulting	12.70	10.39
9	Entertainment—television	12.25	10.78
10	Banks—commercial	11.50	11.48
11	Banks—investment	11.50	11.48
12	Retail—Web	11.30	11.68
13	Credit unions	11.20	11.79
14	Entertainment—music	11.00	12.00
15	Insurance—medical	10.50	12.57
16	Insurance—life	10.00	13.20
17	Stock/commodity brokerage	10.00	13.20
18	Insurance—property and casualty	9.80	13.47
19	Manufacturing—telecommunications	9.75	13.54
20	Telecommunications operations	9.75	13.54
21	Pharmacy chains	9.00	14.67
22	Process control and embedded	9.00	14.67
23	Manufacturing—pharmaceuticals	8.90	14.83
24	Tourism (travel agents, web, etc.)	8.80	15.00
25	Hotels	8.75	15.09

(Continued)

Table 6.18 (*Continued*) U.S. Software Industry Productivity circa 2017

	Industry	*Function Points*	*Work Hours per Function Point*
26	Oil extraction	8.75	15.09
27	Transportation—airlines	8.75	15.09
28	Education—University	8.60	15.35
29	Publishing (books/journals)	8.60	15.35
30	Professional support—medicine	8.55	15.44
31	Government—police	8.50	15.53
32	Professional support—law	8.50	15.53
33	Sports (pro baseball, football, etc.)	8.50	15.53
34	Accounting/financial consultants	8.50	15.53
35	Other industries	8.30	15.90
36	Manufacturing—electronics	8.25	16.00
37	Manufacturing—general	8.25	16.00
38	Wholesale	8.25	16.00
39	Automotive sales	8.00	16.50
40	Hospitals—administration	8.00	16.50
41	Manufacturing—chemicals	8.00	16.50
42	Manufacturing—nautical	8.00	16.50
43	Retail—stores	8.00	16.50
44	Transportation—bus	8.00	16.50
45	Transportation—ship	8.00	16.50
46	Transportation—trains	8.00	16.50
47	Transportation—truck	8.00	16.50
48	Agriculture	7.75	17.03
49	Manufacturing—automotive	7.75	17.03
50	Manufacturing—medical devices	7.75	17.03

(*Continued*)

Table 6.18 (*Continued*) U.S. Software Industry Productivity circa 2017

	Industry	*Function Points*	*Work Hours per Function Point*
51	Education—secondary	7.60	17.37
52	Manufacturing—appliances	7.60	17.37
53	Automotive repairs	7.50	17.60
54	Education—primary	7.50	17.60
55	Games—traditional	7.50	17.60
56	Manufacturing—aircraft	7.25	18.21
57	Public utilities—water	7.25	18.21
58	Real estate—commercial	7.25	18.21
59	Real estate—residential	7.25	18.21
60	Government—intelligence	7.20	18.33
61	Construction	7.10	18.59
62	Food—restaurants	7.00	18.86
63	Government—municipal	7.00	18.86
64	Manufacturing—apparel	7.00	18.86
65	Mining—metals	7.00	18.86
66	Mining—coal	7.00	18.86
67	Public utilities—electricity	7.00	18.86
68	Waste management	7.00	18.86
69	Manufacturing—defense	6.85	19.27
70	Government—military	6.75	19.56
71	Natural gas generation	6.75	19.56
72	Government—county	6.50	20.31
73	Government—federal civilian	6.50	20.31
74	Government—state	6.50	20.31
75	ERP vendors	6.00	22.00
	Averages	8.90	14.83

Table 6.19 U.S. Software Quality Results circa 2017

	Industry	*Defect Potentials per Function Point*	*DRE (%)*	*Delivered Defects per Function Point*
1	Manufacturing—medical devices	4.60	99.50	0.02
2	Government—military	4.70	99.00	0.05
3	Manufacturing—aircraft	4.70	99.00	0.05
4	Smartphone/tablet applications	3.30	98.50	0.05
5	Government—intelligence	4.90	98.50	0.07
6	Software (commercial)	3.50	97.50	0.09
7	Telecommunications operations	4.35	97.50	0.11
8	Manufacturing—defense	4.65	97.50	0.12
9	Manufacturing—telecommunications	4.80	97.50	0.12
10	Process control and embedded	4.90	97.50	0.12
11	Manufacturing—pharmaceuticals	4.55	97.00	0.14
12	Professional support—medicine	4.80	97.00	0.14
13	Transportation—airlines	5.87	97.50	0.15
14	Manufacturing—electronics	4.90	97.00	0.15
15	Banks—commercial	4.15	96.25	0.16
16	Entertainment—films	4.00	96.00	0.16

(Continued)

Table 6.19 (*Continued*) U.S. Software Quality Results circa 2017

	Industry	*Defect Potentials per Function Point*	*DRE (%)*	*Delivered Defects per Function Point*
17	Manufacturing—automotive	4.30	96.25	0.16
18	Retail—web	4.20	96.00	0.17
19	Manufacturing—chemicals	4.80	96.50	0.17
20	Manufacturing—appliances	4.30	96.00	0.17
21	Insurance—life	4.60	96.00	0.18
22	Banks—investment	4.30	95.50	0.19
23	Software (outsourcing)	4.45	95.50	0.20
24	Insurance—property and casualty	4.50	95.50	0.20
25	Pharmacy chains	3.75	94.50	0.21
26	Government—police	4.80	95.50	0.22
27	Insurance—medical	4.80	95.50	0.22
28	Open source development	4.40	95.00	0.22
29	Social networks	4.90	95.50	0.22
30	Games—computer	3.75	94.00	0.23
31	Entertainment—television	4.60	95.00	0.23
32	Transportation—trains	4.70	95.00	0.24
33	Public utilities—electricity	4.80	95.00	0.24
34	Public utilities—water	4.40	94.50	0.24
35	Accounting/financial consultants	3.90	93.50	0.25

(Continued)

Table 6.19 (*Continued*) U.S. Software Quality Results circa 2017

	Industry	*Defect Potentials per Function Point*	*DRE (%)*	*Delivered Defects per Function Point*
36	Professional support—law	4.75	94.50	0.26
37	Credit unions	4.50	94.00	0.27
38	Manufacturing—nautical	4.60	94.00	0.28
39	Transportation—bus	4.60	94.00	0.28
40	Sports (pro baseball, football, etc.)	4.00	93.00	0.28
41	Publishing (books/journals)	4.50	93.50	0.29
42	Manufacturing—apparel	3.00	90.00	0.30
43	Hospitals—administration	4.80	93.00	0.34
44	Transportation—ship	4.30	92.00	0.34
45	Consulting	4.00	91.00	0.36
46	Real estate—commercial	4.00	91.00	0.36
47	Oil extraction	4.15	91.00	0.37
48	Entertainment—music	4.00	90.00	0.40
49	Other industries	4.50	91.00	0.41
50	Natural gas generation	4.80	91.50	0.41
51	Automotive sales	4.75	91.00	0.43
52	Games—traditional	4.00	89.00	0.44
53	Wholesale	4.40	90.00	0.44
54	Education—University	4.50	90.00	0.45
55	Government—municipal	4.80	90.00	0.48

(Continued)

Table 6.19 (*Continued*) U.S. Software Quality Results circa 2017

	Industry	*Defect Potentials per Function Point*	*DRE (%)*	*Delivered Defects per Function Point*
56	Hotels	4.40	89.00	0.48
57	Tourism (travel agents, Web, etc.)	4.90	90.00	0.49
58	Government—state	4.95	90.00	0.50
59	Government—county	4.70	89.00	0.52
60	Retail—stores	5.00	89.50	0.53
61	Stock/commodity brokerage	5.15	89.50	0.54
62	Automotive repairs	4.20	87.00	0.55
63	Real estate—residential	4.80	88.50	0.55
64	Education—primary	4.30	87.00	0.56
65	Education—secondary	4.35	87.00	0.57
66	Manufacturing—general	4.75	88.00	0.57
67	Construction	4.70	87.00	0.61
68	Mining—metals	4.90	87.50	0.61
69	ERP vendors	5.70	89.00	0.63
70	Agriculture	4.90	87.00	0.64
71	Waste management	4.60	86.00	0.64
72	Transportation—truck	4.80	86.50	0.65
73	Government—federal civilian	5.60	88.00	0.67
74	Mining-coal	5.00	86.50	0.68
75	Food—restaurants	4.80	85.50	0.70
	Averages	4.54	92.76	0.33

The new SNAP metric may be useful for quality because the author's preliminary data indicate higher volumes of defects for nonfunctional requirements than for functional requirements, but the author does not yet have enough SNAP quality data to include in this chapter.

Comparing Software Globally

It is important to show how to compare software development productivity and quality internationally because software is a global business. In fact, some applications may have development teams in half a dozen countries.

Countries vary widely in work hours per year, public holidays, and both paid and unpaid overtime. They also vary widely in average compensation and in inflation rates. This chapter discusses the impact of these variations on software productivity.

This portion of the chapter also discusses productivity and quality benchmark variances from country to country. Readers who want more details on national work hours can visit the website of the Organization for Economic Cooperation and Development (OECD.org).

Using OECD data, the 10 countries with the greatest number of work hours per year are presented in Table 6.20.

The OECD data is for all industries and not specifically for software. Software tends to have somewhat longer work hours than many industries and also more unpaid overtime since most software workers are "exempt" and therefore not eligible for paid overtime. However, in some countries, software personnel may be members of trade or craft unions and therefore would receive paid overtime.

Table 6.20 Ten Countries with Longest Work Hours per Year

India	2280.00
Taiwan	2112.00
Mexico	2226.00
China	2232.00
Peru	2,208.00
Colombia	2,112.00
Pakistan	2,112.00
Hong Kong	2,280.00
Thailand	2,016.00
Malaysia	2,304.00

Table 6.21 Ten Countries with Shortest Work Hours per Year

Canada	1710.00
Australia	1728.00
Ireland	1529.00
Spain	1686.00
France	1479.00
Iceland	1706.00
Sweden	1621.00
Norway	1420.00
Germany	1397.00
Netherlands	1381.00

By contrast, the 10 countries with the smallest number of work hours per year are shown in Table 6.21.

In this chapter, let us consider the impact of varying numbers of work hours on a sample benchmark. Assume that exactly the same software is going to be produced in five countries (Table 6.22).

In order to include cost results as well as productivity results, we will also include approximate burdened monthly costs for salaries and benefits (Table 6.23).

Note that costs vary from region to region within countries and also from industry to industry. Therefore, Table 6.23 is only approximate and should be used as hypothetical data only. Do not use this table for actual cost estimates but use actual local corporate data instead.

Table 6.23 assumes burdened costs with factors such as benefits, health care, office space, and other non-personnel topics included.

Table 6.22 Paid and Unpaid Hours per Month in Selected Countries

Country	*Paid Work Hours*	*Unpaid Overtime*	*Total*
China	186	8	196
Russia	145	4	149
United States	132	10	142
Australia	127	0	127
Germany	116	0	116

Table 6.23 Approximate Monthly Burdened Costs

China	$5,000
Russia	$8,000
United States	$10,000
Australia	$9,000
Germany	$11,000

By contrast, Table 6.24 shows the median average salaries for software engineers in 50 countries, taken from Bloomberg.com.

As of 2017, there is a total of 204 de facto countries. Of these, 193 countries are members of the United Nations. Software is probably produced in all of them. Interestingly, the FIFA world cup soccer organization recognizes 209 countries.

There are also about 126 software occupation groups and they all have different compensation levels. Though this chapter is not about occupations, benchmark costs need to include the various occupations listed in Table 6.25.

For large systems, pure programming can be <35% of the total cost. Both benchmarks and cost estimates need to include the full spectrum of software occupation groups.

For a sample project itself, let us assume the parameters listed in Table 6.26 for the United States. Bear in mind that the example assumes exactly the same kind and size of application in every country.

Having set up a sample project to be done in five countries, Table 6.27 shows the amount of human effort and the schedule in calendar months for the five versions.

Note that while the number of work hours is identical in all five counties, the number of months and the schedule in calendar months are quite different due to the number of hours worked per month.

Table 6.28 shows the comparative productivity rates for the five countries using four different metrics: (1) work hours per function point, (2) function points per staff month, (3) work hours per KLOC, and (4) LOC per month.

This table illustrates that hourly and monthly productivity rates are not interchangeable when doing global benchmark comparisons across multiple countries.

Table 6.29 shows cost data and also brings up a chronic problem for software economic studies. Unpaid overtime is often excluded from benchmarks and productivity comparisons, but exerts a fairly large impact. Unpaid overtime shortens project schedules and lowers costs. It should be tracked and included in all software benchmarks.

Table 6.29 also shows that there can be very significant differences between projects where the teams are paid on a monthly basis and projects where the teams are paid on an hourly basis. Obviously, the countries that work many hours each month will have lower costs when compensation is based on months rather than on hours.

Table 6.24 Bloomberg Median Annual Salary for Software Engineers (www.Bloomberg.com)

	Country	*Annual Salary*
1	Switzerland	$104,200
2	Norway	$81,400
3	United States	$76,000
4	Denmark	$71,500
5	Israel	$70,700
6	Australia	$65,900
7	Germany	$63,800
8	Sweden	$61,400
9	New Zealand	$59,600
10	Canada	$57,500
11	United Kingdom	$56,200
12	France	$54,700
13	Ireland	$53,900
14	Netherlands	$53,900
15	Finland	$51,500
16	Austria	$49,300
17	Belgium	$48,400
18	Japan	$40,700
19	Singapore	$38,800
20	Spain	$37,000
21	Italy	$34,900
22	Qatar	$33,400
23	Czech Republic	$33,000
24	Hong Kong	$29,400
25	United Arab Emirates	$29,000

(Continued)

Table 6.24 (*Continued*) Bloomberg Median Annual Salary for Software Engineers (www.Bloomberg.com)

	Country	*Annual Salary*
26	Brazil	$27,300
27	Saudi Arabia	$26,600
28	Greece	$25,300
29	Bulgaria	$25,200
30	Kuwait	$25,200
31	South Africa	$24,000
32	Taiwan	$24,000
33	China	$23,100
34	Poland	$22,600
35	Russia	$22,100
36	Mexico	$22,000
37	Portugal	$21,600
38	Lithuania	$21,000
39	Hungary	$19,200
40	Turkey	$17,100
41	Malaysia	$16,500
42	Argentina	$14,000
43	Ukraine	$11,200
44	Philippines	$7,900
45	Egypt	$7,300
46	Pakistan	$7,200
47	Romania	$6,800
48	Sri Lanka	$6,700
49	Indonesia	$6,400
50	India	$6,200

Table 6.25 Software Benchmark Occupation Groups

1	Architects
2	Business analysts
3	Cost-estimation specialists
4	Database analysts
5	Developers
6	Function point specialists
7	Program librarians
8	Integration/configuration control specialists
9	Project managers
10	Project office staff
11	Software quality assurance (SQA)
12	Technical writers
13	Testers (if separate from developers)
14	Customer support specialists
15	Users (if direct participants)

Table 6.30 shows the overall costs for the project and also costs per function point and costs per KLOC.

At first glance, it might look as though software could migrate to the countries with the lowest labor costs and the largest numbers of work hours per month. However, doing work across multiple time zones and across language barriers is difficult and there are frictional losses in efficiency.

Since the author of this chapter has worked on software projects and collected benchmarks in 27 countries, there are some "frictional" losses in efficiency when doing international software projects. Table 6.31 shows the major topics that can reduce efficiency and raise costs for international projects.

One of the purposes of carrying out benchmark studies is to find out which factors have the greatest impact on software projects for good or ill. Although this chapter concentrates on work hours, that topic is only one of many topics that influence software project results.

Table 6.32 shows a total of 50 factors that are either beneficial and *lower* project work hours or harmful and *raise* project work hours.

Note that the data in Table 6.32 is not symmetrical. Inexperienced managers and clients can do a great deal of harm to software projects that are difficult to overcome by technical factors alone.

Table 6.26 Sample Software Benchmark Parameters

Size in function points	1,000
Size in logical code statements	50,000
Size in KLOC	50.00
Programming language	Java
Development methodology	Agile/scrum
Team experience level	Average
Team size (developers, testers, management)	7.00
Local work hours per month	132.00
Local unpaid overtime per month	0.00
Local paid overtime per month	0.00
Project effort (staff months)	125.00
Project effort (staff hours)	16,500
Team function points per month	16.50
Team work hours per function point	15.00
Net LOC per hour	3.33
Project defect potential per function point	3.75
Project defect potential (all valid defects)	3,750
Project DRE	92.50%
Delivered defects per function point	0.281
Delivered valid defects	281
High-severity defect percentage	12.50%
High-severity defects delivered	35
High-severity defects per function point	.0035
Security flaws delivered	4
Security flaws per function point	.0004

Table 6.27 Comparisons of Five Countries for Hours and Months

	Work Hours	*Hours per Month*	*Work Months*	*Calendar Months*
China	15,000	196	76.53	10.93
Russia	15,000	149	100.67	14.38
USA	15,000	142	105.63	15.09
Australia	15,000	127	118.11	16.87
Germany	15,000	116	129.31	18.47

Table 6.28 Productivity Levels for Five Countries

	Work Hours per FP	*FP per Month*	*Work Hours per KLOC*	*LOC per Month*
China	15.00	13.07	300	653
Russia	15.00	9.93	300	497
USA	15.00	9.47	300	473
Australia	15.00	8.47	300	423
Germany	15.00	7.73	300	387

Table 6.29 Labor Costs Examples for Five Countries

	Monthly Costs	**Paid Hours**	**Unpaid Hours**	**Total Hours**	**Hourly Costs**
China	$5,000.00	186	8	194	$25.77
Russia	$8,000.00	145	4	149	$53.69
USA	$10,000.00	132	10	142	$70.42
Australia	$9,000.00	127	0	127	$70.87
Germany	$11,000.00	116	0	116	$94.83

The implications of this table are that long work hours in countries like India and China do not by themselves guarantee shorter schedules and lower costs. The experience levels of the development teams and the technologies used by those teams have an even greater impact.

International software is common in 2017 but is also a complex issue that has hazards as well as cost benefits. Further, inflation rates in many countries are much

Table 6.30 Software Cost Samples for Five Countries

	Project Cost	*Cost per FP*	*Cost per KLOC*
China	$386,598	$387	$7,732
Russia	$805,369	$805	$16,107
USA	$1,056,338	$1,056	$21,127
Australia	$1,062,992	$1,063	$21,260
Germany	$1,422,414	$1,422	$28,448

Table 6.31 Loss of Efficiency due to International Complications

Efficiency Factors	*Loss (%)*
Poor knowledge of English	–15
Poor local quality control	–15
Exaggerated claims by outsourcer	–15
Falsely optimistic status reports	–15
Communicating requirements changes	–10
More than four time zones	–5
Poor local infrastructure	–5
Long approval cycles	–5
Total	–85

higher than U.S. inflation rates. This means that in a few years current cost differentials will change and the U.S. will benefit due to its comparatively low inflation rates compared to both China and India, which today are the major software outsource countries although there are dozens of others.

Table 6.33 shows a small sample of current inflation rates for 10 countries taken from Web sources.

Of course, inflation rates are not constant values and change frequently. Therefore, current data are needed to estimate actual software projects.

It is obvious that entering into long-range contracts with countries that have high inflation rates will eventually lead to erosion in economic benefits.

A final point about international software should be mentioned but was not illustrated in this short chapter. Many software projects involve multiple countries at the same time. In this case, both project estimation and benchmarks need to be adjusted for all local conditions.

Table 6.32 Impact of Critical Factors on Software Project Work Hours

	Productivity/Quality Factors	*Impact on Work Hours (%)*
1	Certified reusable components	–80
2	Experienced development teams	–65
3	Mashups <1000 function points	–50
4	Effective methodologies for specific project types	–40
5	High-level programming languages	–30
6	Use of inspections for complex systems	–27
7	Use of SEMAT for complex systems	–25
8	Experienced managers	–25
9	Moderate unpaid overtime by teams	–20
10	Low requirements creep	–20
11	Logical, planned architecture for large systems	–20
12	Use of static analysis before testing	–18
13	High CMMI levels	–15
14	Low cyclomatic complexity (<10)	–15
15	Effective project status tracking	–15
16	Effective defect prevention (JAD, Kaizen, etc.)	–15
17	Experienced test teams	–12
18	TSP or RUP >5000 function points	–12
19	Experienced clients	–10
20	SCRUM <1000 function points	–10
21	Agile <1000 function points	–8
22	Effective parametric estimating tools	–7
23	Testing by certified test personnel	–7
24	Formal mathematical test case design	–6
25	Colocated teams	–5

(Continued)

Table 6.32 (*Continued*) Impact of Critical Factors on Software Project Work Hours

	Productivity/Quality Factors	*Impact on Work Hours (%)*
26	Average development approaches	0
27	TSP or RUP <1000 function points	5
28	Testing by developers only	6
29	Informal test case design	8
30	Scrum >5000 function points	8
31	Agile >5000 function points	9
32	Low CMMI levels	10
33	Waterfall >5000 function points	12
34	Ineffective methodologies	15
35	Inexperienced test teams	15
36	Distributed teams: poor communications	15
37	Pair programming	16
38	High cyclomatic complexity (>25)	18
39	Poor status tracking	20
40	Excessive unpaid overtime by team	23
41	Adding personnel to late projects	25
42	Low-level programming languages	25
43	Concurrent maintenance and development tasks	30
44	Inaccurate manual estimates	33
45	Inexperienced development teams	35
46	Chaotic, unplanned architecture for large systems	37
47	Inexperienced clients	40
48	Truncating testing to "meet schedule"	45
49	Inexperienced managers	50
50	High requirements creep: poor change control	60

Table 6.33 Sample of Annual Inflation Rates for 10 Countries

Country	*Annual Inflation (%)*
Venezuela	60.90
Turkey	9.42
Russia	7.40
Brazil	6.50
India	6.50
Japan	3.67
China	2.50
Canada	2.10
United States	1.99
United Kingdom	1.59

It is technically possible to develop software in three countries that are separated by 8 h of time. At the end of each shift, the software work products would be moved electronically to the next country. Thus, continuous development for 24 h a day without overtime is technically possible. However, doing this requires much better architecture, design, and coding practices that are common in 2017.

Other countries also have major variances in work patterns by industry. This is why the author and his company Namcook Analytics LLC use the North American Industry Classification (NAIC) code when collecting benchmark data. It is important to know the industry as well as the country.

Table 6.34 illustrates a sample of the three-digit NAIC codes. (NAIC codes add digits for additional precision, but the three-digit NAIC codes are generally sufficient for software benchmark purposes.)

The author and his colleagues have done benchmarks in over 70 industries so it is important to use NAIC codes to ensure "apples to apples" benchmark comparisons. Of course, some industries are similar from a software standpoint such as various kinds of banks and for that matter banks and insurance.

Readers interested in the details of the NAIC code structure or in the codes for specific subindustries can easily do a Google search on "NAIC codes" and bring up a variety of tools and support for finding NAIC codes of any level of granularity.

For that matter, it is also important to know geographic regions. Major cities such as New York, Tokyo, Geneva, and Beijing have higher costs than rural areas and smaller cities. This is why so many software engineering labs are moving to low-cost regions in every industrialized country. The Namcook benchmarks utilize telephone calling codes for country and regional classifications.

Table 6.34 Three-Digit NAIC Codes for Industry Classification

Industry	*NAIC Code*
Agriculture	111
Agriculture support	115
Mining	212
Utilities	221
Building construction	236
Civil engineering	237
Food production	311
Beverage and tobacco production	312
Textile mills	313
Apparel manufacturing	315
Paper manufacturing	322
Printing and publishing	323
Petroleum manufacturing	324
Chemical manufacturing	325
Plastic and rubber manufacturing	326
Machinery manufacturing	333
Computer manufacturing	334
Air transportation	481
Ship transportation	483
Software publishers	511
Rail transportation	582
Telecommunications	517
Internet service providers	518
Commercial banks	522
Securities and investments	523
Insurance carriers	524

(Continued)

Table 6.34 (*Continued*) Three-Digit NAIC Codes for Industry Classification

Industry	*NAIC Code*
Real estate	531
Custom computer programming	541
Federal reserve banks	601
Education	611
Hospitals	622
Federal government	910
State governments	920
Local governments	930

NAIC, North American Industry Classification codes.

The author and his colleagues have collected benchmark data in 24 countries, so it is important to encode the national sources of the benchmark data. Table 6.35 shows a sample of international country codes, but the full set of over 200 countries is not required in this short chapter.

There are also alphabetic country codes used in Internet names. A few samples of these two-character alphabetic codes include the following:

- AR Argentina
- AT Austria
- AU Australia
- BR Brazil
- CA Canada
- CN China
- DE Germany
- ES Spain
- FR France
- IN India
- MX Mexico
- UK United Kingdom
- US United States

Some benchmark service providers prefer to conceal the identity of the countries where benchmarks are collected, but since software is a global industry, it is better to show countries that provide benchmark data.

Table 6.36 is the next to lasttable in this section on global comparisons. It shows results for 52 countries. The assumptions of the table are 1000 function point

Table 6.35 Samples of Country Codes Used for Benchmarks

Country	*Country Codes*
Argentina	54
Austria	43
Belgium	32
Brazil	55
Bolivia	591
Canada	1
United States	1
Chile	56
China	86
Columbia	57
Costa Rica	506
Cuba	53
Denmark	45
Egypt	20
Finland	358
Germany	49
Hong Kong	852
Iceland	354
India	91
Indonesia	62
Iran	98
Iraq	964
Ireland	353
Israel	972
Italy	39
Japan	81
Malaysia	60
Mexico	52
Netherlands	31
Pakistan	92
Peru	51
Philippines	63
Russia	7

Table 6.36 International Variations due to Local Work Hours per Month

	Countries	OECD National Work Hours per Year	OECD National Work Hours per Month	Software Work Hours per Month	Software Unpaid Overtime per Month	Software Total Hours per Month	FP per Month for 1000 FP (15 WH per Funct. Pt.)
1	India	2280.00	190.00	190.00	12.00	202.00	13.47
2	Taiwan	2112.00	188.00	188.00	10.00	198.00	13.20
3	Mexico	2226.00	185.50	185.50	12.00	197.50	13.17
4	China	2232.00	186.00	186.00	8.00	194.00	12.93
5	Peru	2208.00	184.00	184.00	6.00	190.00	12.67
6	Colombia	2112.00	176.00	176.00	6.00	182.00	12.13
7	Pakistan	2112.00	176.00	176.00	6.00	182.00	12.13
8	Hong Kong	2280.00	190.00	168.15	12.00	180.15	12.01
9	Thailand	2016.00	168.00	168.00	8.00	176.00	11.73
10	Malaysia	2304.00	192.00	169.92	6.00	175.92	11.73
11	Greece	2034.00	169.50	169.50	6.00	175.50	11.70
12	South Africa	2016.00	168.00	168.00	6.00	174.00	11.60
13	Israel	1910.00	159.17	159.17	8.00	167.17	11.14

(Continued)

Table 6.36 (*Continued*) International Variations due to Local Work Hours per Month

	Countries	*OECD National Work Hours per Year*	*OECD National Work Hours per Month*	*Software Work Hours per Month*	*Software Unpaid Overtime per Month*	*Software Total Hours per Month*	*FP per Month for 1000 FP (15 WH per Funct. Pt.)*
14	Vietnam	1920.00	160.00	160.00	6.00	166.00	11.07
15	Philippines	1920.00	160.00	160.00	4.00	164.00	10.93
16	Singapore	2112.00	176.00	155.76	8.00	163.76	10.92
17	Hungary	1956.00	163.00	163.00	6.00	163.00	10.87
18	Poland	1929.00	160.75	160.75	2.00	162.75	10.85
19	Turkey	1877.00	156.42	156.42	4.00	160.42	10.69
20	Brazil	2112.00	176.00	155.76	4.00	159.76	10.65
21	Panama	2112.00	176.00	155.76	4.00	159.76	10.65
22	Chile	2029.00	169.08	149.64	8.00	157.64	10.51
23	Estonia	1889.00	157.42	157.42	0.00	157.42	10.49
24	Japan	1745.00	145.42	145.42	12.00	157.42	10.49
25	Switzerland	2016.00	168.00	148.68	8.00	156.68	10.45
26	Czech Republic	1800.00	150.00	150.00	0.00	150.00	10.00

(Continued)

Table 6.36 (*Continued*) International Variations due to Local Work Hours per Month

	Countries	*OECD National Work Hours per Year*	*OECD National Work Hours per Month*	*Software Work Hours per Month*	*Software Unpaid Overtime per Month*	*Software Total Hours per Month*	*FP per Month for 1000 FP (15 WH per Funct. Pt.)*
27	Russia	1973.00	164.42	145.51	4.00	149.51	9.97
28	Argentina	2016.00	168.00	148.68	0.00	148.68	9.91
29	Korea—South	1656.00	138.00	138.00	6.00	144.00	9.60
30	United States	1790.00	149.17	132.00	6.00	138.00	9.20
31	Saudi Arabia	1920.00	160.00	141.60	0.00	141.60	9.44
32	Portugal	1691.00	140.92	140.92	0.00	140.92	9.39
33	United Kingdom	1654.00	137.83	137.83	2.00	139.83	9.32
34	Finland	1672.00	139.33	139.33	0.00	139.33	9.29
35	Ukraine	1872.00	156.00	138.06	0.00	138.06	9.20
36	Venezuela	1824.00	152.00	134.52	2.00	136.52	9.10
37	Austria	1609.00	134.08	134.08	0.00	134.08	8.94
38	Luxembourg	1609.00	134.08	134.08	0.00	134.08	8.94
39	Italy	1752.00	146.00	129.21	2.00	131.21	8.75

(*Continued*)

Table 6.36 (*Continued*) International Variations due to Local Work Hours per Month

	Countries	*OECD National Work Hours per Year*	*OECD National Work Hours per Month*	*Software Work Hours per Month*	*Software Unpaid Overtime per Month*	*Software Total Hours per Month*	*FP per Month for 1000 FP (15 WH per Funct. Pt.)*
40	Belgium	1574.00	131.17	131.17	0.00	131.17	8.74
41	New Zealand	1739.00	144.92	128.25	2.00	130.25	8.68
42	Denmark	1546.00	128.83	128.83	0.00	128.83	8.59
43	Canada	1710.00	142.50	126.11	2.00	128.11	8.54
44	Australia	1728.00	144.00	127.44	0.00	127.44	8.50
45	Ireland	1529.00	127.42	127.42	0.00	127.42	8.49
46	Spain	1686.00	140.50	124.34	2.00	126.34	8.42
47	France	1479.00	123.25	123.25	0.00	123.25	8.22
48	Iceland	1706.00	142.17	120.00	0.00	120.00	8.00
49	Sweden	1621.00	135.08	119.55	0.00	119.55	7.97
50	Norway	1420.00	118.33	118.33	0.00	118.33	7.89
51	Germany	1397.00	116.42	116.42	0.00	116.42	7.76
52	Netherlands	1381.00	115.08	115.08	0.00	115.08	7.67
	Average	1861.79	155.38	148.21	3.85	151.94	10.13

OECD, Organization for Economic Cooperation and Development.

applications where each country had exactly 15 work hours per function point. This is an artificial assumption, but the purpose of this table is to show how local work hours impact function points per staff month.

As can be seen, the metric function points per month varies widely based on local work hours. The values shown in Table 6.36 are defaults in our estimating tool, but tool users can easily override the defaults and provide their own values for local work hours per month.

Software quality is the most important topic of all, so it has been saved until last. Table 6.37 shows approximate quality levels by country, circa 2017.

As can be noted, there are wide international ranges in software quality levels. Japan has long achieved a good reputation for quality in essentially all products including software. India has more CMMI level 5 organizations than any other country and a national push to achieve high quality in all products.

Unfortunately, sample sizes are small in some countries and indeed some countries don't publish quality data at all. What is interesting to the author is that countries using function point metrics are clustered in the top half. This is because measuring quality is on the critical path to improving quality.

China is a major economic power that wants to succeed in software. But China was late in using function point metrics and even later in measuring quality. However, in recent years, China has been making a strong push to improve software. In fact, Chinese clients have commissioned more studies from the author on software metrics and measurement topics than clients from any other country for the past 5 years.

Unfortunately, China has also been making a strong push in cyber-warfare, which is an important topic but outside the scope of this book.

Probably the main contenders in military cyber warfare in 2017 are as follows:

1. China
2. Russia
3. United States
4. United Kingdom
5. North Korea
6. Iran

It would not be surprising if many of the North Korean missile failures were found to be due to bugs and poor software quality.

Software is a major product in all industrialized countries. Unfortunately, software is also rapidly being weaponized.

For any business or military purpose, software progress will be slow without effective metrics and measurement practices. The five metrics of (1) work hours per function point, (2) function points per month, (3) defect potentials in function points, (4) DRE for all forms of removal including static analysis, and (5) delivered

Table 6.37 Global Software Quality circa 2017

	Countries	*Defect Potential per FP*	*DRE (%)*	*Delivered Defects per FP*
Best Quality				
1	Japan	4.25	96.00	0.17
2	India	4.90	95.50	0.22
3	Finland	4.40	94.50	0.24
4	Switzerland	4.40	94.50	0.24
5	Denmark	4.25	94.00	0.26
6	Israel	5.00	94.80	0.26
7	Sweden	4.45	94.00	0.27
8	Netherlands	4.40	93.50	0.29
9	Hong Kong	4.45	93.50	0.29
Good Quality				
10	Brazil	4.50	93.00	0.32
11	Singapore	4.80	93.40	0.32
12	United Kingdom	4.55	93.00	0.32
13	Malaysia	4.60	93.00	0.32
14	Norway	4.65	93.00	0.33
15	Taiwan	4.90	93.30	0.33
16	Canada	4.70	93.00	0.33
17	Ireland – South	4.75	93.00	0.33
18	Korea – South	4.50	92.50	0.34
19	United States	4.71	92.58	0.35
20	Hungary	4.60	92.40	0.35
21	Mexico	4.70	92.50	0.35
22	Australia	4.85	92.70	0.35
23	Austria	4.65	92.30	0.36

(Continued)

Table 6.37 (*Continued*) Global Software Quality circa 2017

	Countries	*Defect Potential per FP*	*DRE (%)*	*Delivered Defects per FP*
24	Peru	4.80	92.50	0.36
25	Belgium	4.50	91.80	0.37
26	Luxembourg	4.20	91.00	0.38
27	Spain	4.65	91.80	0.38
28	France	4.85	92.00	0.39
Average Quality				
29	Germany	4.90	91.70	0.41
30	Philippines	5.00	91.80	0.41
31	Czech Republic	4.70	91.00	0.42
32	Ireland – North	4.80	91.00	0.43
33	New Zealand	4.60	90.50	0.44
34	Thailand	4.60	90.50	0.44
35	South Africa	4.90	91.00	0.44
36	Italy	4.95	91.00	0.45
37	Poland	4.80	90.70	0.45
38	Kuwait	4.70	90.50	0.45
39	Costa Rica	4.55	90.00	0.46
40	Bolivia	4.60	90.00	0.46
41	Estonia	4.65	90.00	0.47
42	Chile	5.20	91.00	0.47
43	Panama	4.75	90.00	0.48
44	Argentina	4.80	90.00	0.48
45	China	4.45	89.10	0.49
46	Iceland	4.30	88.50	0.49
47	Cuba	4.80	89.50	0.50

(Continued)

Table 6.37 (*Continued*) Global Software Quality circa 2017

	Countries	*Defect Potential per FP*	*DRE (%)*	*Delivered Defects per FP*
48	Bahrain	4.70	89.00	0.52
49	Ukraine	4.95	89.50	0.52
50	Venezuela	4.80	88.90	0.53
51	Portugal	4.85	89.00	0.53
52	Indonesia	4.95	89.00	0.54
53	Viet Nam	4.95	89.00	0.54
54	Jordan	5.00	89.00	0.55
55	Tunisia	5.05	89.00	0.56
56	Colombia	4.70	87.80	0.57
57	Saudi Arabia	5.00	88.50	0.58
58	Bangladesh	4.80	88.00	0.58
59	Greece	4.80	88.00	0.58
60	Algeria	4.90	88.00	0.59
61	Turkey	4.90	88.00	0.59
62	Lebanon	4.75	87.50	0.59
Poor Quality				
63	Syria	5.00	88.00	0.60
64	Pakistan	5.05	88.00	0.61
65	Libya	5.10	88.00	0.61
66	Iraq	5.60	89.00	0.62
67	Burma	4.75	87.00	0.62
68	Korea—North	5.50	88.50	0.63
69	Russia	5.30	89.00	0.58
70	Iran	5.50	88.00	0.66
	Average/sum	4.77	90.86	0.44

defects per function point are the best available metrics for both economic productivity analysis and software quality analysis.

Older and alternate metrics such as LOC, cost per defect, story points, use case points, technical debt, and quite a few others lack ISO standards and some are bad enough to distort reality and conceal progress.

Topics That Cannot Be in Benchmarks due to Laws or Union Regulations

National laws prohibit measuring some topics that are sensitive such as age, gender, and ethnic background. Although few U.S. software personnel are in unions, quite a few software personnel are unionized in Europe, Australia, and some other countries. Union regulations may prohibit measuring individual performance such as productivity.

In any case, benchmark organizations need to know the topics that are out of bounds for most software benchmarks.

The main point of Table 6.38 is that in spite of careful and thorough historical data collection, there are still some topics that impact software projects that cannot easily be studied or included in normal benchmarks.

However, many of the topics do show up during discovery and deposition phases of litigation, so some information is known to consultants who work as expert witnesses in breach of contract litigation. For example, the author has noted during expert witness assignments that management failures are far more common in breach of contract litigation than failures by the technical teams.

The author has observed major project disruptions due to the mass resignation of 65% of a corporate software organization caused by a personnel dispute. This mass resignation impacted more than 30 ongoing software projects. A benchmark study in Australia was canceled due to a strike by software staff. Another strike disrupted a benchmark study in France. A hurricane damaged a data center and shut down computer services for 2 weeks, delaying over a dozen software projects. The author has also noted project delays caused by injunctions and legal issues when a judge ordered the source code of an application frozen due to a legal dispute involving trade secrets. These unusual factors do not occur often and are not normally part of benchmark analysis.

Because of the significant volume of data collected during Namcook benchmarks, the data cannot be self-reported. Namcook consultants work with software development teams in collecting the information shown in this book.

The data collection meetings (either live or by Skype) run from 2 to 4 h per project. The client teams comprise of the project manager and up to four technical personnel. Of course, the clients see the Namcook questionnaire before the data collection sessions so some of the data is already available before the session starts.

Table 6.38 Topics That Cannot Be Studied due to Laws or Policy Restrictions

1. Academic degrees of software team members
2. Academic performance or grades of team personnel
3. Academic performance or grades of project management
4. Academic performance of clients
5. Ages of software team members
6. Genders of software team members
7. Ethnic backgrounds of software team members
8. Appraisal scores of software team members
9. Appraisal scores of project management
10. Academic accomplishments of clients providing requirements
11. Attrition rates of team during project
12. Attrition rates of key software clients during project
13. Bankruptcy or business failure of client
14. Bankruptcy or business failure of vendors
15. Black projects that are highly classified such as cyber-warfare
16. Canceled projects due to corporate embarrassment
17. Competence of clients to set schedules
18. Competence of executives to set schedules
19. Competence of managers to set schedules
20. Divestitures of business units during major software projects
21. Acquisitions of business units during major software projects
22. Legal injunctions or seizure of software assets
23. Litigation against outsource vendors
24. Litigation such as violation of employment agreements
25. Litigation for alleged patent violations
26. Litigation for alleged theft of trade secrets

(Continued)

Table 6.38 (*Continued*) Topics That Cannot Be Studied due to Laws or Policy Restrictions

27. Loss of key personnel due to downsizing or layoffs
28. Loss of key personnel due to illness or personal factors
29. Strikes or slowdowns by team members during major projects
30. Natural disasters (hurricanes, earthquakes, floods, etc.)
31. Prior breach of contract litigation against vendor building software
32. Reverse appraisal scores of project managers
33. Union membership of software personnel
34. Unpaid overtime by software personnel during project
35. Wastage, or time spent on fixing bugs or canceled projects

The interviews are designed to correct "leaks" in client historical data and provide a solid empirical base for future projects of the same size and type.

Examples of Three Software Benchmarks

This section of Chapter 6 contains three sample benchmarks produced for a client in the telecommunications industry. The three are all for central office telecommunications switching systems. One is a best-case example, one is average, and the third is a worst-case example. We are often commissioned to provide such sets of three benchmarks for the same size and type of software.

Usually benchmark data for major projects are provided verbally in meetings supported by PowerPoint slides. Paper reports using text and spreadsheets are also provided to the clients, but are secondary and used for extended study and process improvements.

We normally present output data to the clients and don't bother to present the inputs because the clients have already seen them when the data was collected. But in this book where readers have not seen the benchmark inputs, they are provided first to provide a context for the output data.

Note that all benchmarks in this book use even numbers of function points such as 10,000. In real life, projects are almost never even numbers in terms of size. But the author has a proprietary size conversion tool that converts applications to any target size. This is not easy because quality and productivity don't change at the same rate as size itself. The three benchmarks are all shown here as 10,000 function points, but in real life they were 9,800, 10,650, and 11,280 function points in size.

To speed up benchmarks and compensate for missing data, we use a technique the author calls "synthetic benchmarks." For example, if a client does not have data on topics such as unpaid overtime or specialist effort such as technical writers or quality assurance, we run our SRM estimation tool and ask the client if our results seem acceptable. Usually they are, so we use SRM estimate data to fill in holes and gaps in client data.

We have some other benchmark tools that also speed things up. Our SRM tool can convert story points or use case points to function points in <30 s. If the client uses LOC, we can convert that too. If the client has no size data at all we can use the SRM quick sizing tool and generate function point size in <5 min.

It still takes several hours to collect our benchmark data, but without our high-speed tools the same data might take several days.

Because of the large total volume of benchmark data we use, they are subset into specific sections with explanatory text in this book.

Apologies to readers for showing benchmark data in 8-point type. But with three columns of data, there is no other way to put the data onto a page in a portrait format. It is technically possible to publish the data in landscape format, but the author does not enjoy turning books sideways unless there is no other alternative.

Readers will probably think that we collect possibly too much data. The author's view is that big software projects need about the same level of data as collected by physicians when doing thorough annual medical examinations. Nobody says that doctors collect too much data, except perhaps insurance companies.

Software has suffered too long with little or no quantitative data. It is time to take software quality and software economic analysis seriously and collect the pertinent data points that impact quality, schedules, and costs.

The ultimate purpose of our benchmarks is not just to look backward at historical projects, but rather to use the benchmark information to avoid future problems and make future projects faster and more reliable.

Benchmark Section 0: Executive Summary

Because of the large amount of information provided with our benchmarks, we always start with an executive summary that contains highlights data.

As can be seen, the executive summary provides only the tip of the iceberg in terms of data. Even so, the data are important to C-level executives.

Part 0 Benchmark Executive Summary

	Benchmark 1	Benchmark 2	Benchmark 3
	Best Case	Average Case	Worst Case
	TSP/Personal Software Process (PSP)	Agile with scrum	Waterfall development
	10,000 function points	10,000 function points	10,000 function points
	1,200 SNAP points	1,500 SNAP points	1,800 SNAP points
	426,667 LOC	533,333 LOC	1,067,000 LOC
	427 KLOC	533 KLOC	1,067 KLOC
	Erlang	C++	CHILL
Security Level	*Confidential*	*Confidential*	*Confidential*
Nondisclosure agreement	Yes	Yes	Yes
Executive Summary	*Best-Case Results*	*Average-Case Results*	*Worst-Case Results*
Schedule months	22.19	23.71	29.90
Schedule slippage months	–1.25	0.71	7.71
Schedule slip (%)	–5.63	2.99	25.79
Project risk percentage (%)	9.39	24.78	–63.37
Development $ per function point	$1,214.26	$1,828.66	$4,901.58
Defect potentials	17,605	22,498	59,190
DRE (%)	99.99	99.91	99.50
Delivered defects	13	131	1,651
High-severity defects	2	20	297
Security defects	0	3	45
Customer satisfaction score (%)	97.98	81.74	35.29

Benchmark Section 1: Input Data

Because the benchmark input data are usually collected at least a week prior to the benchmark report being delivered, we usually don't present inputs to the clients, but instead provide them as an appendix to the written report. However, since this is a book and many readers are not familiar with software benchmarks, it seems best to show them first before showing outputs.

Benchmark users need to know need to know certain basic facts about the project including but not limited to the following:

1. Project size in function points
2. Project class
3. Project type
4. Country of origin
5. Salary and burden costs of project team
6. Additional costs such as travel and legal fees
7. Methodologies used
8. Programming languages used
9. Project start data
10. Planned or actual delivery date

These essential benchmark topics are supported by additional kinds of information, much of which is optional.

Part 1 Benchmark Input Data

Project Identification				
	Project name	Telecom 1	Telecom 2	Telecom 3
	Project description	Telecom switching	Telecom switching	Telecom switching
	Industry code (NAIC codes)	517 Telecom	517 Telecom	517 Telecom
	Country location (optional)	Belgium	United States	Germany
	City location (optional)	Nondisclosure	Nondisclosure	Nondisclosure
	State location (optional)	Nondisclosure	Nondisclosure	Nondisclosure
	Organization (optional)	AAAA Corporation	BBBB Corporation	CCCC Corporation

(Continued)

Part 1 (*Continued*) Benchmark Input Data

	Manager (optional)	John Doe	Jame Does	Alice Doe
	Benchmark collected by	Namcook Analytics	Namcook Analytics	Namcook Analytics
	Data provided by (requiredl)	Capers Jones	Capers Jones	Capers Jones
Key Project Dates				
	Current date (MM/DD/YY)	9/10/2013	9/10/2013	9/10/2013
	Project start date (MM/DD/YY)	2/3/2009	2/3/2009	2/3/2009
	Planned delivery date (MM/DD/YY)	1/17/2011	1/10/2013	12/24/2010
	Actual delivery date (MM/DD/YY)	11/30/2010	1/15/2011	7/19/2011
	Planned schedule in calendar months	23.44	23.00	22.19
	Actual schedule in calendar months	22.19	23.71	29.90
	Plan vs. actual schedule (plus or minus)	−1.25	0.71	7.71
Work and Staffing Inputs				
	In-house/outsourced/mixed	In-house	In-house	In-house
	U.S. default for work hours per month	132	132	132
	Project benchmark work hours per month	136	132	128
	Percentage of normal U.S work month (%)	103.03	100.00	96.97
	Unpaid overtime per month	0	0	0
	Paid overtime per month	0	0	0

(*Continued*)

Part 1 (*Continued*) Benchmark Input Data

	Schedule impact	0	0	0
	Project staffing goals	3	3	3
		Average	Average	Average
Cost Inputs				
	Funding source (optional)	Corporate R&D	Corporate R&D	Corporate R&D
	Currency	Euros converted to U.S. Dollars	U.S. Dollars	Euros converted to U.S. Dollars
Management Consultant Fees				
	Burden rate percentage	$0	$0	$0
	Loaded daily consulting cost	$5,000	$2,500	$1,800
	Loaded hourly consulting costs	$625.00	$312.50	$225.00
	Development monthly labor cost	$10,000	$10,000	$10,000
	Burden rate percentage (%)	50	50	50
	Loaded monthly development cost	$15,000	$15,000	$15,000
	Loaded hourly development cost	$113.64	$113.64	$113.64
	Maintenance monthly labor cost	$7,500	$7,500	$7,500
	Burden rate percentage (%)	50	50	50
	Loaded monthly maintenance cost	$11,250	$11,250	$11,250
	Loaded hourly maintenance cost	$85.23	$85.23	$85.23
	Customer support labor costs	$3,500	$3,500	$3,500

(Continued)

Part 1 (*Continued*) Benchmark Input Data

	Burden rate percentage (%)	35.00	35.00	35.00
	Loaded monthly support cost	$4,725	$4,725	$4,725
	Loaded hourly support cost	$35.80	$35.80	$35.80
	User monthly labor cost	Not included	Not included	Not included
	Burden rate percentage	Not included	Not included	Not included
	Loaded monthly user cost	Not included	Not included	Not included
Additional Costs (if any)				
	Project travel costs (to clients)	$350,000	$425,000	$825,000
	Project travel costs (other labs)	$200,000	$300,000	$600,000
	Patent filing fees	$275,000	$0	$0
	Management consulting risk analysis	$125,000	$0	$0
	Management consulting project turnaround	$0	$50,000	$675,000
	Litigation costs as plaintiff	$0	$295,000	$0
	Litigation costs as defendant	$0	$375,000	$550,000
	Total additional costs	$950,000	$1,445,000	$2,650,000
	Additional costs per function point	$95.00	$144.50	$265.00
	Additional costs per KLOC	$2,194.00	$2,711.07	$2,483.60
Value Inputs (if known)				
	Direct revenues	$125,000,000	$75,000,000	$50,000,000
	Indirect revenues	$133,750,000	$56,250,000	$20,000,000
	Cost reductions	$35,000,000	$20,000,000	($10,000,000)

(*Continued*)

Part 1 (*Continued*) Benchmark Input Data

	Total value	$293,750,000	$151,250,000	$60,000,000
	Total value per function point	$29,375.00	$15,125.00	$6,000.00
	Development cost	$13,093,875.98	$19,733,423.23	$51,670,755
	5-year M&E cost	$5,237,550.39	$13,813,396.26	$59,421,368
	Total cost of ownership (TCO)	$18,331,426.38	$33,546,819.49	$111,092,122
	TCO per function point	$1,833.14	$3,354.68	$11,109
	Project value total	$293,750,000	$151,250,000	$60,000,000
	Project TCO	$18,331,426	$33,546,819	$111,092,122
	Project return on investment (ROI)	$275,418,574	$117,703,181	($51,092,122)
	$ value per $ expense	$16.02	$4.51	–$5.19
Project Taxonomy Inputs				
1	Project nature—new application	10	10	10
2	Project scope—corporate system	130	130	130
3	Project class—external; bundled with hardware	130	130	130
4	Project type —telecommunications	140	140	140
5	Secondary type—public switching	140.1	140.1	140.1
6	Hardware platform—IBM mainframe	120	120	120
7	Software Platform—IBM operating system	80	80	80

(*Continued*)

Part 1 (*Continued*) Benchmark Input Data

Complexity Inputs				
8	Problem complexity	5.00	5.00	5.00
9	Code complexity	3.00	5.00	6.00
10	Data complexity	6.00	6.00	6.00
	Average complexity	4.67	5.33	5.67
		Below average	Average	Above average
Methodology/Goals Inputs				
	CMMI Level	5	3	1
	Methodology	TSP/PSP 9	Agile/Scrum 8	Waterfall 29
	Project goals—average staffing levels	3	3	3
Experience Inputs				
1	Client experience	1.00	3.00	5.00
2	Development team experience	1.00	3.00	5.00
3	Test team experience	1.00	3.00	5.00
4	SQA team experience	1.00	3.00	5.00
5	Maintenance team experience	1.00	3.00	5.00
6	Support team experience	1.00	3.00	5.00
7	Technical communications experience	1.00	3.00	5.00
8	Project management experience	1.00	3.00	5.00
	Experience average	1.00	3.00	5.00
		Very experienced	Average experience	Inexperienced

(*Continued*)

Part 1 (*Continued*) Benchmark Input Data

Programming Language Inputs				
Programming Languages (New Applications)				
	Number of languages in application (1 to 10)	1	1	1
	Primary language	Erlang	C++	CHILL
	Primary language level	7.50	6.00	3.00
	LOC per function point	42.67	53.33	106.7
	Percentage of application (%)	100	100	100
	Percentage of reused code and other deliverables (%)	26.02	22.50	9.53

Readers may wonder why many of the input data points are in numeric form such as "5" or "10." The answer is that numeric data are needed because sometimes mathematical calculations are performed on the answers.

We use many of the same input topics for cost estimation and therefore need to have numeric values.

For example, the team-experience input values run from 1 to 5 with the following meanings:

1. Expert with significant experience
2. Above-average experience
3. Average experience
4. Below-average experience
5. Novice with little experience

Readers should not forget that benchmark data may be collected for several hundred projects within a large company and for many thousands of projects within specific industries. In order to carry out statistical analysis of the benchmark data, most of the input information needs to be recorded numerically.

Benchmark Section 2: Output Data

Normally benchmark outputs are provided to clients in face-to-face meetings or webinars using PowerPoint slides. Spreadsheet and text data are provided too, but are intended for more thoughtful long-range analysis than the initial presentation.

The first segment of our benchmark outputs shows the client the effort and costs of collecting the benchmark data. These three samples have benchmark costs between $15,000 and $22,000.

Part 2 Benchmark Outputs

Detailed Benchmark Data	*TSP/PSP*	*Agile with Scrum*	*Waterfall Development*
Project Cost-Driver Overview	**TSP/PSP Cost Drivers**	**Agile Project Drivers**	**Waterfall Cost Drivers**
	1) The cost of requirements creep	1) The cost of meetings	1) The cost of fixing bugs
	2) The cost of programming	2) The cost of fixing bugs	2) The cost of security flaws
	3) The cost of paperwork	3) The cost of programming	3) The cost of paperwork
	4) The cost of meetings	4) The cost of requirements creep	4) The cost of requirements creep
	5) The cost of fixing bugs	5) The cost of paperwork	5) The cost of programming
	6) The cost of management	6) The cost of management	6) The cost of meetings
	7) The cost of security flaws	7) The cost of security flaws	7) The cost of management
Benchmark Collection Cost Analysis	*TSP/PSP*	*Agile with Scrum*	*Waterfall Development*
Benchmark Client Effort			
Client Benchmark Data Collection			
Executive hours—data collection	1.00	1.00	1.00
Management hours—data collection	2.00	4.00	3.00
Team hours—data collection	8.00	12.00	9.00
Total hours—data collection	11.00	17.00	13.00

(Continued)

Part 2 (*Continued*) Benchmark Outputs

Client Benchmark Data Reporting			
Executive hours—benchmark data report	12.00	12.00	12.00
Management hours—benchmark data report	4.00	4.00	4.00
Team hours—benchmark data report	16.00	16.00	16.00
Total hours—data reporting	32.00	32.00	32.00
Total client benchmark hours	43.00	49.00	45.00
Benchmark Consulting Effort			
Benchmark consulting modeling with SRM	1.00	1.50	1.00
Benchmark consultant hours—collection	2.50	3.50	3.00
Benchmark consultant hours—risk analysis	1.50	2.50	3.50
Benchmark consultant hours—data analysis	6.00	7.00	7.50
Benchmark consultant hours—prep report	6.00	14.00	15.00
Benchmark consultant hours—presentations preparation	3.00	3.00	3.00
Benchmark consultant hours—present data	4.00	4.00	5.00
Total consulting hours	24.00	35.50	38.00
Total benchmark effort	47.00	84.50	83.00

(*Continued*)

Part 2 (*Continued*) Benchmark Outputs

Client benchmark collection costs	$1,250	$1,932	$1,477
Client benchmark report costs	$3,636	$3,636	$3,636
Consulting costs	$10,800	$15,975	$17,100
Total benchmark costs	$15,687	$21,543	$22,214
Benchmark costs per function point	$1.57	$2.15	$2.22
Benchmark costs per KLOC	$36.76	$40.39	$22.08

As can be seen, the best-case example was the cheaper and easier to collect data than the worst-case example. There are several reasons for this and one of them is that the best-case projects have better data and use tools such as the Automated Project Office (APO) that makes the data readily available.

The worst-case projects seldom have complete data so a lot of digging is needed on the part of both benchmark consultants and client benchmark team members. Worst-case projects usually spend many hours of unpaid overtime but seldom measure this key factor.

Sizing and Other Quantitative Topics Including Risk Assessments

All benchmarks need size in function points. We also collect size in terms of logical code statements too. In addition, we collect data on numbers and kinds of paper documents produced.

Readers may wonder why paperwork is important, but for defense software projects paperwork is the #1 cost driver and is more expensive than source code itself. Military standards lead to the creation of enormous document sets about three times larger than for civilian projects of the same size and type.

Size and Requirements Creep

Size at End of Requirements				
	New application size (function points)	9,037	8,412	7,523
	SNAP points	1,084	1,260	1,440

(*Continued*)

(*Continued*) Size and Requirements Creep

	New application size (logical code)	385,609	448,612	802,704
Size at Delivery				
	Total requirements creep (%)	9.63	15.88	24.77
	Size at delivery (function points)	10,000	10,000	10,000
	SNAP points	1,200	1,500	18,000
	Size at delivery (logical code)	426,667	533,333	1,067,000
	Size at delivery (KLOC)	426.67	533.33	1,067.00
	Percentage reuse (%)	26.02	16.25	9.53
		TSP/PSP	*Agile with Scrum*	*Waterfall Development*
Major Software Project Risks				
1	Cancellation of project (%)	10.41	26.47	38.26
2	Executive dissatisfaction with progress (%)	12.80	29.40	94.90
3	Negative ROI (%)	11.65	33.53	62.14
4	Cost overrun (%)	15.75	29.12	56.59
5	Schedule slip (%)	16.21	35.30	84.35
6	Unhappy customers (%)	4.00	36.00	68.00
7	Low team morale (%)	4.50	16.30	77.60
8	High warranty repair costs (%)	6.74	17.63	76.45
9	Litigation (for outsourced work) (%)	5.30	11.65	17.64
10	Competition releasing superior product (%)	6.50	12.44	57.80
	Average risks (%)	9.39	24.78	63.37

(*Continued*)

(*Continued*) Size and Requirements Creep

	Documentation Size Benchmarks (pages)	*TSP Pages*	*Agile Pages*	*Waterfall Pages*
1	Requirements	2,230	758	2,497
2	Architecture	376	125	414
3	Initial design	2,753	826	3,358
4	Detail design	5,368	939	6,818
5	Test plans	1,158	463	695
6	Development plans	550	40	165
7	Cost estimates	376	188	207
8	User manuals	2,214	1,993	2,126
9	HELP text	1,964	1,846	1,689
10	Courses	1,450	1,160	1,508
11	Status reports	776	854	1,319
12	Change requests	1,867	2,128	2,464
13	Bug reports	14,084	17,999	47,352
	Total	35,166	29,319	70,612
	Total pages per function point	3.52	2.93	7.06
	Total pages per KLOC	82.42	54.97	66.18
	Total diagrams/graphics in documents	4,118	3,867	3,182
	Total words in documents	17,582,868	14,659,479	35,305,709
	Total words per function point	1,758.29	1,465.95	3,530.57
	Total words per LOC	41.21	27.49	33.09
	Work hours per page	1.05	0.95	1.00
	Paperwork months for all documents	271.50	211.01	551.65

(Continued)

(*Continued*) Size and Requirements Creep

	Paperwork hours for all documents	36,924.02	27,853.01	70,611.42
	Document hours per function point	3.69	2.79	7.06
	Documentation costs	$4,196,046	$3,165,216	$8,024,282
	Documentation costs per function point	$419.60	$316.52	$802.43
	Document percentage of total costs (%)	34.56	17.31	16.37
	Project Meetings/ Communications	*TSP/PSP Meetings*	*Agile Meetings*	*Waterfall Meetings*
	Conference calls	42	33	55
	Client meetings	26	368	25
	Architecture/design meetings	14	13	22
	Team status meetings	78	356	90
	Executive status meetings/phase reviews	18	19	28
	Problem analysis meetings	7	7	23
	Total meetings	184	796	244
		TSP/PSP Participants	*Agile Participants*	*Waterfall Participants*
	Conference calls	6	7	10
	Client meetings	5	3	7
	Architecture/design meetings	11	10	14
	Team status meetings	27	45	71
	Executive status meetings/phase reviews	9	8	15
	Problem analysis meetings	6	8	26

(*Continued*)

(*Continued*) Size and Requirements Creep

	Total participants	65	81	143
	Average participants	11	14	24
		TSP/PSP Meeting Hours	*Agile Meeting Hours*	*Waterfall Meeting Hours*
	Conference calls	0.60	0.45	0.70
	Client meetings	2.70	1.30	3.80
	Architecture/design meetings	4.00	3.00	5.00
	Team status meetings	2.00	2.65	3.00
	Executive status meetings/phase reviews	4.00	4.00	5.00
	Problem analysis meetings	5.50	6.25	12.00
	Total duration hours	18.80	17.65	29.50
	Average duration hours	3.13	2.94	4.92
		TSP/PSP Meeting Hours	*Agile Meeting Hours*	*Waterfall Meeting Hours*
	Conference calls	152.78	107.11	404.79
	Client meetings	377.03	1,228.25	614.66
	Architecture/design meetings	631.42	402.30	1,531.74
	Team status meetings	4,249.92	42,401.96	19,116.16
	Executive status meetings/phase reviews	647.61	607.01	2,130.35
	Problem analysis meetings	213.71	365.73	7,183.08
	Total duration hours	6,272.47	45,112.37	30,980.78
	Conference calls	$17,362	$12,172	$46,000

(Continued)

(*Continued*) Size and Requirements Creep

		TSP/PSP Meeting Total Costs	*Agile Meeting Total Costs*	*Waterfall Meeting Total Costs*
	Client meetings	$42,846	$139,578	$69,850
	Architecture/design meetings	$71,754	$45,718	$174,067
	Team status meetings	$482,961	$4,818,559	$2,172,361
	Executive status meetings/phase reviews	$73,594	$68,981	$242,093
	Problem analysis meetings	$24,286	$41,562	$816,286
	Total meeting costs	$712,803	$5,126,570	$3,520,656
	Meeting $ per function point	$71.28	$512.66	$352.07
	Meeting percentage of total costs (%)	5.87	28.03	7.18

As can be observed, both paper documents and meeting costs are expensive. For agile projects, meeting costs can actually approach 30% of total application costs and may be more expensive than code itself.

Software Quality Benchmark Data

For the entire software industry, the #1cost driver is that of finding and fixing bugs. These costs include pretest inspections and static analysis, all forms of testing, and post-release bug repairs. Therefore, all software benchmarks should be very accurate on defect numbers and defect repair costs.

One topic needs to be explained. When we provide benchmark data on individual projects, the data is usually in multiple columns and side by side. But for showing three projects on the same page, the author had to change the format to a vertical format.

Usually our quality data looks like this: defect potential, removal efficiency, delivered defects.

Unfortunately, in this book, to get everything on the same page for three separate projects, the author had to change the format to the following:

Defect potential
Removal efficiency
Delivered defects

Quality and Defect Potentials (Requirements + Design + Code + Documents + Bad Fixes)				
		Best Case	*Average Case*	*Worst Case*
		TSP	*Agile with Scrum*	*Waterfall Development*
Per Function Point				
1	Requirement defects	0.41	0.39	1.33
2	Design defects	0.60	0.67	0.90
3	Code defects	0.66	1.04	3.04
4	Document defects	0.07	0.07	0.08
5	Bad fixes (secondary defects)	0.02	0.07	0.57
	Defect Potentials	1.76	2.24	5.92
		Below average for telecom	*Average for telecom*	*Well above average for telecom*
	Defect potentials (per KLOC—logical statements)	41.26	42.18	55.47
	Application Defect Potentials	*Best Case*	*Average Case*	*Worst Case*
1	Requirement defects	4,086	3,942	13,277
2	Design defects	6,037	6,675	9,003
3	Code defects	6,629	10,448	30,399
4	Document defects	674	738	810
5	Bad fixes (secondary defects)	179	696	5,700
	Total defect potential of application	17,605	22,498	59,190
	DRE	*Defects Remaining*	*Defects Remaining*	*Defects Remaining*
1	JAD (%)	27.00	27.00	0.00
	Defects remaining	12,851	16,424	59,190

2	QFD (%)	30.00	0.00	0.00
	Defects remaining	8,996	16,424	59,190
3	Prototype (%)	30.00	30.00	24.00
	Defects remaining	6,297	11,497	44,984
4	Desk check (%)	27.00	27.00	23.00
	Defects remaining	4,597	8,393	34,638
5	Pair programming (%)	0.00	0.00	15.00
	Defects remaining	4,597	8,393	29,442
6	Static analysis (%)	55.00	55.00	55.00
	Defects remaining	2,069	3,777	15,587
7	Formal inspections (%)	93.00	65.00	0.00
	Defects remaining	145	1,322	15,587
8	Unit test (%)	32.00	30.00	28.00
	Defects remaining	98	925	11,223
9	Function test (%)	35.00	33.00	31.00
	Defects remaining	64	620	7,744
10	Regression test (%)	14.00	13.00	12.00
	Defects remaining	55	539	6,814
11	Component test (%)	32.00	32.00	32.00
	Defects remaining	37	367	4,634
12	Performance test (%)	14.00	14.00	14.00
	Defects remaining	32	315	3,985
13	System test (%)	36.00	36.00	36.00
	Defects remaining	21	202	2,550
14	Platform test (%)	22.00	22.00	22.00
	Defects remaining	16	157	1,989
15	Acceptance test (%)	17.00	17.00	17.00
	Defects remaining	13	131	1,651

	Delivered defects	13	131	1,651
	Defects per function point	0.0013	0.0131	0.1651
	Defects per KLOC	0.0313	0.2501	1.5475
	High-severity defects	2	20	297
	Security flaws	0	3	45
	DRE (%)	99.99	99.91	99.50
	Defect removal work months	302.19	424.27	1,082.17
	Defect removal work hours	41,097	56,004	138,518
	Defect removal hours per function point	4.11	5.60	8.20
	Defect removal costs	$4,670,314	$6,364,257	$15,741,199
	Defect removal costs per function point	$467.03	$636.43	$1,574.12
	Defect removal percentage of total costs (%)	38.46	40.27	32.11
	Test coverage (%) (approximate)	96.50	80.76	74.20
	Test cases (approximate)	39,811	36,308	34,041
	Test case errors (approximate)	995	1,071	1,004
	Maximum cyclomatic complexity	10	16	37
	Reliability (days to initial defect)	30.00	21.24	3.60
	Mean time between failures (days)	248.00	183.00	23.00
	Stabilization period (months) (months to approach zero defects)	0.72	5.83	22.69
	Error-prone modules (>5 indicates quality malpractice)	0	1	4
	Customer satisfaction (%)	97.98	81.74	35.29

This format is harder to read and understand but is necessary to show three separate projects on the same page.

Note that the final data point for quality is customer satisfaction. IBM discovered in the 1960s that customer satisfaction had a strong correlation to product quality. In fact, IBM's attention to product quality is one of the key reasons why IBM was able to grow faster than its competitors in the early days of the computer era.

Schedule, Effort, and Cost Benchmark Data

We come now to the heart of software benchmarks: project schedules, staffing, effort, and costs. As it happens, C-level execs tend to be more interested in schedules than any other topic, although they should be interested in quality.

For internal software projects with local users we also include user costs in our benchmark data collection. However, these are three telecommunications switching systems so the "users" are all indirect users who interact with the software when making phone calls. In other words, user costs are irrelevant since none of the users are local or part of the software construction.

Software schedules are tricky to measure due to "overlap." There is no true waterfall in software. Design starts before requirements are done, and coding starts before design is done. This is the normal situation for software.

Development Schedule, Effort, and Costs (From Requirements through Delivery to Clients)				
		TSP	*Agile with Scrum*	*Waterfall Development*
	Average monthly cost	$15,000	$15,000	$15,000
	Average hourly cost	$113.64	$113.64	$113.64
Overall Project				
	Development schedule (months)	22.19	23.71	29.90
	Staff (technical+management)	36.49	51.41	109.29
	Development effort (staff months)	809.51	1,219.10	3,267.72
	Development costs	$12,142,633	$18,286,560	$49,015,807
	Development $ per function point	$1,214.26	$1,828.66	$4,901.58

	Development Work Months	*Work Months*	*Work Months*	*Work Months*
1	Requirements	36.98	46.57	64.97
2	Design	59.17	74.52	103.96
3	Coding	333.43	545.49	1,741.47
4	Testing	200.06	327.29	1,044.88
5	Documentation	26.90	33.87	47.25
6	Quality assurance	34.64	42.34	57.28
7	Management	118.34	149.03	207.91
	Totals	809.52	1,219.10	3,267.72
Normalized Productivity Data				
	IFPUG function points per month	12.35	8.20	3.06
	IFPUG work hours per function point	11.01	16.09	41.83
	Logical LOC per month	527.07	437.48	326.53
	Development Activity Schedules	*Schedule Months*	*Schedule Months*	*Schedule Months*
1	Requirements	2.57	2.56	2.64
2	Design	3.11	3.10	3.19
3	Coding	9.64	10.26	13.18
4	Testing	6.54	6.97	8.96
5	Documentation	3.61	3.63	3.77
6	Quality assurance	5.71	5.59	5.65
7	Management	22.19	23.71	29.90
	Totals	31.17	32.11	37.39
	Overlap percentage (%)	71.20	73.8	80.0
	Overlapped net schedule	22.19	23.71	29.90
	Probable schedule without reuse	26.62	27.27	33.19

	Development Activity Staffing	*Development Staffing*	*Development Staffing*	*Development Staffing*
1	Requirements	14.42	18.17	24.58
2	Design	19.06	24.07	32.60
3	Coding	34.58	53.17	132.14
4	Testing	30.57	46.97	116.65
5	Documentation	7.46	9.34	12.54
6	Quality assurance	6.07	7.57	10.13
7	Management	6.83	8.53	11.44
	Totals	36.49	51.41	109.29
	Development Activity Costs	*Development Costs*	*Development Costs*	*Development Costs*
1	Requirements	$554,721	$698,581	$974,601
2	Design	$887,553	$1,117,730	$1,559,362
3	Coding	$5,001,405	$8,182,284	$26,121,978
4	Testing	$3,000,843	$4,909,370	$15,673,187
5	Documentation	$403,433	$508,059	$708,801
6	Quality assurance	$519,573	$635,074	$859,153
7	Management	$1,775,106	$2,235,461	$3,118,725
	Totals	$12,142,633	$18,286,560	$49,015,807
	Cost per function point	$1,214.26	$1,828.66	$4,901.58
	Cost per logical code statement	$28.46	$34.29	$45.94
8	Additional costs (travel, consulting, etc.)	$950,000	$1,445,000	$2,650,000
	Project grand total	*$13,093,876*	*$19,733,423*	*$51,670,755*
	Grand total per function point	$1,309.39	$1,973.34	$5,167.08
	Grand total per LOC	$30.69	$37.00	$48.43

	Development Productivity 1	*Work Hours per Function Point*	*Work Hours per Function Point*	*Work Hours per Function Point*
1	Requirements	0.50	0.61	0.83
2	Design	0.80	0.98	1.33
3	Coding	4.53	7.20	22.29
4	Testing	2.72	4.32	13.37
5	Documentation	0.37	0.45	0.60
6	Quality assurance	0.47	0.56	0.73
7	Management	1.61	1.97	2.66
	Total with reusable materials	11.01	16.09	41.83
	Totals without reusable materials	14.44	20.30	46.21
	Development Productivity 2	*Function Points per Month*	*Function Points per Month*	*Function Points per Month*
1	Requirements	270.41	214.72	153.91
2	Design	169.00	134.20	96.19
3	Coding	29.99	18.33	5.74
4	Testing	49.99	30.55	9.57
5	Documentation	371.81	295.24	211.62
6	Quality assurance	288.70	236.19	174.59
7	Management	84.50	67.10	48.10
	Total with reusable materials	12.35	8.20	3.06
	Totals without reusable materials	9.42	6.50	2.77
	Project Occupations and Specialists	*Normal Full-Time Staffing*	*Normal Full-Time Staffing*	*Normal Full-Time Staffing*
1	Programmers	28.7	43.5	106.7

2	Testers	25.5	38.6	94.3
3	Designers	12.4	15.6	21.2
4	Business analysts	12.4	15.6	21.2
5	Technical writers	5.6	7.0	9.4
6	Quality assurance	4.9	6.1	8.1
7	First-line managers	5.6	7.4	14.7
8	Database administration	2.7	3.9	8.2
9	Project office staff	2.4	3.3	7.1
10	Administrative support	2.7	3.9	8.2
11	Configuration control	1.6	2.2	4.6
12	Project librarians	1.3	1.8	3.8
13	Second-line managers	1.1	1.5	2.9
14	Estimation specialists	0.9	1.3	2.7
15	Architects	0.6	0.9	1.9
16	Security specialists	1.0	0.5	1.1
17	Performance specialists	1.0	0.5	1.1
18	Function point counters	1.0	0.5	1.1
19	Human factors specialists	1.0	0.5	1.1
20	Third-line managers	0.3	0.4	0.7

Note that a study funded by AT&T on software occupation groups turned up a total of 126 different kinds of occupations working on software. In our benchmarks, we normally only show 20 of these occupations unless there is some compelling reason to show others.

It is an unfortunate fact that many companies do not record software occupations. Our study for AT&T included IBM, Texas Instruments, the U.S. Navy, and about a dozen other large organizations. Not a single person of the human resource departments knew how many software workers were employed, much less their occupations. Many HR groups used "member of the technical staff" as a generic job title that could include dozens of different kinds of specialists.

In order to get accurate data on software staff numbers and occupation groups, we had to interview line managers in various business units. HR departments were essentially worthless for demographic data. This raises an interesting question in how the Department of Commerce is able to publish software demographic data since they tend to depend on HR groups and never go on site as we do.

Benchmark Section 3: Technology Stack Evaluation

In addition to collecting quantitative data, we also evaluate the tools, methods, and technologies used for the project.

We use a scale that runs from 1 to 10, with 10 as the maximum positive value. Usually C-level executives are very interested in these evaluations. They can also be used in possible litigation, although projects with good technology stacks usually don't have any litigation.

Part 3 Technology Stack Evaluation and Scoring

Scoring Totals				
	>300	Excellent		
	225–299	Good		
	175–225	Average		
	125–175	Marginal		
	<125	Poor		
	<75	Failure close to 100%		
	Technology Stack Scoring	*Best Weighted Scores*	*Average Weighted Scores*	*Worst Weighted Scores*
1	Benchmarks (validated historical data from similar projects)	10.00	0.00	0.00
2	Defect potential <2.5	10.00	0.00	0.00
3	DRE >99%	10.00	0.00	0.00
4	Estimates: parametric estimation tools	10.00	0.00	0.00
5	Estimates: TCO cost estimates	10.00	0.00	0.00
6	Formal and early quality predictions	10.00	0.00	0.00

(Continued)

Part 3 (*Continued*) Technology Stack Evaluation and Scoring

7	Formal and early risk abatement	10.00	0.00	0.00
8	Inspection of all critical deliverables	10.00	0.00	0.00
9	Metrics: defect potential measures	10.00	0.00	0.00
10	Metrics: DRE measures	10.00	10.00	0.00
11	Metrics: SRM pattern-matching sizing	10.00	10.00	0.00
12	Pre-requirements risk analysis	10.00	0.00	0.00
13	Static analysis of all source code	10.00	10.00	0.00
14	Metrics: IFPUG function points	0.00	9.50	9.50
15	SEMAT usage on project	9.50	0.00	0.00
16	APO	9.00	0.00	0.00
17	Estimates: activity-based cost estimates	9.00	9.00	9.00
18	Estimates: cost of quality (COQ) estimates	9.00	9.00	9.00
19	Metrics: COSMIC function points	0.00	0.00	0.00
20	Metrics: defect detection efficiency (DDE) measures	9.00	0.00	0.00
21	Metrics: FISMA function points	0.00	0.00	0.00
22	Metrics: NESMA function points	0.00	0.00	0.00
23	Reusable test materials	0.00	0.00	0.00
24	Test coverage >96%	9.00	0.00	0.00
25	Test coverage tools used	9.00	9.00	9.00
26	Accurate cost tracking	8.50	8.50	0.00
27	Accurate defect tracking	8.50	8.50	0.00
28	Accurate status tracking	8.50	0.00	0.00
29	DRE >95%	8.50	8.50	8.50
30	Methods: TSP/PSP	8.50	0.00	0.00
31	Methods: hybrid: (agile/TSP)	0.00	8.50	0.00
32	Requirements change tracking	0.00	0.00	0.00
33	Reusable designs	8.50	0.00	0.00

(*Continued*)

Part 3 (*Continued*) Technology Stack Evaluation and Scoring

34	Reusable requirements	0.00	0.00	0.00
35	Reusable source code	8.50	8.50	8.50
36	Automated requirements modeling	8.00	0.00	0.00
37	Mathematical test case design	8.00	0.00	0.00
38	Methods: RUP	0.00	0.00	0.00
39	Metrics: cyclomatic complexity tools	8.00	8.00	8.00
40	Requirements change control board	8.00	8.00	8.00
41	Reusable architecture	8.00	8.00	0.00
42	Reusable user documents	8.00	8.00	0.00
43	CMMI 5	0.00	0.00	0.00
44	Methods: Agile <1000 function points	0.00	0.00	0.00
45	Methods: correctness proofs—automated	0.00	0.00	0.00
46	Methods: hybrid (waterfall/agile)	0.00	0.00	0.00
47	Methods: quality function deployment (QFD)	0.00	0.00	0.00
48	Metrics: unadjusted function points	0.00	0.00	0.00
49	Six sigma for software	0.00	0.00	0.00
50	Automated testing	7.00	7.00	7.00
51	DRE >90%	0.00	0.00	7.00
52	Methods: joint application design (JAD)	0.00	0.00	0.00
53	Metrics: COQ measures	0.00	0.00	0.00
54	Static analysis of text requirements	0.00	0.00	0.00
55	CMMI 4	0.00	0.00	0.00
56	Maintenance: data mining	0.00	0.00	0.00
57	Methods: DevOps	0.00	0.00	0.00
58	Metrics: automated function points	6.00	6.00	6.00
59	Metrics: earned value analysis (EVA)	0.00	0.00	0.00

(*Continued*)

Part 3 (*Continued*) Technology Stack Evaluation and Scoring

60	Metrics: FOG/Flesch readability scores	0.00	0.00	0.00
61	Methods: extreme programming	0.00	0.00	0.00
62	Methods: service-oriented models	0.00	0.00	0.00
63	Metrics: mark II function points	0.00	0.00	0.00
64	CMMI 3	5.00	5.00	5.00
65	Methods: continuous development	0.00	0.00	0.00
66	Methods: Kanban/Kaizen	0.00	0.00	0.00
67	Metrics: story point metrics	0.00	0.00	0.00
68	Methods: iterative	0.00	0.00	0.00
69	Methods: iterative development	0.00	0.00	0.00
70	Methods: spiral development	0.00	0.00	0.00
71	Metrics: technical debt measures	4.50	4.50	0.00
72	Certified quality assurance personnel	4.00	4.00	0.00
73	Certified test personnel	4.00	4.00	0.00
74	Defect potential 2.5–4.9	4.00	4.00	4.00
75	ISO risk standards	0.00	0.00	0.00
76	Maintenance: ITIL	0.00	0.00	0.00
77	Metrics: use case metrics	0.00	0.00	0.00
78	CMMI 2	0.00	0.00	0.00
79	ISO quality standards	3.00	3.00	0.00
80	Requirements modeling—manual	0.00	0.00	0.00
81	Estimates: phase-based cost estimates	0.00	0.00	0.00
82	Test coverage <90%	0.00	0.00	0.00
83	Testing by untrained developers	0.00	0.00	0.00
84	CMMI 0 (not used)	0.00	0.00	0.00
85	Methods: correctness proofs—manual	0.00	0.00	0.00
86	Benchmarks (unvalidated self-reported benchmarks)	0.00	0.00	0.00
87	CMMI 1	0.00	0.00	–2.00

(*Continued*)

Part 3 (*Continued*) Technology Stack Evaluation and Scoring

88	Methods: agile >5000 function points	0.00	–3.00	0.00
89	Cyclomatic complexity >20	0.00	0.00	–4.00
90	Methods: waterfall development	0.00	0.00	–4.50
91	Test coverage not used	0.00	0.00	–4.50
92	Methods: pair programming	0.00	0.00	–5.00
93	Metrics: cost per defect metrics	0.00	0.00	–5.00
94	Inaccurate cost tracking	0.00	0.00	–5.50
95	Estimates: manual estimation >250 function points	0.00	0.00	0.00
96	Inaccurate defect tracking	0.00	0.00	–6.00
97	Inaccurate status tracking	0.00	0.00	–7.00
98	Metrics: no function point measures	0.00	0.00	0.00
99	DRE <85%	0.00	0.00	0.00
100	Defect potential >5	0.00	0.00	–8.00
101	Metrics: no productivity measures	0.00	0.00	0.00
102	Methods: cowboy development	0.00	0.00	0.00
103	Metrics: LOC for economic study	0.00	0.00	0.00
104	Metrics: no quality measures	0.00	0.00	0.00
105	No static analysis of source code	0.00	0.00	–10.00
	Scoring sum	338.50	175.50	37.00
	Scoring average	8.26	7.02	1.54

As is evident, the best-case project had quite an impressive set of beneficial technologies and essentially zero harmful technologies. The worst-case project was almost a total reversal with many harmful technologies and few good ones.

Benchmark Section 4: Topics Observed During Benchmark Process

When we do on-site benchmarks, we tend to have many discussions with clients about how they build software. These are informal and not included in benchmark quantitative data. However, we like to show clients these observations.

Part 4 Reasons for Best-, Average-, and Worst-Case Results

From regression studies.			
Data takes about 4–6 h.			
Analysis of the collected data requires 8–16 h.			
	Best Case	*Average Case*	*Worst Case*
	TSP/PSP	*Agile with Scrum*	*Waterfall Development*
	Specific Project Factors	*Specific Project Factors*	*Specific Project Factors*
	Software reuse of >26%	Software reuse of 18%	Software reuse of <10%
	Expert teams	Average teams	Inexperienced teams
	Effective methodology	Fair methodology	Inadequate methodology
	Erlang programming language	C++ programming language	CHILL programming language
	Formal inspections	Informal inspections	No inspections
	Full static analysis	Partial static analysis	No static analysis
	CMMI level 5	CMMI level 3	CMMI level 1
	Work week of 136 h	Work week of 132 h	Work week of 128 h

(Continued)

Part 4 (*Continued*) Reasons for Best-, Average-, and Worst-Case Results

	Telecom quality is a key factor	Telecom quality is a key factor	Telecom quality is a key factor
	Solo programming	Solo programming	Pair programming (very expensive)
	Project structure	*Project structure*	*Project structure*
	Effective decomposition into discrete, buildable components	Partial decomposition but many structural problems	Complex structure does not allow effective decomposition
	Maximum Cyclomatic Complexity<10	*Maximum Cyclomatic Complexity < 20*	*Maximum Cyclomatic Complexity > 50*
	Certified reusable materials >25%	Uncertified reusable materials	Unreliable reusable materials
	Unplanned requirements <0.25% per month	Unplanned requirements >0.8% per month	Unplanned requirements >1.5% per month
	Very few security vulnerabilities <0.01 per function point	Some security vulnerabilities >0.15 per function point	Many security vulnerabilities >0.3 per function point
	Zero error-prone modules	At least one error-prone module	Multiple error-prone modules

(*Continued*)

Part 4 (*Continued*) Reasons for Best-, Average-, and Worst-Case Results

	Good maintainability (5 stars)	Fair maintainability (3 stars)	Poor maintainability (1 star)
	Static analysis of legacy code	Partial static analysis of legacy code	No static analysis of legacy code
	Full renovation of legacy code before enhancements	Partial renovation of legacy code before enhancements	No renovation of legacy code before enhancements
	Team Education	*Team Education*	*Team Education*
	Effective curriculum planning	Marginal curriculum planning	No curriculum planning
	Annual Training	*Annual Training*	*Annual Training*
	Executives=>3 days	Executives=>±2 days	Executives=±0 days
	Managers=>5 days	Managers=±3 days	Managers=<2 days
	Developers=>5 days	Developers=±3 days	Developers=<2 days
	Maintainers=>5 days	Maintainers=±3 days	Maintainers=<2 days
	Testers=>5 days	Testers =±3 days	Testers=<2 days
	SQA=>5 days	SQA =±3 days	SQA=<2 days

(*Continued*)

Part 4 (*Continued*) Reasons for Best-, Average-, and Worst-Case Results

	Webinars:=>8 per year	Webinars:=±4 per year	Webinars=±2 per year
	Certification for Key Positions	*Certification for Key Positions*	*Certification for Key Positions*
	Managers=>30%	Managers=±15%	Managers=<10%
	Developers=>20%	Developers=±15%	Developers=<10
	Maintainers=15%	Maintainers=±12%	Maintainers:=<10%
	Testers=>50%	Testers=±20%	Testers=<15%
	SQA=>25%	SQA=±15%	SQA=<10%
	Significant annual training $	Average annual training $	Inadequate annual training $
	Executives >$10,000	Executives >$5,000	Executives <$2,500
	Managers: >$3,000	Managers: ±$2,500	Managers: <$1,000
	Developers >$3,500	Developers: ±$2,000	Developers: <$1,500
	Maintainers: >$3,000	Maintainers: ±$1,500	Maintainers: <$1000
	Testers: >$3,000	Testers:±$2,500	Testers: <$1,500
	SQA: >$1,500	SQA: ±$1,000	SQA: <$500
	Total: >$25,000	Total: ±$14,500	Total: <$8,000

(*Continued*)

Part 4 (*Continued*) Reasons for Best-, Average-, and Worst-Case Results

	Average:=$4,167	Average:=$2,416	Average: <$1,100
	Percentage:=172%	Percentage:=100%	Percentage:=56%
	Predevelopment Risk Studies	*Predevelopment Risk Studies*	*Predevelopment Risk Studies*
	Formal value analysis adjusted for risks	Informal value analysis	Informal, inadequate value analysis
	Early risk analysis	Late risk analysis	Not used
	Early risk solutions	Late risk solutions	No risk solutions
	Continuous risk monitoring	Intermittent risk monitoring	No risk monitoring
	Rapid project corrections	Slow project corrections	Inadequate project corrections
	Automated cost estimates	Formal manual cost estimates	Informal manual cost estimates
	Automated schedule estimates	Manual schedule estimates	Inaccurate manual schedule estimates
	Formal, accurate progress tracking	Inaccurate progress tracking	Grossly inaccurate progress tracking
	Formal, accurate cost tracking	Inaccurate cost tracking	Grossly inaccurate cost tracking

(*Continued*)

Part 4 (*Continued*) Reasons for Best-, Average-, and Worst-Case Results

	Leakage from cost data <5%	Leakage from cost data <25%	Leakage from cost data >60%
	Leakage from quality data <2%	Leakage from quality data <10%	Leakage from quality data >75%
	Effective "dashboard" status monitoring	Marginally effective status monitoring	Ineffective or erroneous status monitoring
	Formal benchmark comparisons: Internal	Partial benchmark comparisons: Internal	No internal benchmark data available
	Formal benchmark comparisons: external (ISBSG or others)	Partial benchmark data from unknown sources	No external benchmark data utilized
	Formal change control	Informal change control	Casual change control
	Effective governance	Partial governance	Inadequate governance
	Accurate Estimates and Tracking	Marginal Estimates and Tracking	Poor Estimates and Tracking
	Automated cost estimates	Formal manual cost estimates	Informal manual estimates
	Automated quality estimates	Partial manual quality estimates	Not used

(*Continued*)

Part 4 (*Continued*) Reasons for Best-, Average-, and Worst-Case Results

	Automated risk estimates	Manual risks estimates, if any	Not used
	Defect tracking: from project start	*Defect tracking: from function test*	*Defect tracking: from delivery to customers*
	Automated test coverage measurements	Partial test coverage measurements	Not used
	Full cyclomatic complexity measures	Partial cyclomatic complexity measures	Not used
	Function point measures (100%)	Partial use of function points (small projects only)	Not used
	Full quality measures	Partial quality measures	Partial quality measures
	Defect potentials	Not used	Not used
	DDE	Not used	Not used
	DRE	Not used	Not used
	Defect severity	Defect severity	Defect severity
	Duplicate defects	Duplicate defects	Not used
	Test coverage for all tests	Test coverage for all tests	Not used

(*Continued*)

Part 4 (*Continued*) Reasons for Best-, Average-, and Worst-Case Results

	Bad-fix injection	Not used	Not used
	Invalid defects	Not used	Not used
	Defects by origin	Not used	Not used
	False positives	Not used	Not used
	Defect repair cost	Defect repair cost	Defect repair cost
	Defect repair time	Defect repair time	Defect repair time
	Root causes	Not used	Not used
	Adherence to ISO standards	Adherence to ISO standards	Adherence to ISO standards
	Adherence to ISO 9126 quality standards	Partial adherence to ISO 9126	Not used
	Adherence to ISO 14764 maintainability standards	Partial adherence to ISO 14764	Not used
	Adherence to ISO 14143 functional sizing	No adherence to ISO 14143	Not used
	Adherence to ISO 12207 software life cycle	Partial adherence to ISO 12207	Not used

(*Continued*)

Part 4 (*Continued*) Reasons for Best-, Average-, and Worst-Case Results

	Effective Requirement Analysis	*Average Requirement Analysis*	*Ineffective Requirements Analysis*
	JAD	Embedded users due to agile	Informal interviews with users
	QFD	Not used	Not used
	Kaizen	Not used	Not used
	Six Sigma	Not used	Not used
	Effective Methodologies	*Popular Methodologies*	*Traditional Methodologies*
	Methods chosen based on technical evaluation and empirical data	Methods chosen by popularity with partial evaluation of results	Methods chosen because "that is always the way we have done software" without any checks
	CMMI 4 or 5	CMMI 2 or 3	CMMI 1
	PSP/TSP	Agile	Waterfall
	RUP	XP	Cowboy
	XP	Hybrid	Partial, ineffective agile
	Hybrid	Uncertified reuse	Hazardous uncertified reuse

(Continued)

Part 4 (*Continued*) Reasons for Best-, Average-, and Worst-Case Results

	Certified reuse model-based development	Pair programming (< 500 function points)	Pair programming (< 500 function points)
	Pretest Defect Removal	*Pretest Defect Removal*	*Pretest Defect Removal*
	Certified SQA	Informal SQA	Informal or no SQA
	Static analysis—text	Not used	Not used
	FOG index—req. and des.	Not used	Not used
	Requirements inspections	Not used	Not used
	Design inspections	Not used	Not used
	Architecture inspections	Not used	Not used
	Refactoring	Informal refactoring	Occasional refactoring
	Static analysis—code	Partial use of static analysis	Partial use of static analysis
	Code inspections	Partial use of code inspections	Not used
	Test inspections	Not used	Not used
	Formal Testing by Experts	*Informal Testing by Developers*	*Informal Testing by Developers*
	Certified test teams	Some certification of testers	No certification of testers

(*Continued*)

Part 4 (*Continued*) Reasons for Best-, Average-, and Worst-Case Results

	Formal test case design mathematical models	Informal test case design	Informal test case design
	Formal, effective test library control <10 builds during test	Fairly effective test library control <15 builds during test	Marginal test library controls <25 builds during test
	Risk-based tests	Some risk-based tests	Not used
	Many reusable tests	Some reusable tests	Not used
	Unit test-1	Unit test-1	Unit test-1
	Function test-2	Function test-1	Function test-1
	Regression test-2	Regression test-1	Regression test-1
	Performance test-2	Performance test-2	Not used
	Security test-2	Not used	Not used
	Component test-2	Component test-1	Not used
	Usability test-2	Usability test-1	Not used
	Platform test-3	Platform test-3	Platform test-3
	System test-2	System test-2	System test-2
	Beta test-3	Beta test-3	Beta test-3
	Acceptance test-3	Acceptance test-3	Acceptance test-3

(*Continued*)

Part 4 (*Continued*) Reasons for Best-, Average-, and Worst-Case Results

	(1=tests by developers)	(1=tests by developers)	(1=tests by developers)
	(2=tests by certified test experts)	(2=tests by certified test experts)	(2=tests by certified test experts)
	(3=tests by customers)	(3=tests by customers)	(3=tests by customers)
	Test coverage analysis (percentage of code executed during testing; risks covered by testing; paths covered by testing; inputs covered by testing; outputs covered by testing)	Test coverage: execution paths only	Not used
	Automated Tools for Key Tasks	*Automated Tools for Key Tasks*	*Automated Tools for Key Tasks*
1	Project planning	Project planning	Project planning
2	Project sizing (all deliverables)	No automated tools	No automated tools
3	Cost estimates	Partial automation	No automated tools
4	Project quality estimates	No automated tools	No automated tools
5	Project maintenance estimates	Partial automation	No automated tools
6	Project customer support estimates	Partial automation	No automated tools
7	Project tracking—milestones	Partial automation	No automated tools
8	Project tracking—requirements creep	No automated tools	No automated tools

(*Continued*)

Part 4 (*Continued*) Reasons for Best-, Average-, and Worst-Case Results

9	Risk analysis	No automated tools	No automated tools
10	Project cost tracking	Cost tracking	Cost tracking
11	Project configuration control	Partial automation	No automated tools
12	Defect tracking	Defect tracking	Defect tracking
13	FOG index—text	No automated tools	No automated tools
14	Static analysis—text	No automated tools	No automated tools
15	Static analysis—code	Static analysis—code	No automated tools
16	Test library support tools	Test library support tools	Test library support tools
17	Automated test tools	Automated test tools	Automated test tools
18	Legacy code analysis	No automated tools	No automated tools
19	Code structure analysis	No automated tools	No automated tools
20	Code coverage analysis	Code coverage analysis	No automated tools
21	Code update and change analysis	No automated tools	No automated tools
22	Customer satisfaction tracking	Customer satisfaction tracking	Customer satisfaction tracking
23	Customer support tracking	Customer support tracking	Customer support tracking
24	Warranty and repair tracking	Warranty and repair tracking	Warranty and repair tracking
25	Project governance	Partial automation	No automated tools

These are usually the final parts of our development benchmark output data. Some clients, about 15% of total clients, also commission benchmarks for maintenance and enhancements. We usually do 3 years of maintenance and enhancement benchmarks. This means, of course, that the projects have been operational for at least 3 years.

It is hard to get good data longer than 3 years before the current year. Maintenance and enhancement benchmark data is not shown here since it is about as large and complex as the development data.

We need maintenance and enhancement data to calculate TCO. However, since our SRM tool predicts 3 years of maintenance, we can produce synthetic data from SRM for TCO calculations.

Benchmark Section 5: Maintenance, Enhancement, and Support Benchmarks

We capture effort for enhancements, maintenance (bug repairs), and customer support. We also capture numbers of bugs per year and also number of "incidents." An incident is a query or contract from a customer about a software product, but it may not be a bug report.

For internal software projects with only a few users, customer support is not very pricey. But for high-volume commercial software such as Microsoft Office or Apple applications with millions of users, customer support is very pricey indeed, assuming the software company decides to do support reasonably well.

Summary and Conclusions on Global Benchmarks

Software is a major industry in more than 200 countries. Software outsourcing is a major subindustry in more than 50 countries. It is important to understand the impacts of local work hours and local cost structures. It is also important to know inflation rates because these will degrade the long-term benefits of international outsource agreements.

Software is probably the largest and the most important industry to lack effective data on economic productivity and quality. Good benchmarks are a step in the direction of improving both software quality and software economic performance.

This book discusses the key factors that are known to influence software on a global basis. There are many other factors that impact software, but national work patterns and unpaid overtime patterns need to be included in all software benchmark studies and should technology stacks.

Function point metrics are the best available metrics for productivity and also for normalizing quality data. The metrics set of (1) local work hours per month, (2) work hours per function point, (3) function points per month, (4) defect potentials in function points, (5) DRE, and (6) delivered defects per function point are the combinations used in this book, and are also the best available metrics as of 2017. These can be combined to create COQ and TCO values for applications. Growth in function points over time also needs to be included. SNAP points for nonfunctional requirements may be useful in the future but lack data in 2017.

Older metrics such as story points, LOC, use case points, velocity, code defect density, and cost per defect all have serious mathematical flaws and all distort reality. There are no accurate or effective benchmarks in these metrics. Technical debt is an interesting metaphor but not an effective metric in 2017 because it has no standard definition and every company evaluates it differently. It also omits major quality factors such as consequential damages to clients and litigation costs for poor quality, both of which are common and expensive.

The author hopes that in the future all software projects <1000 function points in size will use parametric estimation tools before starting and create formal benchmarks data collections using function point metrics when completed.

Chapter 7

Advancing Software Benchmark Technology

This chapter discusses some topics that are not current in benchmarks in the calendar year 2017. Some of these appear to add value to benchmarks and hopefully will become common in the future.

Synthetic Benchmarks Using Parametric Estimation to Improve Speed and Accuracy

Since the outputs of a software parametric estimation tool are usually more accurate than what are called historical data, why not use a parametric estimation tool to help improve benchmark measurement accuracy? This is the essential idea of parametric benchmarks.

This new way of collecting historical data involves high-speed and low-cost function point analysis combined with using parametric estimates to help validate historical data. This new method is called "parametric benchmarks." This new method collects quantitative benchmark data as described in this paper. The same method can be used for collecting qualitative assessment data, but that is a separate topic and has not been discussed here.

As mentioned previously, normal function point analysis proceeds at an average rate of about 400 function points per day. The author's high-speed function point method can size more than 30 applications per day of any size when used by experienced personnel.

Assuming each of these 30 applications averages 400 function points in size, the daily rate would be 12,000 function points per day as opposed to manual sizing of only 400 function points per day. This method is embedded in a tool called Software Risk Master (SRM). With SRM the speed of function point sizing drops to about 90 s and the cost drops to pennies per function point.

Thus, the new high-speed sizing method eliminates the high cost and rather slow speed of normal function point analysis.

Since remote self-reported benchmarks have gaps and omissions, and on-site benchmarks are expensive, there is a new form of benchmark data collection that combines very high accuracy with very low cost.

Using this new method, the benchmark questionnaire will be based on the input questionnaire to a parametric estimation tool such as SRM. The questionnaire would be available to benchmark clients.

Once the initial input questions have been answered, the tool would generate both the function point size of the application and also provide a detailed estimate of the schedules, effort, staffing, and costs of the project using an activity-based cost structure. These estimates include the elements that are often omitted from benchmarks such as unpaid overtime and management costs.

With the SRM tool as an example, sizing an application and generating a full estimate normally takes between 4 and 7 min.

Having a full estimate available as a basis for benchmark data collection will prompt users to include all relevant information and not ignore topics such as unpaid overtime or project management.

When used in measurement mode, the estimation tool will allow users to perform any of these four actions for every activity displayed:

1. *Accept* the predicted value if it matches historical data.
2. *Modify* the predicted value so that it matches historical data exactly.
3. *Inform* the estimation/measurement tool that certain predicted activities were not in fact performed (i.e., inspections, static analysis, etc.) by specifying *not used.*
4. *Inform* the estimation/measurement tool that a new kind of activity not included in the standard chart of account was performed by specifying *add activity.*

This form of benchmark would only take about 7 min to complete the questionnaire. Assuming that three people discussed the project before attempting to answer the questions, the net effort would probably be three people at 15 min each, or a total of 45 min per project. At internal costs of $50 per hour, this would cost about $37.50.

By merely scrolling through the predicted values from the estimate and accepting or modifying them, very accurate historical data could be gathered. Even better, this form of measurement could help to calibrate the estimates themselves. When used in predictive mode, SRM displays a set of software benchmark parameters, i.e., staff, effort, schedule, and costs for each activity (Table 7.1).

Table 7.1 SRM in Benchmark Prediction Mode

Activities	*Staff*	*Effort Months*	*Schedule Months*	*Project Costs*
Requirements	2.05	3.91	1.91	$39,127
Design	2.57	6.26	2.43	$62,604
Coding	5.39	35.08	6.51	$350,812
Testing	4.41	19.65	4.46	$196,455
Documentation	1.28	2.85	2.23	$28,456
Quality assurance	1.12	3.56	3.17	$35,570
Management	1.21	12.52	13.73	$125,208
Totals	6.11	83.82	20.71	$838,232
Gross schedule months			20.71	
Overlap (%)			66.29	
Predicted delivery schedule and date			13.73	

When used in measurement mode, the SRM tool has a "micrometer" function that allows very high precision measurements to better than 1% precision. The micrometer mode uses three special parameters, assignment scopes, production rates, and overlap.

The assignment scope is the amount of work assigned to one person. The production rate is the amount of work performed in a fixed time interval such as 1 month. The overlap is the amount of an activity still not finished when the next activity begins. These three parameters are used in measurement mode to achieve very high precision (Table 7.2).

By adjusting the assignment scope and production rate parameters (and the third parameter for schedule overlaps), historical project data can be matched exactly.

It would still be desirable to have some external validation of the benchmark data before it is accepted. This might consist of a short telephone call with a benchmark consultant or possibly an exchange of e-mails or an online chat.

This new method of parametric benchmarks simultaneously lowers the high costs of function point analysis and speeds up the collection of data, while providing standard data elements to users in order to help avoid common omissions.

The new parametric benchmark method combines several attractive features, including ease of use, low cost, and rapid results. Parametric benchmarks might easily top 100 projects per month or 1200 per year in the United States, and perhaps 300 per month or 3600 projects per year globally.

Table 7.2 SRM in Parametric Measurement Mode

Activities	*Assignment Scope*	*Production Rate*	*Staff*	*Effort Months*	*Schedule Months*
Requirements	483	192	2.05	3.91	1.91
Design	362	120	2.57	6.26	2.43
Coding	8,186	1,140	5.39	35.08	6.51
Testing	10,232	2,036	4.41	19.65	4.46
Documentation	965	264	1.28	2.85	2.23
Quality assurance	1,207	211	1.12	3.56	3.17
Management	1,062	60	1.21	12.52	13.73
Totals	123	11.93	6.11	83.82	20.71
Gross schedule months					20.71
Overlap (%)					66.29
Predicted delivery months					13.73

Measurement accuracy cannot achieve 100% precision or capture every single cost element. Too many unpredictable events such as layoffs of stakeholders, weather delays, illness of key personnel, and other transient phenomena can degrade measurement precision.

Measuring Brand New Tools, Methods, and Programming Languages

New software development methodologies occur about every 8 months. The author's current collection of software methodologies circa 2016 has 70 named methodologies. These include Agile, container development DevOps, Extreme Programming, mashups, object-oriented programming, the Rational Unified Process (RUP), the Team Software Process, iterative development, waterfall development, V-model development, Prince2, Merise, and many others. In addition, there are hundreds of hybrid variations that use portions of several methods.

New programming languages have been coming out at a rate of about one per month since the 1970s. As of 2017, there are more than 3000 programming languages.

New tools have been coming at a rate of about one per week for the past 15 years. As of 2017, there are 25 major categories of software management and technical tools, and a total tool inventory that probably tops 5000 commercial tools.

Capturing data on brand new methods and tools is a challenge for software benchmark providers. For one thing, brand new tools or languages have no historical data of any kind and sometimes no users other than the original developers.

The solution developed by the author is to provide general "boxes" for various kinds of tools and methods. For example, a partial list of project management tool categories in alphabetical order includes the following:

1. Cost estimation tools
2. Cost tracking tools
3. Defect tracking tools
4. Human resource assessment tools
5. Project management tools
6. Quality estimation tools
7. Project tracking tools

A partial list of software development tools in alphabetical order includes the following:

1. Assemblers
2. Compilers
3. Configuration control
4. Debuggers
5. Inspection support
6. Integration
7. Interpreters
8. Static analysis

Users will select the appropriate category and then enter the actual name of the tool or languages used.

For example, if you are using a brand new programming language called "JavaDecafe," you would select compilers as the tool category.

When brand new tools or languages are being measured for the very first time, it is not possible to predict their impact with high precision. It is usually necessary to collect empirical data on perhaps a dozen uses of the new tool or language to form a preliminary judgment. Of course, if a new programming language is just another variant of C or Java, you can make reasonable assumptions as to its impact. But if it is truly new and has unique features, they need to be measured.

Analysis of the results of new tools, languages, and methods is of necessity a quite complicated activity that requires careful collection of data.

As an example, a new tool from a group called IntegraNova is entering the U.S. market. The company is headquartered in Spain and its current clients are in Europe. The tool is a requirements modeling engine that feeds into an application generator.

It would be useful for IntegraNova to know the productivity and quality results from using this tool in a U.S. context. From European data it is possible to compare results against other methods such as Agile and RUP. However, until there are U.S. projects and careful measurements, the results are speculative rather than definitive.

Executive Interest Levels in Software Benchmark Types

This book primarily deals with individual project benchmarks for software project productivity and quality levels. There are many other kinds of benchmarks as well. Table 7.3 shows the levels of interest in all types of benchmarks based on interviews with executives in client companies.

As can be observed, all levels and types of executives are interested in benchmarks. However, chief technology officers have the widest range of interest, as might be expected from the nature of CTO work.

CEOs are more interested in competitive and large-scale benchmarks than in specific project benchmarks, with the important exception of very high interest levels in applications that might be of strategic importance or might cost more than $10,000,000.

Software Benchmark Providers

The Web is the best source for additional information on benchmarks. Google searches on "software benchmarks," "software productivity," "international inflation rates," and "work hours by country" will turn up scores of government and corporate reports (Table 7.4).

It is evident that there are many sources of software benchmarks in the United States and Europe. India, China, Japan, other Eastern countries, and the Middle East have fewer choices. Brazil has ti Metricas (www.Metricas.com/BR), but the rest of Central and South America have sparse benchmark providers.

Summary and Conclusions on Estimation, Benchmarks, and Quantified Results

Software estimation is simple in concept, but difficult and complex in reality. The difficulty and complexity required for successful estimates exceeds the capabilities of most software project managers to produce effective manual estimates.

The commercial parametric software estimating tools can outperform human estimates in terms of accuracy, and always in terms of speed and cost-effectiveness. However, no method of estimation is totally error-free. The current "best practice"

Table 7.3 Executive Interests

	Software Benchmark Types	*CEO Interest*	*CFO Interest*	*CIO Interest*	*CTO Interest*
1	Competitive practices within industry	10	10	10	10
2	Project failure rates (size, methods)	10	10	10	10
3	Development schedules	10	10	10	10
4	Outsource contract success/ failure	10	10	10	10
5	Project risks	10	10	10	10
6	Capability maturity model integrated (CMMI) assessments	9	9	9	10
7	Security attacks (number, type)	10	10	10	10
8	Return on investment (ROI)	10	10	10	10
9	Total cost of ownership (TCO)	10	10	10	10
10	Customer satisfaction	10	10	10	10
11	Employee morale	10	9	10	10
12	Data quality	10	10	10	10
13	Customer support benchmarks	9	9	10	10
14	Cost of quality (COQ)	10	10	10	10
15	Best practices—requirements	6	9	9	10
16	Development costs	10	10	10	10
17	Team attrition rates	7	8	10	10
18	ISO standards certification	9	9	9	9
19	Best practices—maintenance	8	9	10	10
20	Best practices—test efficiency	8	7	9	10
21	Best practices—defect prevention	5	6	8	10
22	Best practices—pretest defects	6	6	9	10

(*Continued*)

Table 7.3 (*Continued*) Executive Interests

	Software Benchmark Types	*CEO Interest*	*CFO Interest*	*CIO Interest*	*CTO Interest*
23	Technical debt	8	8	9	9
24	Productivity—project	8	8	9	10
25	Coding speed	7	8	8	8
26	Code quality (only code—nothing else)	7	7	10	10
27	Cost per defect (caution: unreliable)	6	7	10	10
28	Application sizes by type	8	8	8	9
29	Enhancement costs	8	10	10	8
30	Maintenance costs (annual)	8	9	9	9
31	Team morale	8	8	8	8
32	Litigation—patent infringements	10	10	10	10
33	Best practices—design	5	5	9	10
34	Litigation—intellectual property	10	10	10	10
35	Portfolio maintenance costs	10	10	10	10
36	Team compensation level	10	9	9	7
37	Litigation—breach of contract	10	10	10	10
38	Earned value analysis (EVA)	5	7	7	10
39	Methodology comparisons	5	6	9	10
40	Test coverage benchmarks	5	6	9	8
41	Productivity—activity	4	6	8	10
42	Application types	7	7	7	9
43	Litigation—employment contracts	7	8	8	8
44	Tool suites used	4	4	7	10

(*Continued*)

Table 7.3 (*Continued*) Executive Interests

	Software Benchmark Types	*CEO Interest*	*CFO Interest*	*CIO Interest*	*CTO Interest*
45	CMMI levels within industries	6	5	5	9
46	Standards benchmarks	5	5	6	8
47	Country productivity	8	5	8	9
48	Database size	9	8	8	9
49	Industry productivity	10	7	6	9
50	Metrics used	3	3	7	9
51	Programming languages	3	3	6	8
52	Application class by taxonomy	4	5	6	8
53	Certification benchmarks	3	3	6	7
54	Cyclomatic complexity benchmarks	1	1	5	7
55	SNAP nonfunctional size metrics	3	4	5	8
	Totals	412	421	475	513
		7.49	7.65	8.64	9.33

Table 7.4 Software Benchmark Providers (Listed in Alphabetic Order)

1	4SUM Partners	www.4sumpartners.com
2	Bureau of Labor Statistics, Dept. of Commerce	www.bls.gov
3	Capers Jones (Namcook Analytics LLC)	www.namcook.com
4	CAST Software	www.castsoftware.com
5	Congressional Cyber Security Caucus	cybercaucus.langevin.house.gov
6	Construx	www.construx.com
7	COSMIC function points	www.cosmicon.com

(*Continued*)

Table 7.4 (*Continued*) Software Benchmark Providers (Listed in Alphabetic Order)

8	Cyber Security and Information Systems	www.CSIAC.org
9	David Consulting Group	www.davidconsultinggroup.com
10	Forrester Research	www.forrester.com
11	Galorath Incorporated	www.galorath.com
12	Gartner Group	www.gartner.com
13	German Computer Society	http://metrics.cs.uni-magdeburg.de/
14	Hoovers Guides to Business	www.hoovers.com
15	Howard Rubin	www.rubinworldwide.com
16	IDC	www.IDC.com
17	ISBSG Limited	www.isbsg.org
18	ITMPI	www.itmpi.org
19	Jerry Luftman (Stevens Institute)	http://howe.stevens.edu/index.php?id=14
20	Level 4 Ventures	www.level4ventures.com
21	Namcook Analytics LLC	www.namcook.com
22	Price Systems	www.pricesystems.com
23	Process Fusion	URL no longer available
24	Q/P Management Group	www.qpmg.com
25	QuantiMetrics	www.quantimetrics.net
26	Quantitative Software Management (QSM)	www.qsm.com
27	RBCS, Inc.	www.rbcs-us.com
28	Reifer Consultants LLC	www.reifer.com
29	SANS Institute	www.SANS.org
30	Software Benchmarking Organization (SBO)	www.sw-benchmark.org

(*Continued*)

Table 7.4 (*Continued*) Software Benchmark Providers (Listed in Alphabetic Order)

31	Software Engineering Institute (SEI)	www.sei.cmu.edu
32	Software Improvement Group (SIG)	www.sig.eu
33	Software Productivity Research	www.SPR.com
34	Standish Group	www.standishgroup.com
35	Strassmann, Paul	www.strassmann.com
36	System Verification Associates LLC	http://sysverif.com
37	Test Maturity Model Integrated	www.experimentus.com

for software cost estimation is to use a combination of software cost estimating tools coupled with software project management tools, under the careful guidance of experienced software project managers and estimating specialists.

Software benchmarks are troubling due to the "gaps" in historical data caused by inadequate tracking tools and methods. A majority of U.S. companies utilize substantial amounts of unpaid overtime but do not record it. Software projects may have dozens of specialists but benchmarks may exclude many such as business analysts, project office staff, technical writers, quality assurance, and the like. Benchmarks should report 100% of occupation groups and 100% of effort.

Among the author's clients, their self-reported benchmarks exclude over 60% of software occupations and capture only about 37% of total costs. These partial benchmarks are worthless for serious economic analysis.

References and Readings on Software Costs

Boehm, B.; *Software Engineering Economics*; Prentice Hall, Englewood Cliffs, NJ; 1981; 900 pages.

Brooks, F.; *The Mythical Man-Month*; Addison-Wesley, Reading, MA; 1995; 295 pages.

Cohn, M.; *Agile Estimating and Planning*; Prentice Hall, Englewood Cliffs, NJ; 2005; ISBN 0131479415.

Conte, S.D., Dunsmore, H.E., and Shen, V.Y.; *Software Engineering Models and Metrics*; The Benjamin Cummings Publishing Company, Menlo Park, CA; 1986; ISBN 0-8053-2162-4; 396 pages.

DeMarco, T.; *Controlling Software Projects*; Yourdon Press, New York; 1982; ISBN 0-917072-32-4; 284 pages.

DeMarco, T.; *Why Does Software Cost So Much*? Dorset House Press, New York; 1995; ISBN 0-932633-34-X; 237 pages.

Gack, G.; *Managing the Black Hole: The Executive's Guide to Project Risk*; The Business Expert Publisher; Thomson, GA; 2010; ISBN 10: 1-935602-01-2.

Galorath, D.D. and Evans, M.W.; *Software Sizing, Estimation, and Risk Management*; Auerbach Publications, New York; 2006.

Garmus, D. and Herron, D.; *Measuring the Software Process: A Practical Guide to Functional Measurement*; Prentice Hall, Englewood Cliffs, NJ; 1995.

Garmus, D. and Herron, D.; *Function Point Analysis*; Addison Wesley Longman, Boston, MA; 1996.

Hill, P.R; *Practical Software Project Estimation*; McGraw Hill; 2010.

Jones, C.; *A Ten-Year Retrospective of the ITT Programming Technology Center*; Software Productivity Research, Burlington, MA; 1988.

Jones, C.; *Software Productivity and Quality Today: The Worldwide Perspective*; Information Systems Management Group; 1993a; ISBN-156909-001-7; 200 pages.

Jones, C.; *Critical Problems in Software Measurement*; Information Systems Management Group; 1993b; ISBN 1-56909-000-9; 195 pages.

Jones, C.; *Assessment and Control of Software Risks*; Prentice Hall; 1994; ISBN 0-13-741406-4; 711 pages.

Jones, C.; *Patterns of Software System Failure and Success*; International Thomson Computer Press, Boston, MA; 1995; 250 pages; ISBN 1-850-32804-8; 292 pages.

Jones, C.; *Software Quality: Analysis and Guidelines for Success*; International Thomson Computer Press, Boston, MA; 1997a; ISBN 1-85032-876-6; 492 pages.

Jones, C.; *The Economics of Object-Oriented Software*; SPR Technical Report; Software Productivity Research, Burlington, MA; 1997b; 22 pages.

Jones, C.; Software project management practices: Failure versus success; *Crosstalk*; 19, 6; 2006; pp. 4–8.

Jones, C.; *Software Estimating Methods for Large Projects*; Crosstalk; 2005.

Jones, C.; *Estimating Software Costs*; 2nd edition; McGraw Hill, New York; 2007; 700 pages.

Jones, C; *Applied Software Measurement*; 3rd edition; McGraw Hill; 2008; ISBN 978-0-07-150244-3; 662 pages.

Jones, C.; *Software Engineering Best Practices*; McGraw Hill, New York; 2010; ISBN 978-0-07-162161-8; 660 pages.

Jones, C.; *The Technical and Social History of Software Engineering*; Addison Wesley; 2014.

Jones, C.; *Samples of Software Risk Master (SRM) Estimating Assumptions*; Namcook Analytics LLC; 2015.

Jones, C.; *New Directions in Software Management; Information Systems Management Group*; ISBN 1-56909-009-2; 150 pages.

Jones, C. and Bonsignour, O.; *The Economics of Software Quality*; Addison Wesley, Boston, MA; 2011; ISBN 978-0-13-258220-9; 587 pages.

Kan, S.H.; *Metrics and Models in Software Quality Engineering*; 2nd edition; Addison Wesley Longman, Boston, MA; 2003; ISBN 0-201-72915-6; 528 pages.

Kemerer, C.F.; An empirical validation of software cost estimation models; *Communications of the ACM*; 30; 1987; pp. 416–429.

Kemerer, C.F.; Reliability of function point measurement: A field experiment; *Communications of the ACM*; 36; 1993; pp. 85–97.

Laird, L.M. and Brennan, C.M; *Software Measurement and Estimation: A Practical Approach*; John Wiley & Sons, Hoboken, NJ; 2006; ISBN 0-471-67622-5; 255 pages.

Love, T.; *Object Lessons*; SIGS Books, New York; 1993; ISBN 0-9627477 3-4; 266 pages.

McConnell, S.; *Software Estimating: Demystifying the Black Art*; Microsoft Press, Redmund, WA; 2006.

Mills, H.; *Software Productivity*; Dorset House Press, New York; 1998; ISBN 0-932633-10-2; 288 pages.

Park, R.E. et al; Software cost and schedule estimating: A process improvement initiative; Technical Report CMU/SEI 94-SR-03; Software Engineering Institute, Pittsburgh, PA; May 1994.

Park, R.E. et al; Checklists and criteria for evaluating the costs and schedule estimating capabilities of software organizations; Technical Report CMU/SEI 95-SR-005; Software Engineering Institute, Pittsburgh, PA; January 1995.

Paulk, M. et al; *The Capability Maturity Model; Guidelines for Improving the Software Process*; Addison Wesley, Reading, MA; 1995; ISBN 0-201-54664-7; 439 pages.

Pressman, R.; *Software Engineering: A Practitioner's Approach*; McGraw Hill, New York; 1982.

Roetzheim, W.H. and Beasley, R.A.; *Best Practices in Software Cost and Schedule Estimation*; Prentice Hall, Saddle River, NJ; 1998.

Royce, W.E.; *Software Project Management: A Unified Framework*; Addison Wesley, Reading, MA; 1999.

Rubin, H.; *Software Benchmark Studies for 1997; Howard Rubin Associates*, Pound Ridge, NY; 1997.

St-Pierre, D.; Maya, M.; Abran, A., and Desharnais, J.-M.; Full function points: Function point extensions for real-time software, concepts and definitions; University of Quebec. Software Engineering Laboratory in Applied Metrics (SELAM); TR 1997-03; 1997; 18 pages.

Strassmann, P.; *The Squandered Computer*; The Information Economics Press, New Canaan, CT; 1997; ISBN 0-9620413-1-9; 426 pages.

Stutzke, R.D.; *Estimating Software Intensive Systems*; Addison Wesley, Boston, MA; 2005.

Yourdon, E.; *Death March: The Complete Software Developer's Guide to Surviving "Mission Impossible" Projects*; Prentice Hall, Upper Saddle River, NJ; 1997; ISBN 0-13-748310-4; 218 pages.

Index

Note: Page numbers followed by "*f*" and "*t*" refer to figures and tables.

D

G

Q

R

S

T